Advances in the Use of Beneficial Microorganisms to Improve Nutritional and Functional Properties of Fermented Foods

Advances in the Use of Beneficial Microorganisms to Improve Nutritional and Functional Properties of Fermented Foods

Editors

Carlo Giuseppe Rizzello
Palmira De Bellis

MDPI

Basel • Beijing • Wuhan • Barcelona • Belgrade • Novi Sad • Cluj • Manchester

Editors
Carlo Giuseppe Rizzello
Sapienza University of Rome
Rome, Italy

Palmira De Bellis
National Research Council
(CNR)
Bari, Italy

Editorial Office
MDPI
St. Alban-Anlage 66
4052 Basel, Switzerland

This is a reprint of articles from the Special Issue published online in the open access journal *Foods* (ISSN 2304-8158) (available at: https://www.mdpi.com/journal/foods/special_issues/Beneficial_Microorganisms).

For citation purposes, cite each article independently as indicated on the article page online and as indicated below:

Lastname, A.A.; Lastname, B.B. Article Title. *Journal Name* **Year**, *Volume Number*, Page Range.

ISBN 978-3-0365-9969-4 (Hbk)
ISBN 978-3-0365-9970-0 (PDF)
doi.org/10.3390/books978-3-0365-9970-0

Contents

About the Editors

Carlo Giuseppe Rizzello

Carlo Giuseppe Rizzello has been a Full Professor in Food Microbiology at La Sapienza University of Rome from 2020. Before, he was Associate Professor in Food Microbiology at University of Bari. He is the author of 172 peer-reviewed articles (total citations, approx. 10677; h-index, 67 reported on Scopus) and of 23 book chapters and 13 patents (with more than 25 extensions reported on google patents). He is co-Editor of the books: "Biotecnologia dei prodotti lievitati da Forno" (Casa Editrice Ambrosiana) and "Basic Methods and Protocols on Sourdough" (Springer Nature). His area of expertise includes lactic acid bacteria, nutritional and functional aspects of fermented foods, food biotechnology. From 2022, he has been vice-president of the SIMTREA, Italian Society of Agro-Food And Environmental Microbiology.

Palmira De Bellis

Palmira De Bellis graduated in Agriculture Sciences and received her PhD at the University of Bari (Italy). Since 2011, she has been a Researcher at the National Research Council (CNR) - Institute of Sciences of Food Production (ISPA), Bari. Her main research area is food microbiology with expertise on the molecular microbiology and biotechnology to improve nutritional and functional value of foods. She is the co-author of over 70 papers.

foods

Editorial

Advances in the Use of Beneficial Microorganisms to Improve Nutritional and Functional Properties of Fermented Foods

Palmira De Bellis [1,*] and Carlo Giuseppe Rizzello [2]

1 Institute of Sciences of Food Production (ISPA), National Research Council (CNR), Via G. Amendola 122/O, 70126 Bari, Italy
2 Department of Environmental Biology, Sapienza University of Rome, Piazzale Aldo Moro 5, 00185 Rome, Italy; carlogiuseppe.rizzello@uniroma1.it
* Correspondence: mirella.debellis@ispa.cnr.it; Tel.: +39-080-5929459; Fax: +39-080-5929374

The World Health Organization [1] has highlighted the need to improve the nutritional and functional characteristics of foods and beverages in order to enhance quality of life and prevent chronic diseases. Many foods often present critical issues such as high glycemic response, low biological protein value, high salt and fat concentrations, a lack of functional compounds such as fibers and polyphenols, and the presence of ingredients associated with hypersensitivity reactions. The use of beneficial microorganisms such as lactic acid bacteria (LAB) is an excellent strategy to improve the nutritional and functional properties of foods via the synthesis of bioactive compounds or through the degradation of antinutritional factors. In recent years, many microorganisms with metabolic characteristics useful for improving traditional and novel fermented foods have been identified, and the relationships between their application and fermented food quality, safety, and health promoting features have been elucidated.

In this Special Issue, an overview is provided of the latest scientific evidence on improving the nutritional and functional properties of food resulting from the use of beneficial microorganisms. Recent increases in scientific efforts have resulted in the developed of both new and traditional fermented products that combine the beneficial characteristics of microbial fermentation with the nutritional properties of animal- and plant-derived matrices.

In detail, bioprocessing techniques (sprouting and sourdough fermentation) applied to nonwheat grains, such as barley and lentil, can produce ingredients able to improve the technological and nutritional properties of fortified wheat breads. The modulation of fermentation parameters (native or sprouted cereal and legume flour, DY, and temperature) can lead to the production of a dextran-enriched sourdough that is suitable for the production of bread with enhanced nutritional quality (low HI and pGI), functionality (high soluble and total fiber content), and sensory attributes [2]. Some aspects related to the production of EPS by a *Weissella cibaria* strain were explored in depth by De Bellis et al. [3]. The strain selected as a high-EPS producer in the presence of sucrose was used to produce an EPS-enriched sourdough suitable for use as a fat replacer in baked goods [4]. The authors characterized the EPS produced by *W. cibaria* C43-11 and investigated the possible genetic regulatory elements responsible for the modulation of *dextransucrase (dsr)* gene expression [3].

Sourdough and bread with improved features were developed using fermented water extracts from Asian pears and Assam tea leaves with co-cultures of *Lactiplantibacillus plantarum* and *Saccharomyces cerevisiae* strains as starter cultures. In particular, a sourdough bread supplemented with fermented water had 10% less sugar, 12% higher dietary fiber, and two to three times higher total phenolic content and antioxidant activity than common sourdough bread [5].

Included in this Special Issue, a study addressed the microbial diversity associated with different flatbread doughs leavened with traditional brewer's yeast or type II sourdough.

Citation: De Bellis, P.; Rizzello, C.G. Advances in the Use of Beneficial Microorganisms to Improve Nutritional and Functional Properties of Fermented Foods. *Foods* **2024**, *13*, 155. https://doi.org/10.3390/foods13010155

Received: 15 December 2023
Accepted: 23 December 2023
Published: 2 January 2024

The latter was started with a strain of *Leuconostoc citreum*, which was selected for the production of "yeast-free" bread. The authors explored, using an in-depth metagenetic approach, the bacterial and fungal microbiota of the dough, highlighting how the use of a microbial starter profoundly influenced the composition of the microbiota of the dough which was directly responsible for the product quality [6].

The promising results reported through several studies indicate a possible market exploitation of innovative fermented beverages. In particular, He et al. [7] found that oats are a suitable fermentation substrate. They evaluated the effects of fermentation on the content of the bioactive components of oats, such as β-glucan, polyphenols, flavonoids, and volatile compounds. The results differed depending on the strains used in oat fermentation, indicating that oats can be used for the production of fermented beverages, providing alternative products for vegetarians and those who are lactose-intolerant.

Demarinis et al. [8] identified a suitable combination of starters and legume grains that can be used to produce a legume-based milk substitute containing high concentrations of free peptides and amino acids. These lupin- and pea-based beverages may also be carriers of probiotic LAB.

Soy drinks are an alternative to milk, especially for those who are allergic to milk proteins, lactose-intolerant, or follow a vegan diet. However, the bean flavor of soy beverages is the main factor limiting consumer acceptance. Sun et al. [9] proposed the use of a Chinese indigenous edible and medicinal fungus, *Naematelia aurantialba*, to produce a fermented soybean beverage with improved health-benefitting characteristics and a weaker beany taste compared with typical soy beverages, which are essential for its commercialization [9].

Researches have focused on the production of new fermented foods with functional properties, the consumer demand for which is increasing. Canonico et al. [10] produced a craft beer with functional properties through the conjunction of a legume-fortified beer with functional yeasts. In particular, this craft beer was fortified with hydrolyzed red lentils using selected nonconventional yeasts (*Lachancea thermotolerans* and *Kazachstania unispora*) through both pure and mixed fermentation.

The management of food waste and by-products is crucial for the agri-food industry both for managing the costs relating to disposal and to meet current environmental regulations. Cheese whey represents the main by-product of the dairy industry; however, this whey is also a source of functional and bioactive compounds, in particular, proteins and peptides. A biotechnological protocol aimed at the valorization of exhausted ricotta whey enriched the substrate in bioactive peptides, thus creating a supplement to produce a functional ricotta [11].

Finally, Ha et al. [12] performed the fermentation of *Tenebrio molitor* larvae by *Cordyceps militaris* mycelium to increase the contents of nutrients such as total protein, total fiber, β-glucan, and cordycepin contents. The proposed method could allow the bio-fortification of several foods, which is crucial for populations that do not have a proper and balanced diet.

This Special Issue also includes an in-depth review on the potential of LAB as biopreservatives through the production of various antifungal metabolites [13]. Foods are highly susceptible to spoilage due to numerous fungi, which cause economic and production losses and represent a toxicological risk to humans through the production of mycotoxins. Furthermore, consumers are increasingly demanding healthier foods that are prepared with the minimal use of chemical preservatives. Therefore, the use of LAB as biopreservatives is among the safest strategies to inhibit the growth of fungi in foods, resulting in a lengthening of their shelf life.

In summary, this Special Issue provides insights into the recent application of beneficial microorganisms to improve nutritional and functional properties of various fermented foods. We hope that readers will find this Special Issue interesting.

Author Contributions: Writing—original draft preparation, P.D.B.; writing—review and editing, C.G.R. and P.D.B. All authors have read and agreed to the published version of the manuscript.

Funding: This study received no external funding.

Conflicts of Interest: The authors declare no conflicts of interest.

References

1. WHO World Health Organization. Healthy Diet. Fact Sheet N°394. 2018. Available online: http://www.who.int/mediacentre/factsheets/fs394/en/ (accessed on 30 October 2023).
2. Perri, G.; Rizzello, C.G.; Ampollini, M.; Celano, G.; Coda, R.; Gobbetti, M.; De Angelis, M.; Calasso, M. Bioprocessing of Barley and Lentil Grains to Obtain In Situ Synthesis of Exopolysaccharides and Composite Wheat Bread with Improved Texture and Health Properties. *Foods* **2021**, *10*, 1489. [CrossRef] [PubMed]
3. De Bellis, P.; Ferrara, M.; Bavaro, A.R.; Linsalata, V.; Di Biase, M.; Musio, B.; Gallo, V.; Mulè, G.; Valerio, F. Characterization of Dextran Produced by the Food-Related Strain *Weissella cibaria* C43-11 and of the Relevant Dextransucrase Gene. *Foods* **2022**, *11*, 2819. [CrossRef] [PubMed]
4. Bavaro, A.R.; Di Biase, M.; Conte, A.; Lonigro, S.L.; Caputo, L.; Cedola, A.; Del Nobile, M.A.; Logrieco, A.F.; Lavermicocca, P.; Valerio, F. *Weissella cibaria* short-fermented liquid sourdoughs based on quinoa or amaranth flours as fat replacer in focaccia bread formulation. *Int. J. Food Sci. Technol.* **2021**, *56*, 3197–3208. [CrossRef]
5. Supasil, R.; Suttisansanee, U.; Santivarangkna, C.; Tangsuphoom, N.; Khemthong, C.; Chupeerach, C.; On-nom, N. Improvement of Sourdough and Bread Qualities by Fermented Water of Asian Pears and Assam Tea Leaves with Co-Cultures of *Lactiplantibacillus plantarum* and *Saccharomyces cerevisiae*. *Foods* **2022**, *11*, 2071. [CrossRef] [PubMed]
6. Ferrara, M.; Sisto, A.; Mulè, G.; Lavermicocca, P.; De Bellis, P. Metagenetic Analysis for Microbial Characterization of Focaccia Doughs Obtained by Using Two Different Starters: Traditional Baker's Yeast and a Selected *Leuconostoc citreum* Strain. *Foods* **2021**, *10*, 1189. [CrossRef] [PubMed]
7. He, Z.; Zhang, H.; Wang, T.; Wang, R.; Luo, X. Effects of Five Different Lactic Acid Bacteria on Bioactive Components and Volatile Compounds of Oat. *Foods* **2022**, *11*, 3230. [CrossRef] [PubMed]
8. Demarinis, C.; Verni, M.; Pinto, L.; Rizzello, C.G.; Baruzzi, F. Use of Selected Lactic Acid Bacteria for the Fermentation of Legume-Based Water Extracts. *Foods* **2022**, *11*, 3346. [CrossRef] [PubMed]
9. Sun, T.; Jiang, H.; Yang, K.; Li, X.; Wang, S.; Yao, H.; Wang, R.; Li, S.; Gu, Y.; Lei, P.; et al. Nutritional Function and Flavor Evaluation of a New Soybean Beverage Based on *Naematelia aurantialba* Fermentation. *Foods* **2022**, *11*, 272. [CrossRef] [PubMed]
10. Canonico, L.; Agarbati, A.; Zannini, E.; Ciani, M.; Comitini, F. Lentil Fortification and Non-Conventional Yeasts as Strategy to Enhance Functionality and Aroma Profile of Craft Beer. *Foods* **2022**, *11*, 2787. [CrossRef] [PubMed]
11. Pontonio, E.; Montemurro, M.; De Gennaro, G.V.; Miceli, V.; Rizzello, C.G. Antihypertensive Peptides from Ultrafiltration and Fermentation of the Ricotta Cheese Exhausted Whey: Design and Characterization of a Functional Ricotta Cheese. *Foods* **2021**, *10*, 2573. [CrossRef]
12. Ha, N.-I.; Mun, S.-K.; Im, S.-B.; Jang, H.-Y.; Jeong, H.-G.; Kang, K.-Y.; Park, K.-W.; Seo, K.-S.; Ban, S.-E.; Kim, K.-J.; et al. Changes in Functionality of *Tenebrio molitor* Larvae Fermented by *Cordyceps militaris* Mycelia. *Foods* **2022**, *11*, 2477. [CrossRef] [PubMed]
13. Nasrollahzadeh, A.; Mokhtari, S.; Khomeiri, M.; Saris, P.E.J. Antifungal Preservation of Food by Lactic Acid Bacteria. *Foods* **2022**, *11*, 395. [CrossRef] [PubMed]

foods

Article

Bioprocessing of Barley and Lentil Grains to Obtain In Situ Synthesis of Exopolysaccharides and Composite Wheat Bread with Improved Texture and Health Properties

Giuseppe Perri [1], Carlo Giuseppe Rizzello [2], Marco Ampollini [3], Giuseppe Celano [1], Rossana Coda [4,5], Marco Gobbetti [6], Maria De Angelis [1] and Maria Calasso [1,*]

1 Department of Soil, Plant and Food Science, University of Bari Aldo Moro, 70126 Bari, Italy; giuseppe.perri483@virgilio.it (G.P.); giuseppe.celano@uniba.it (G.C.); maria.deangelis@uniba.it (M.D.A.)
2 Department of Environmental Biology, University of Roma La Sapienza, 00185 Roma, Italy; carlogiuseppe.rizzello@uniroma1.it
3 Puratos Italia S.r.l., 43122 Parma, Italy; mampollinI@puratos.com
4 Department of Food and Nutrition, University of Helsinki, 00100 Helsinki, Finland; rossana.coda@helsinki.fi
5 Helsinki Institute of Sustainability Science, Department of Food and Nutrition, University of Helsinki, 00100 Helsinki, Finland
6 Faculty of Science and Technology, Free University of Bozen, 39100 Bozen, Italy; marco.gobbetti@unibz.it
* Correspondence: maria.calasso@uniba.it; Tel.: +39-080-544-2948

Citation: Perri, G.; Rizzello, C.G.; Ampollini, M.; Celano, G.; Coda, R.; Gobbetti, M.; De Angelis, M.; Calasso, M. Bioprocessing of Barley and Lentil Grains to Obtain In Situ Synthesis of Exopolysaccharides and Composite Wheat Bread with Improved Texture and Health Properties. *Foods* **2021**, *10*, 1489. https://doi.org/10.3390/foods10071489

Academic Editor: Claudia Monika Haros

Received: 24 May 2021
Accepted: 23 June 2021
Published: 27 June 2021

Abstract: A comprehensive study into the potential of bioprocessing techniques (sprouting and sourdough fermentation) for improving the technological and nutritional properties of wheat breads produced using barley and lentil grains was undertaken. Dextran biosynthesis in situ during fermentation of native or sprouted barley flour (B or SB) alone or by mixing SB flour with native or sprouted lentil flour (SB-L or SB-SL) by *Weissella paramesenteroides* SLA5, *Weissella confusa* SLA4, *Leuconostoc pseudomesenteroides* DSM 20193 or *Weissella confusa* DSM 20194 was assessed. The acidification and the viscosity increase during 24 h of fermentation with and without 16% sucrose (on flour weight), to promote the dextran synthesis, were followed. After the selection of the fermentation parameters, the bioprocessing was carried out by using *Leuconostoc pseudomesenteroides* DSM 20193 (the best LAB dextran producer, up to 2.7% of flour weight) and a mixture of SB-SL (30:70% *w*/*w*) grains, enabling also the decrease in the raffinose family oligosaccharides. Then, the SB-SL sourdoughs containing dextran or control were mixed with the wheat flour (30% of the final dough) and leavened with baker's yeast before baking. The use of dextran-containing sourdough allowed the production of bread with structural improvements, compared to the control sourdough bread. Compared to a baker's yeast bread, it also markedly reduced the predicted glycemic index, increased the soluble (1.26% of dry matter) and total fibers (3.76% of dry matter) content, giving peculiar and appreciable sensory attributes.

Keywords: barley; bioprocessing; baked goods; dextran; fibers; germination; glycemic index; lactic acid bacteria; lentil; sourdough

1. Introduction

Bioprocessing of cereals and legumes by using microbial inoculants, with and without the use of commercial enzymes, along with lactic acid bacteria (LAB) fermentation, are performed to improve their technological and functional properties, nutritional value, and consumer acceptability [1–5]. As one of the oldest and natural biotechnologies, sourdough fermentation by LAB offers multiple benefits for bread producers and consumers enhancing the overall quality of baked goods [3,6]. The pro-health effects of bioprocessing for making traditional and novel baked goods includes the microbial metabolites as well as the ability to affecting the levels and the bioavailability of several bioactive compounds, the capability to degrade anti-nutritional factors, the improvement of protein digestibility and the reduction of the glycemic index [6–10]. Among the well-known benefits of LAB fermentation on baked-food properties [11,12], some LAB strains have the ability to synthesize exopolysaccharides (EPS) [13–15]. EPS are polysaccharides commercially used in the food industry as emulsifiers, stabilizers, thickeners, gelling agents as well as for moisture retention [16,17]. In the last decades, the use of EPS-producing starters has received increasing interest from the bakery and cereal global industry since the hydrocolloidal nature of these carbohydrate polymers provides a natural replacement for commercial ones such as hydroxypropyl methylcellulose (HPMC) [18,19]. For several EPS, prebiotic effects have also been described [20]. Moreover, for EPS has been reported additional effects, such as anti-inflammatory, antitumor, or antioxidant properties [21] Among the EPS, the homopolysaccharides dextran and levan are those of relevance for the bakery industry [18,19,22].

α-D-glucans (dextran, mutan, alternan, and reuteran) are homopolysaccharides produced by extracellular glucansucrases, using sucrose as the substrate [23]. Dextran can be produced in situ in fermented products by LAB (e.g., *Leuconostoc* and *Weissella* spp.) or acetic acid bacteria and it has been generally recognized as safe (GRAS) by the Food and Drug Administration. Dextran has been successfully produced in cereals, pseudocereals and legumes by fermentation with selected LAB strains to obtain breads with an enhanced quality and prolonged shelf-life [24–28].

Nowadays, several studies have established the use of germinated grains as means to innovate and obtain foods with improved nutritional quality and content of bioactive compounds [29–32]. The potential of grain germination as an effective, low-cost and sustainable practice to enhance the levels of functional compounds and healthy properties, but also the digestibility, bioavailability and palatability of grains, has been highlighted before (for review see Benincasa et al. 2019) [31].

Germinated grains can be considered and labelled as malted or sprouted whole grain because of containing all the original bran, germ, and endosperm [31]. In the last decade, several composite breads containing sprouted grains have been developed and beyond the wheat [30], minor cereals such as barley [33], pseudocereals [33], and legumes [34,35] have been investigated as germinated ingredients for making functional bread. Although the addition of flour from sprouted grains to wheat bread improves its nutritional value, some detrimental effects on bread rheology and flavor can be observed [6,35], with respect to the conventional counterpart. To overcome this technological drawback, EPS can be used since they have been proven to be good texture modifiers [36,37].

In our previous study [35], we have demonstrated that the addition of 30% *w*/*w* sourdoughs from lentil and sprouted lentil flours enriched with bacterial dextran in white bread increased the fiber content, specific volume and decreased crumb hardness and staling rate compared to wheat control bread, without negative effects on its sensory characteristics.

In this study, we aimed to assess the potential of the bioprocessing of barley (*Hordeum vulgare*) and lentil (*Lens culinaris*) flours by lactic acid bacteria to obtain a sourdough enriched with dextran useful to maximize the texture and health-promoting properties of composite barley–lentil–wheat breads. The main reasons for utilization of barley and lentil grains in food are the health benefits deriving from their peculiar nutritional composition, resulting in wanted functional attributes and improved bread quality. In particular, studies have shown dietary fibers to have many health benefits. Barley is an excellent source of dietary fibers, especially β-glucan, known to have several physiological functions, including the improvement of lipid metabolism and the increase in satiety [38–41]. Lentils is a legume commonly consumed worldwide, particularly in the Mediterranean area, which have high protein content and low caloric value, contain phytochemicals and present antioxidant properties [42]. Despite their large use as food ingredients, the nutritional quality of lentils and derived products may be decreased by the content of ANFs [6]. Germination (sprouting) and fermentation of barley and lentil flours with the simultaneous in situ production of EPS are a potential opportunity to reduce the quality losses. The positive effect of sprouted barley and sprouted lentil flours as a nutritious and functional ingredient has been shown in previous studies [32,34,43].

To this aim, the suitability of native or sprouted barley alone or a blend of sprouted barley with native or sprouted lentil grains as a substrate for dextran synthesis by *Weissella paramesenteroides* SLA5, *Weissella confusa* SLA4, *Leuconostoc pseudomesenteroides* DSM 20193 or *Weissella confusa* DSM 20194, previously shown as good dextran producers [26,27,35,36,44], was assessed. The acidification and the viscosity increase during 24 h of fermentation with and without added sucrose were followed. The best sourdoughs were characterized and used for the manufacture of laboratory-scale composite wheat breads. The breads were evaluated for their rheological, nutritional and sensory properties.

2. Materials and Methods

2.1. Materials

The ingredients used in this study included barley grains (*Hordeum vulgare*, Caporal Grani s.a.s.) (carbohydrate 78.8% on dry matter (on d.m.), fibers 13.5% on d.m., protein (N × 5.70) 13.6% on d.m., fat 1.5% on d.m., moisture 11.1%), lentil grains (*Lens culinaris*, Caporal Grani s.a.s.) (carbohydrate 50.4% on d.m., protein 30.0% on d.m., fat 0.66% on d.m., fibers 23% on d.m., moisture 11%), wheat flour (*Triticum aestivum*, commercial wheat flour type "0", Puratos Italia s.a.s., protein 13% on d.m., fat 1.9% on d.m., fiber 2.3% on d.m., moisture 13.6%), fresh yeast (Puratos Italia), sucrose (Sigma Aldrich, St. Louis, MO, USA) and salt. Barley and lentil grains were sprouted according to the protocol described by Montemurro et al. [33]. Briefly, whole grains were prior disinfected by submersion in 1.25% *w/v* NaClO (seed:water ratio 1:5 *w/v*) for 30 min at room temperature, washed 20–30 min under tap water and then soaked in water at 16.5 °C for 24 h. Afterwards, grains were placed in a germination system (BioSnacky, Biokosma GmbH, Konstanz, Germany) in the dark until they began to sprout (rootlets length correspondent to ca. $\frac{3}{4}$ of the seed length). After germination, sprouted grains, including the rootlets which were not separated, were washed with distilled water and dried in experimental conditions comparable to those used for industrial malting of barley [33]. Flours were obtained from native and sprouted grains by a laboratory mill (IKA-Werke M20 GMBH, and Co. KG, Staufen, Germany). After milling, all the flours were sieved (mesh size 500 μm) to remove the coarse fraction and stored under vacuum until further analysis.

2.2. Liquid Sourdough Fermentation

Weissella paramesenteroides SLA5 and *Weissella confusa* SLA4, belonging to the Culture Collection of the Department of Soil, Plant and Food Sciences (University of Bari Aldo Moro, Italy) previously isolated from sprouted lentil flour [45], *Leuconostoc pseudomesenteroides* DSM 20193 and *Weissella confusa* DSM 20194 from DSMZ-German Collection of Microorganisms and Cell Cultures GmbH (Braunschweig, Germany), were used in this study as starters for preliminary dough fermentation. The strains were selected based on their confirmed dextran-producing capacity using 58.4 mM sucrose as a carbon source [26,27,35,44] and pro-technological properties and sensory characteristics on native or sprouted lentil-based substrates [35]. *Lactiplantibacillus plantarum* (formerly *Lactobacillus plantarum*) DPPMAB24W (culture collection of DiSSPA, University of Bari), was used as a non-EPS-producing control [44]. All LAB strains were maintained in 20% (*v/v*) glycerol as frozen stocks at −20 °C and routinely propagated in de Man Rogosa and Sharpe (MRS) broth at 30 °C (Oxoid, Basingstoke, Hampshire, England).

Selected strains were used as a single starter for fermentation of preliminary formulations of liquid doughs obtained by native or sprouted barley flour (B and SB, respectively) alone or by mixing SB flour with native or sprouted lentil flour (L or SL, respectively) in 40:60 and 30:70 ratios. The tested dough yield (DY, dough weight × 100/flour weight) were 500 for barley or sprouted barley dough alone and 333, 350, 400, 450 or 500 for the blends (Table S1). To support in situ formation of EPS, sucrose-supplemented (5% on dough weight, corresponding to 16% flour weight, f.w.) fermentations were carried out (namely EPS-positive sourdoughs, EPS POS) [36]. For each condition, an EPS-negative sourdough (namely EPS NEG) was prepared with the same starter strain but without sucrose addition [46]. For each formulation, a control dough (CT) without sucrose and without inoculum and a non-EPS-producing control dough (named B24W) added of sucrose and started by *L. plantarum* DPPMAB24W were prepared as described above [44]. Fermentations were carried out at 20 and 25 °C for 24 h. All the doughs were prepared in triplicate in sterile beakers using tap water and mixed manually for 5 min. Before and immediately after fermentation, samples were collected, stored at 4 °C and analyzed within 2 h. All the analyses were carried out in duplicate for each batch of sourdough (a total of six analyses for each type of sourdough). When used for sourdough fermentation LAB cells were cultivated in MRS broth supplemented with 58.4 mM sucrose overnight centrifuged (10,000× *g* for 10 min), washed in 50 mM phosphate buffer pH 7.0 (twice) and re-suspended in the water used for making the dough at the initial cell density of ca. 7 log cfu/g [44].

2.3. pH and Viscosity Measurement

The pH values of the sourdoughs were measured using a food pH meter equipped with a penetration probe (Model HI-99161, Hanna Instruments, Woonsocket, RI, USA) Viscosity of sourdoughs was measured before and after fermentation at 20 °C with a RheolabQC rheometer (Anton Paar, Graz, Austria), using 60 g of each dough mixed thoroughly. Viscosity was performed under different shear rates, from 2 to 100 1/s (up and down sweeps) [44] and the viscosity values at the shear rate of 100 1/s were compared Based on pH and viscosity [27], the DY of 333, the temperature of 20 °C, the SB-L and SB-SL flour blends at a 30:70 ratio, were chosen to produce sourdoughs fermented by *L. pseudomesenteroides* DSM 20193, which were further characterized.

2.4. Enumeration of Cultivable Bacteria and Yeasts

Enumeration of cultivable bacteria and yeasts was carried out according to methods previously described [47]. For each dough, aliquots of 20 g were added to 180 mL of sterile sodium chloride solution (0.9%, *w*/*v*), homogenized with a Stomacher for 180 sec and serially diluted. Appropriate dilutions were plated in selective culture media and supplements purchased from Oxoid (Basingstoke, Hampshire, United Kingdom) Total mesophilic aerobic microorganisms were enumerated using Plate Count Agar (PCA) media after incubating at 30 °C for 48 h under aerobic condition. LAB were estimated using modified MRS, containing 28 mM maltose, 5%, *v/v* fresh yeast extract, pH 5.6, and supplemented with cycloheximide (0.1 g/L), incubating the plates under anaerobiosis (AnaeroJar and AnaeroGen, Oxoid) at 30 °C for 48 h. *Enterobacteriaceae* were enumerated using a Violet Red Bile Glucose Agar (VRBGA) medium and plates were incubated at 37 °C for 24 h. Yeast cells were enumerated by using Wort agar supplemented with chloramphenicol (0.1 g/L), incubating the plates at 30 °C for 48 h. To confirm the microbiological counts, representative colonies from each medium were analyzed for morphology, motility, Gram staining reaction and catalase test.

2.5. Determination of Dextran, Sugars and Organic Acids

Before and after fermentation, the amount of dextran was determined by an enzyme-assisted method based on the enzymatic activity of the dextranase from *Chaetomium erraticum* (10,000 nkat/g) (Sigma-Aldrich, Darmstadt, Germany) and α-glucosidase from *Aspergillus niger* (1000 kat/g) (Megazyme, Ireland), as previously described by Katina et al. [37]. After freeze-drying, removal of free sugars and short oligosaccharides and inactivation of the enzymes, glucose in the sourdough supernatants was analyzed by high performance anion exchange chromatography with pulsed amperometric detection (HPAEC-PAD), using glucose (Merck, Darmstadt, Germany) as standard and 2-deoxy-D-galactose (Sigma-Aldrich,

Dorset, UK) as the internal standard for quantification [37]. Results were calculated as the sum of anhydro-glucose using a corrector factor of 0.90.

For sugar analysis, freeze-dried sourdough before and after fermentation was treated to inactivate enzymes and microbes [35,44] and to remove any polymeric molecules. The samples were analyzed by HPAEC-PAD as reported by Xu et al. [43]. Results are expressed as % on a flour weight basis.

The total titratable acidity (TTA) was measured on 10 g of dough diluted with 90 mL of sterile sodium chloride (0.9% *w*/*v*) solution titrated with 0.1 mol/L NaOH until pH achieved 8.3. TTA was expressed as the total NaOH amount (mL).

Organic acid analysis (lactic acid and acetic acid contents) in the extracts from sourdoughs before and after fermentation was performed using commercial kits, K-DLATE and K-ACET (Megazyme, Wicklow, Ireland) kits. The quotient of fermentation (FQ) was calculated as the molar ratio between lactic and acetic acids.

2.6. Bread Making Trials

Experimental breads were manufactured at the pilot plant of Puratos Italia (Ceparana, La Spezia, Italy). Three types of breads were prepared: control wheat bread manufactured using wheat flour fermented by baker's yeast alone (CWB); sourdough wheat bread (SWB) manufactured using sprouted barley–sprouted lentil sourdough (SB-SL SWB EPS NEG); sourdough wheat bread manufactured using dextran-containing sprouted barley–sprouted lentil sourdough (SB-SL SWB EPS POS). Selected sourdoughs were used in baking at 30% of the dough weight corresponding to a 15% of wheat flour substitution. The selection of the percentage of replacement was based on a calculation (nutritional composition) to obtain a 3% dietary fibers content which allows the nutrition claim "source of fibers" [48].

The required amount of water for the breads was previously determined by a Brabender farinograph (Brabender GmbH & Co. KG, Duisburg, Germany). The amount of flour and water was the same in CWB and SWB breads (DY 162). All breads were manufactured according to a two-stage protocol, which is routinely used in artisanal and industrial bakeries [47]. In stage I, the SB-SL blended flours obtained from sprouted barley and lentil grains were fermented with *L. pseudomesenteroides* DSM 20193 for 24 h at 20 °C in sucrose-supplemented or not fermentation as described before. In stage II, SB-SL EPS NEG or EPS POS sourdoughs were mixed with all ingredients (wheat flour, water and baker's yeast) in a mixer vessel (Sottoriva S.p.a Group). Baker's yeast was added at a percentage of 1.1% *w*/*w*. The doughs were divided into pieces of uniform weight (500 g), rested in pans for 20 min at 25 °C and relative humidity (RH) of 75% and leavened in a fermentation chamber (Zucchelli Forni S.p.a) for 60 min at 30 °C and RH 85%. All types of breads were baked at 220 °C for 30 min in a rotating rack oven (Zucchelli Forni S.p.a). Five replicates for each type of bread were carried out on two different days. All the analyses were carried out in duplicate for three replicates of bread (a total of six analyses for each type of bread). The resulting breads were allowed to cool to room temperature (25 °C) for 2 h before analyses and weighing.

2.7. Bread Technological Characterization

Dextran-containing SWB (EPS POS) were compared to EPS-negative SWB (EPS NEG) and to the control breads [36]. After cooling, loaves were weighed, and loaf volume measured by millet-seed displacement [49]. The specific volume of the bread was calculated by dividing the loaf volume (mL) by the corresponding loaf weight (g). The percent bake loss of the breads was also evaluated (% bake loss = (dough weight−bread weight) * 100/dough weight). Bread was packed in polypropylene micro perforated bags and stored for 7 days at room temperature. For each bread, slices (2 cm) were used for texture profile analysis (TPA) after days 1 and 7 of storage at room temperature using a texture analyzer (TA, TA-XT2i, Stable Micro Systems Ltd., UK) fitted with a cylinder probe (diameter 36 mm) [50]. Data were recorded using a TPA analyzer Stable Micro Systems software exponent (version 5.0.9.0). All measurements were performed in triplicate by two compression cycles, test speed of 1 mm/s, and 30% compression [51]. The texture properties determined were: hardness (maximum peak force), springiness (ratio of a product's original height) cohesiveness (the area of work during the second compression divided by the area of work during the first compression) and resilience (ratio of the first decompression area to the first compression area). The effect of sourdough on the staling rate was studied as the increase in hardness (staling rate = (hardness [day 7−day 1]/days of storage)) after storage for 7 days [27]. The characterization of the crumb structure was performed on two bread slices taken from the center of different loaves. The gas cell number of the breadcrumbs was evaluated 24 h after production using image analysis technology [47]. Images of slices of breads (control, SB-SL SWB EPS NEG and SB-SL SWB EPS POS) were scanned full-scale at 300 dots per inch using an image scanner (Amersham Pharmacia Biotech, Uppsala, Sweden). A threshold method was used for differentiating gas cells [52], and the images were analyzed in grey scale (0–255) using the UTHSCSA ImageTool program (Version 2.0, University of Texas Health Science Centre, San Antonio, Texas, available by anonymous FTP from maxrad6.uthscsa.edu). Analysis was carried out on two sub-images with a resolution of 500 × 500 pixels (field of view) selected from within the bread slice.

2.8. Breads Nutritional Characterization

The in vitro starch hydrolysis index (HI) was determined on each type of bread by an enzyme-assisted procedure that mimicked the in vivo digestion [53]. For this analysis, aliquots of breads containing 1 g of starch were subjected to enzymatic digestion. The released glucose content was determined in each sample using a glucose oxidase kit (Megazyme International, Bray, Co., Wicklow, Ireland). Data are expressed as the % of potentially available starch hydrolyzed after 180 min. The predicted glycemic index (pGI) value was then calculated using the equation: $pGI = 0.549 \times HI + 39.71$ [54] with white wheat bread as a reference (HI = 100).

Total (TDF) and insoluble (IDF) dietary fiber were determined by method AOAC 2011.25 [55]. Soluble (SDF) dietary fiber was calculated as a difference between TDF and IDF according to Tobaruela et al. [55].

2.9. Volatile Organic Compounds Profile of Breads

Evaluation of volatile organic compounds (VOCs) was carried out on a Clarus 680 gas chromatography (Perkin Elmer, Beaconsfield UK) equipped with a Rtx-Wax column (30 m × 0.25 mm i.d., 0.25 μm film thickness) (Restek Superchrom, Milano, Italy) coupled to a single-quadrupole mass spectrometer Clarus SQ8MS (Perkin Elmer). The SPME-GC–MS (solid phase micro-extraction gas chromatography–mass spectrometry) protocol and the identification of volatile compounds were performed according to previous reports, with minor modifications [35,56,57]. An amount of 0.750 g of crushed bread (crumb and crust) samples were placed into 20 mL glass vials and added with 10 μL of 4-methyl-2-pentanol (final concentration of 33 mg/L), as the internal standard. A PAL COMBI-xt autosampler (CTC CombiPAL, CTC Analytics AG, Zwingen, Switzerland) was used to standardize the extraction procedure. The samples were then equilibrated for 10 min at 60 °C. The SPME divinylbenzene/carboxen/polydimethylsiloxane (DVB/CARB/PDMS) fiber (Supelco, Bellefonte, PA, USA) was exposed to the sample headspace for 50 min and finally the fiber was inserted into the injection port of the GC at 230 °C to be thermally desorbed and to separate the head space volatile organic compounds. The temperature program was: 35 °C for 8 min, then programmed at 4 °C/min to 60 °C, at 6 °C/min to 160 °C, and finally at 20 °C/min to 200 °C, which was maintained for 15 min. Injections were carried out in splitless mode, and helium (1 mL/min) was used as the carrier gas. The single-quadrupole MS was used to detect the different compounds. The source and transfer line temperatures were 250 and 230 °C, respectively. The MS detector system operated in scan mode with a mass-to-charge ratio interval 35 to 300 Da [57]. Each chromatogram was analyzed for peak identification by comparing (i) retention times with those of pure compounds for HPLC (Sigma-Aldrich, St. Louis, MO, USA) and (ii) experimental mass spectra with those of the

National Institute of Standards and Technology database (NIST/EPA/NIH Mass Spectral Library with Search Program, data version NIST 05, software version 2.0d). A peak area threshold of >1,000,000 and a match criterion of >85% were used for VOC identification followed by manual visual inspection of the fragment patterns when required. Quantitative data for the compounds identified were obtained by the interpolation of the relative areas versus the internal standard area.

2.10. Bread Sensory Analysis

Sensory analysis of experimental breads after 2–6 h of cooling was carried out by 10 trained panelists (5 males and 5 females, mean age: 30 years, range: 18–54 years) [47]. After a roundtable discussion about the sensory attributes, 15 were selected as the most frequently recognized by all the assessors, which were included in a panel score sheet for the quantitative evaluation using a scale from 0 to 10, with 10 the highest score. Visual and tactual perception (color of crust and crumb, elasticity, consistency, friability), taste (acidic taste, sweetness, salty, legume flavor, bitter flavor), smell perception (acidic odor, caramel-like odor), chewing (chewiness, wetness), and overall aroma were chosen as attributes to characterize the breads. A quarter of each piece of bread sample (including crust and crumb) (1.5 cm thick) were served in random order on a plastic plate encoded with an alpha-numeric code and evaluated by all panelists. Final scores for each attribute were calculated as the means of the data collected during the evaluation.

2.11. Statistical Analysis

Experimental data were subjected to analysis of variance (ANOVA); pair-comparison of treatment means was achieved using Tukey's procedure at $p < 0.05$, using the statistical software Statistica 7.0 for Windows. For each bread, the measured physicochemical (pH, TTA, VOCs), technological, nutritional and sensory characteristic data were used as variables for the Principal Component Analysis (PCA).

3. Results and Discussion

3.1. pH and Viscosity in Preliminary Liquid Sourdough Formulations

Preliminarily, different doughs were produced using native barley (B) or sprouted barley (SB) flour or blend of SB with native (L) or sprouted lentil (SL) flour at different ratio (40:60 or 30:70% *w*/*w*) during sucrose-supplemented (5% on dough weight, corresponding to 16% flour weight, f.w.) or without sucrose addition fermentation. The germination protocols used in this study to obtain sprouted flours from barley and lentil grains were set-up in previous studies [33,35,45]. The doughs were singly inoculated with the dextran-producer *W. confusa* SLA4, *W. paramesenteroides* SLA5, *L. pseudomesenteroides* DSM 20193 and *W. confusa* DSM 20194 strains [26,27,35,36,44], and by *L. plantarum* DPPMAB24W, a dextran not producer strain [44], and fermented for 24 h at 20 or 25 °C. These trials were performed to select the best strains and parameters leading the significant increase in the viscosity in the fermented doughs. The pH and viscosity were analyzed to show the influence of acid and dextran production on the rheological properties of the sourdoughs. Viscosity formation during fermentation indicates the presence of large molecules with water binding properties and correlates with in situ EPS synthesis [18,19]. After 24 h of fermentation, the values of pH became lower than ca. 4.3 in almost all the sourdoughs (Table S2). All the sourdoughs exhibited a shear thinning behavior. The viscosity of all sourdoughs increased after 24 h of fermentation compared to the *L. plantarum* DPPMAB24W fermented doughs. Both the *Weissella* spp. and *L. pseudomesenteroides* sucrose-supplemented sourdoughs exhibited significantly higher viscosities than their EPS-negative counterparts. When fermented with selected strains, sourdoughs containing only native or sprouted barley flour showed low viscosity increase, presumably due to low dextran production, and, therefore, were not further considered. This effect might be due to the high content of maltose or other sugar acceptors which might have favored the formation of low-molecular mass oligosaccharides from sucrose rather than dextran, as typically occurs in cereal substrate, leading to a low viscosity formation [18,19,58,59] and/or to the low pH reached in those sourdoughs (pH average ca. 3.78), also affecting viscosity.

The viscosity of the blends of sprouted barley and native (SB-L) or sprouted lentil (SB-SL) was ca.0.3 Pa*s before fermentation, with similar values between EPS POS and EPS NEG doughs, while ranged from 0.15 to 0.96 Pa *s in EPS NEG and from 0.17 to 3.33 Pa*s in EPS POS sourdoughs after fermentation, with the highest viscosity recorded for SB-SL EPS POS indicating the thickening ability of EPS [44] and the successful outcome of the mixture cereal/legume. Blend of SB-L and SB-SL flours at 30:70% *w/w* ratio and fermented by *L. pseudomesenteroides* DSM 20193 at 20 °C per 24 h, DY333, showed the highest viscosity (2.90 and 3.30 Pa/s for SB-L and SB-SL sourdoughs, respectively) (Table S2). Hence, SB-L and SB-SL blends at 30:70% *w/w* ratio fermented by *L. pseudomesenteroides* DSM 20193 at 20 °C per 24 h, DY333, were selected and further characterized.

3.2. Microbial Growth and Acidification in the Liquid Sourdough Formulations

Microbial profiles of selected sourdoughs changed after the fermentation, but no significant differences were observed between EPS POS and EPS NEG sourdoughs. The initial cell count of lactic acid bacteria and total mesophilic aerobic bacteria in all sourdoughs was approximately 6.7 log cfu/g (Table 1).

In agreement with earlier studies [35], LAB cell density increased ca 2.5 log cycle after 24 h of fermentation, independently of sucrose presence, indicating that SB-L and SB-SL are good substrates for the growth of the strain used. The mesophilic bacteria count in SB-L sourdough was slightly ($p < 0.05$) lower than SB-SL sourdoughs after 24 h Before fermentation, *Enterobacteriaceae* were higher than 5 log cfu/g in SB-SL sourdoughs, while were ca. 2 log cfu/g in all SB-L sourdoughs. In all cases, any significant difference (p <0.05) among EPS POS and EPS NEG sourdough was observed. After 24 h, *Enterobacteriaceae* grew up to four log cycles. The highest increase was observed in SB-L sourdoughs. The initial cell density of yeasts in all sourdoughs was lower than 3 log cfu/g. In the SB-SL sourdoughs, the yeast cell density was 1 log cfu/g higher than SB-L doughs. After fermentation, yeasts were not detected. Before fermentation, the pH of all the doughs was around 6.15; it decreased by ca. 1.7 units after 24 h of fermentation, reaching values in the range from 4.4 to 4.6 (Table 1). The fermentation drove to an increase in the TTA in all the sourdough types. TTA values varied from 11.6 to 14.2 mL, with the highest value for SB-SL EPS POS. Fermentation caused an increase in the concentration of both lactic and acetic acids (Table 1). Lactic acid concentration ranged from 18 to 22.0 mmol/kg and was slightly higher in both SB-SL sourdoughs compared to SB-L doughs. The highest value for acetic acid was found in SB-SL EPS POS. The concentration of lactic and acetic acid in sourdoughs plays a central role in the flavor and taste of sourdough bread. Sucrose addition, through the liberation of fructose, facilitated the production of acetic acid while lactic acid concentration decreased, which was also represented by the fermentation quotient (FQ). It has been described that *Leuconostoc* spp. can reduce the released fructose to mannitol, taking part at the acetic acid formation [60]. FQ was calculated as the molar ratio between lactic acid and acetic acid, and it ranged between 2.59 in SB-SL EPS POS and 4.11 in SB-L EPS NEG, indicating that the addition of sucrose lead to a lower FQ.

Table 1. Microbial growth and acidification [1] of control (EPS NEG) and sucrose-supplemented (16% of flour weight) doughs (DY 333) before and after 24 h of fermentation at 20 °C started by *Leuconostoc pseudomesenteroides* DSM 20193 (initial cell density of ca. 7 log cfu/g). The table shows microbial cell density (log cfu/g) of LAB, total mesophilic bacteria, *Enterobacteriaceae* and yeasts, pH, TTA (mL), lactic acid and acetic acid concentrations (mmol/Kg dough) and their ratios (FQ). SB-L EPS NEG, sprouted barley-raw lentil dough (30:70 ratio); SB-L EPS POS, sucrose-supplemented sprouted barley-raw lentil dough (30:70 ratio); SB-SL EPS NEG, sprouted barley–sprouted lentil dough (30:70 ratio); SB-SL EPS POS, sucrose-supplemented sprouted barley–sprouted lentil dough (30:70 ratio).

Sample Code	Lactic Acid Bacteria	ΔLog [a]	Total Mesophilic Bacteria	*Enterobacteriaceae*	Yeasts	pH	TTA	Lactic Acid	Acetic Acid	FQ [b]
					T0 h					
SB-L EPS NEG	6.81 ± 0.107 a		6.62 ± 0.117 b	1.77 ± 0.097 d	<1 c	6.27 ± 0.113 a	4.18 ± 0.093 d	n.d. [c]	n.d.	-
SB-L EPS POS	6.84 ± 0.058 a		6.53 ± 0.143 b	1.53 ± 0.176 d	1.14 ± 0.176 b	6.22 ± 0.044 a	4.57 ± 0.117 d	n.d.	n.d.	-
SB-SL EPS NEG	6.78 ± 0.192 a		6.74 ± 0.117 b	5.44 ± 0.097 b	2.35 ± 0.087 a	6.07 ± 0.095 a	5.4 ± 0.135 c	n.d.	n.d.	-
SB-SL EPS POS	6.71 ± 0.144 a		6.72 ± 0.033 b	5.53 ± 0.176 b	2.29 ± 0.014 a	6.15 ± 0.085 a	5.6 ± 0.110 c	n.d.	n.d.	-
					T24 h					
SB-L EPS NEG	9.51 ± 0.044 a	2.71 ± 0.035 a	8.87 ± 0.111 a	4.81 ± 0.063 c	<1 c	4.62 ± 0.115 b	11.6 ± 1.241 b	18.00 ± 0.10 c	4.38 ± 0.10 c	4.11± 0.18
SB-L EPS POS	9.35 ± 0.148 a	2.51 ± 0.055 a	9.03 ± 0.201 a	5.01 ± 0.212 c	<1 c	4.53± 0.271 b	12.4 ± 0.115 b	17.20 ± 0.72 c	5.45 ± 0.22 b	3.16 ± 0.13
SB-SL EPS NEG	9.43 ± 0.152 a	2.65 ± 0.082 a	9.14 ± 0.257 a	5.81 ± 0.091 a	<1 c	4.44 ± 0.095 bc	13.6 ± 0.241 ab	22.00 ± 0.10 a	5.40 ± 0.10 b	4.07 ± 0.18
SB-SL EPS POS	9.55 ± 0.074 a	2.84 ± 0.112 a	9.23 ± 0.151 a	6.01 ± 0.151 a	<1 c	4.42 ± 0.045 bc	14.2 ± 0.115 a	20.20 ± 0.10 b	7.80 ± 0.22 a	2.59 ± 0.13

[1] Data are mean values ± standard deviation. a–d, means within a column with different letters are significantly different ($p < 0.05$). [a] The increase in LAB cell density after 24 h of fermentation. [b] FQ, fermentation quotient. [c] n.d., not detected.

3.3. Sugars and Dextran Content in the Liquid Sourdough Formulations

The amount of extractable free sugars in all sourdoughs before and after 24 h of fermentation is shown in Figure 1.

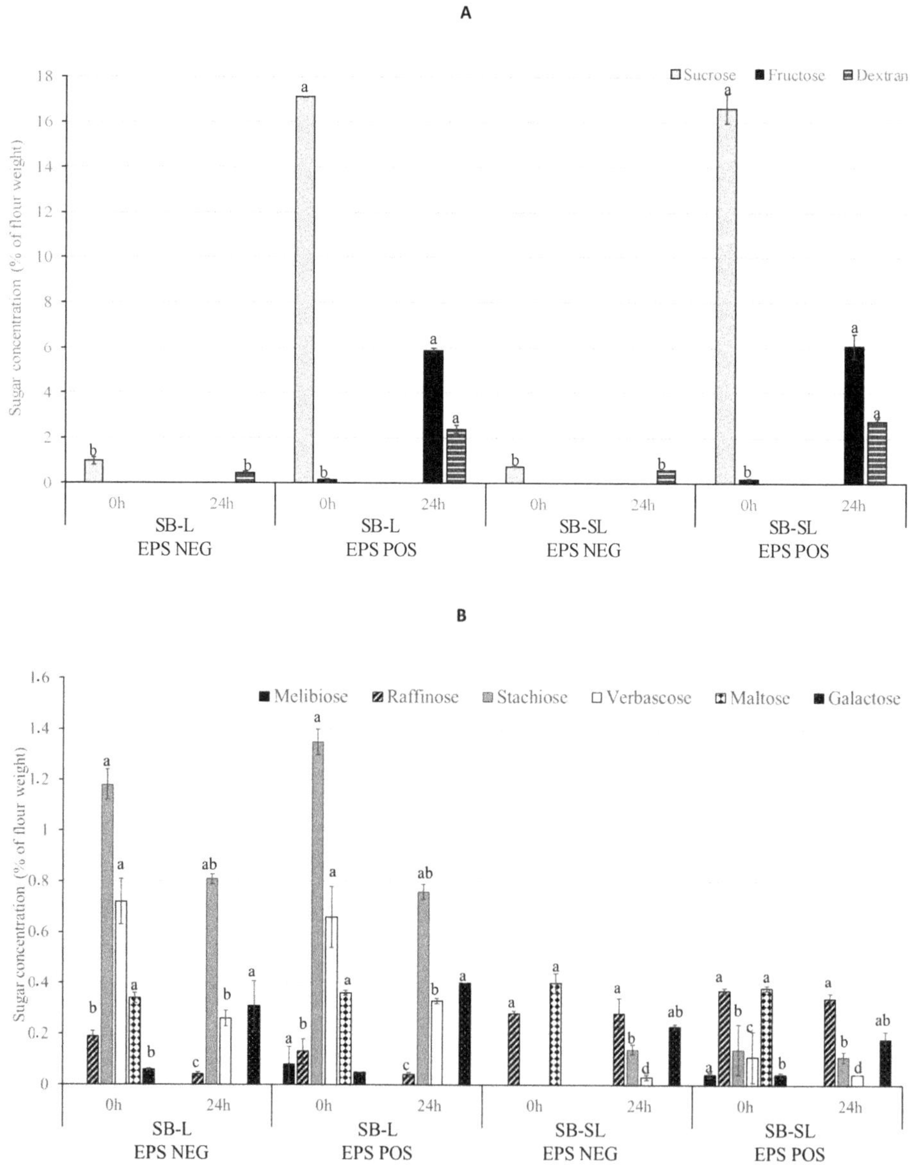

Figure 1. Sugars and dextran concentration (% of flour weight) in control and sucrose-supplemented (16% of flour weight) doughs (DY 333) before and after 24 h of fermentation at 20 °C started by *Leuconostoc pseudomesenteroides* DSM 20193 (initial cell density of ca. 7 log cfu/g). Panel (**A**) shows sucrose, fructose and dextran concentration; panel (**B**) shows melibiose, raffinose, stachyose, verbascose, maltose and galactose concentration. SB-L EPS NEG, sprouted barley-native lentil dough (30:70 ratio); SB-L EPS POS, sucrose-supplemented sprouted barley-native lentil dough (30:70 ratio); SB-SL EPS NEG, sprouted barley–sprouted lentil dough (30:70 ratio); SB-SL EPS POS, sucrose-supplemented sprouted barley–sprouted lentil dough (30:70 ratio). For each sugar, bars with different superscript letters differ significantly ($p < 0.05$).

SB-L and SB-SL doughs naturally contained ca. 0.97–0.74% f.w. of sucrose and 0.34–0.4% f.w. of maltose, respectively, (Figure 1A,B) and the supplementation of sucrose (5% on dough weight, corresponding to 16% f.w.) was necessary to obtain enough EPS yield by *L. pseudomesenteroides* DSM 20193 [15]. After fermentation, sucrose was totally used by *L. pseudomesenteroides* DSM 20193. The low amount of dextran formed in EPS NEG doughs (0.43–0.58% flour basis) may be formed from the sucrose existent in the native flour. *L. pseudomesenteroides* DSM 20193 in SB-L and SB-SL sucrose enriched doughs produced up to 2.7% f.w. of dextran, which is less than the theoretical. Based on sucrose addition, up to 8% f.w. of dextran could be formed. Previously, dextran-forming *L. pseudomesenteroides* DSM 20193 produced 3.6% of dextran on a wet weight base in faba bean sourdough upon the addition of 25% f.w. sucrose [44,46]. The yield of EPS is resultant of sucrose content, the presence of acceptors molecules, such as maltose, and the starter strain and growth conditions, amongst others [61]. Based on the sugar analysis, glucose released from added sucrose was utilized by *L. pseudomesenteroides* DSM 20193 only partly for dextran production. The glucansucrase acted on sucrose-synthesizing glucan and liberating fructose, as confirmed by the residual fructose in all sucrose-containing doughs, in the range from 5.89% to 6.09% [60].

Together with the increase in viscosity, the capacity of *L. pseudomesenteroides* DSM 20193 to reduce different Raffinose Family Oligosaccharides (RFO), such as raffinose, stachyose and verbascose, during fermentation of the blend of SB-L and SB-SL flours was investigated, since they are one of the main limits to the use of legumes in animal and human nutrition [62]. RFOs were detected at different extends in all doughs before fermentation (Figure 1B), without significant difference among EPS NEG and EPS POS samples. Raffinose, as expected, was not affected by germination and was contained in both the blends [51]. Raffinose cannot be broken down by human digestive enzymes but can be utilized by anaerobic bacteria in the large intestine, thus, causing the production of flatus gases and gastrointestinal discomfort. Nevertheless, stachyose and verbascose concentration was very low in SB-SL compared to SB-L (Figure 1B), which can be attributed to the more intensive effect of different endogenous α-galactosidases during sprouting. RFOs can be enzymatically hydrolyzed by LAB during fermentation [63,64], thus, increasing product digestibility and reducing digestive discomfort [65]. Fermentation with *L. pseudomesenteroides* DSM 20193 decreased the RFOs content in both SB-L and SB-SL sourdoughs and particularly in the SB-L flour sourdough. Fermentation significantly increased ($p < 0.05$) galactose, a degradation product of RFO by α galactosidase, in all doughs.

3.4. Breads Characterization

Considering the higher in situ production of dextran and the balanced acid production, SB-SL sourdoughs were selected for subsequent baking trials. The content of dextran synthesized in situ in SB-SL sourdough by *L. pseudomesenteroides* DSM 20193 was 2.73% of flour basis and consequently the correspondent final breads contained 0.36% flour basis of dextran, which is in the range (0.1–2%) of the number of commercial hydrocolloids such as CMCHPMC, GG and κ-CAR applied in baking [66]. Fermentation and bread making were carried out applying the process parameters typical for type I sourdough fermentation [47]. The addition of dextran-containing sourdoughs into bread significantly affected the technological features compared to the control bread (Figure 2).

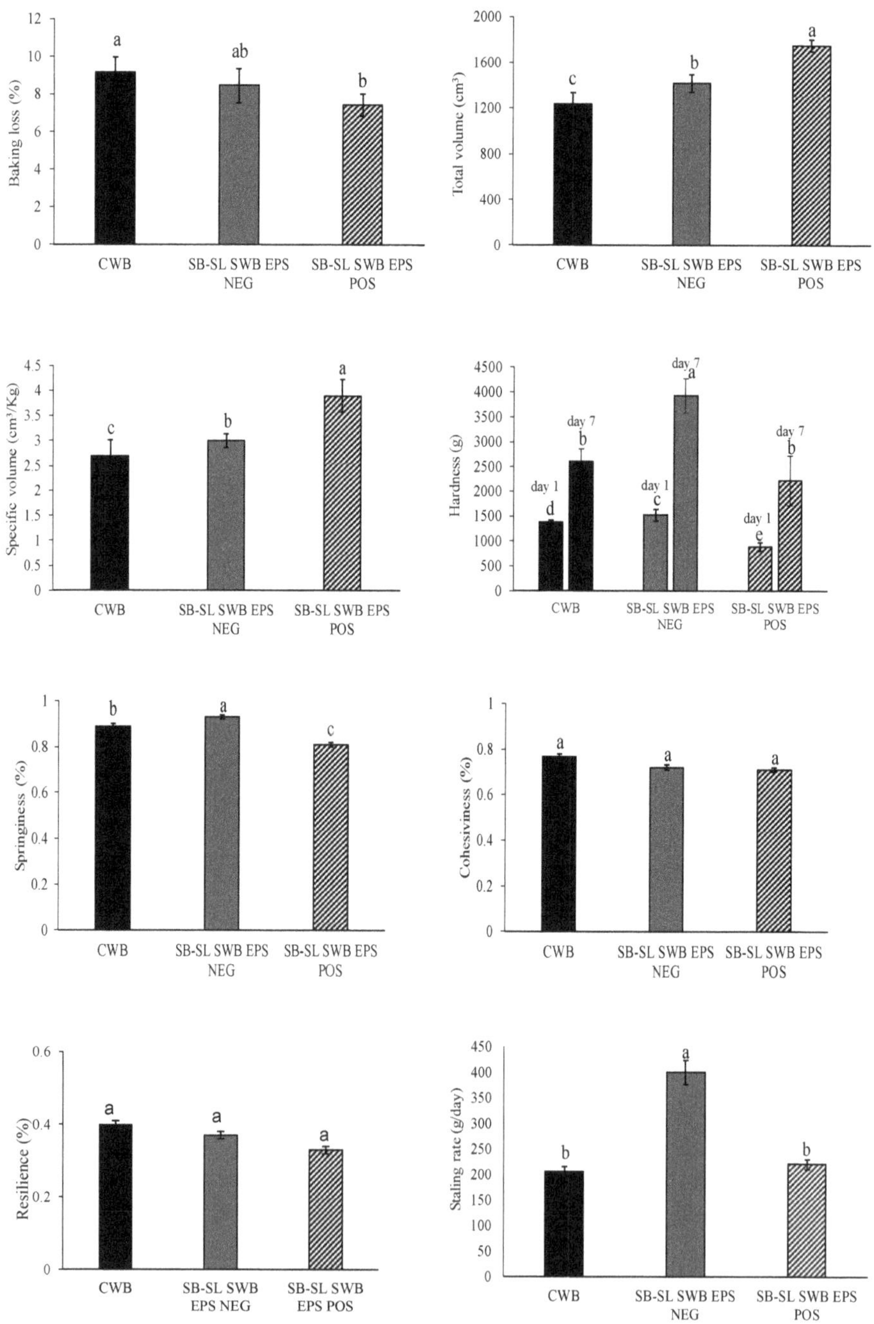

Figure 2. Technological characteristics of control and sourdough breads (DY 162). CWB, control wheat bread started with baker's yeast; SB-SL SWB EPS NEG, bread containing wheat flour added with 30% (*w*/*w*) sprouted barley–sprouted lentil (30:70% *w*/*w*) sourdough fermented by *Leuconostoc pseudomesenteroides* DSM 20193; SB-SL SWB EPS POS, bread containing wheat flour added to 30% (*w*/*w*) dextran-containing sprouted barley–sprouted lentil (30:70% *w*/*w*) sourdough fermented by *L. pseudomesenteroides* DSM 20193. The initial cell density of the strain was ca. 7 log cfu/g. The sourdough was fermented at 20 °C for 24 h. For each bread, bars with different superscript letters differ significantly ($p < 0.05$).

Compared to CWB, the substitution of wheat with dextran-enriched or EPS NEG SB-SL fermented doughs, resulted in a significant ($p < 0.05$) increase in the bread volume accompanied by a decrease in baking loss (37% of baking loss in SB-SL SWB EPS POS bread), leading to a superior quality of the wheat bread. The use of dextran-enriched sourdough increases the specific volumes compared to breads fortified with the corresponding EPS NEG counterpart. The presence of dextran significantly decreased the hardness of the composite bread. After 1 day, the hardness of bread crumb was significantly ($p < 0.05$) lower in SB-SL SWB EPS POS (892 $\pm$ 79 g) when compared to CWB and their respective negative EPS NEG counterparts. Furthermore, the hardness of SB-SL SWB EPS NEG was higher ($p < 0.05$) than CWB. After 7 days of storage, the hardness of dextran-enriched SWB (2218 g) was lower but not significantly different (p >0.05) compared to CWB (2610 g), while strongly higher was the hardness of SB-SLS WB EPS NEG (3927 g) (Figure 2). It is known that the use of grains different from wheat or legumes markedly affect the properties of baked goods [6], usually leading to a weak dough structure and baking quality, decreased bread volume and elasticity of the crumb and to an increased hardness of the loaves [38,67]. Springiness is how well a product physically springs back after it has been deformed. A lower value of springiness was observed for SB-SL SWB compared to CWB. Cohesiveness is how well the bread withstands a second deformation relative to its resistance under the first deformation. Resilience is how well a product fights to regain its original height. Moreover, for these two parameters, SB-SLSWB EPS POS showed the lowest value (Figure 2). The staling rate, as determined by TPA analysis, was affected ($p < 0.05$) by the addition of sourdoughs and followed the order SB-SL SWB EPS NEG > SB-SL SWB EPS POS > CWB. Bread staling involves changes in both crumb and crust and is a complex and physicochemical irreversible phenomenon involving starch amylopectin recrystallization and water redistribution [68]. The anti-staling effect of hydrocolloids, such as dextran, could be related to dextran polymers competing for water, so that less water molecules are available for the development of amylopectin crystallites [69]. Similar findings were reported in recent works studying the effect of the addition of dextran to faba bean–wheat, pearl millet–wheat and lentil-or sprouted lentil–wheat composite bread [26,27,35]. Image analysis technology was performed on crumb grain of bread slices after 24 h of storage to provide a more detailed view of the bread texture, Digital images were pre-processed to detect crumb cell total area by a binary conversion (black/white pixels). The gas cell-total area (corresponding to the black pixel ratio) of the breads containing sourdough were significantly ($p < 0.05$) higher than CWB. Crumb cell detection of bread slice portions showed that no significant difference in the mean area of gas cells could be observed between SB-SL SWB EPS NEG and SB-SL SWB EPS POS, having values of 56.4 $\pm$ 0.11 and 59.1 $\pm$ 0.08% pixels, respectively. The pH of the CWB bread crumb after baking was 5.92 $\pm$ 0.111, higher ($p < 0.05$) than both the sourdough breadcrumbs (4.68 $\pm$ 0.202 and 4.72 $\pm$ 0.137 for EPS POS and EPS NEG, respectively). All doughs containing sourdoughs presented TTA values significantly higher ($p < 0.05$) (13.6 $\pm$ 0.303 and 15.2 $\pm$ 0.093 mL NaOH 0.1N for SB-SL SWB EPS NEG and SB-SL SWB EPS POS breads, respectively) than the control bread (5.20 $\pm$ 0.404).

3.5. Dietary Fiber and Starch Hydrolysis Index

Regarding the nutritional value, the incorporation of 30% EPS NEG and dextran-containing SB-SL sourdoughs led to a significant increase in total dietary fibers (TDF) compared to CWB, which may be beneficial for consumers' health (Figure 3).

The dextran-containing SB-SL bread had the highest content of SDF compared to the corresponding EPS NEG and CWB breads which can be due to the dextran and oligosaccharides produced. According to EC Regulation No 1924/2006 [48] on nutrition and health claims on food products, the composite grain breads can be labelled as "source of fiber", since containing at least 3 g of fiber/100 g of bread.

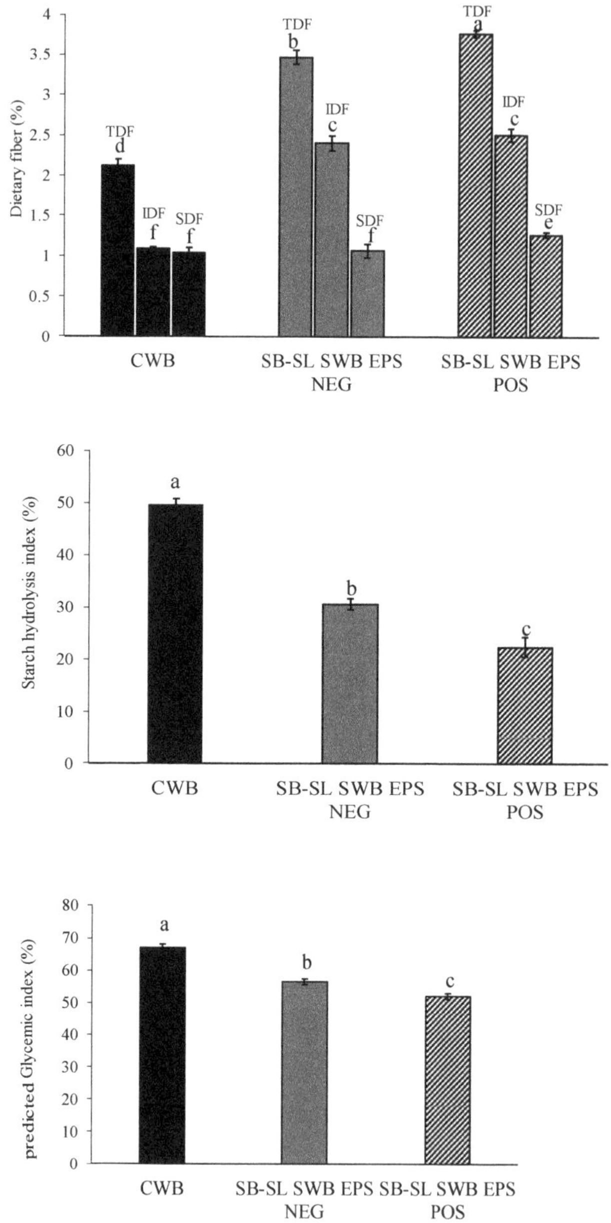

Figure 3. Nutritional characteristics of control and sourdough breads (DY 162). CWB, control wheat bread started with baker's yeast; SB-SL SWB EPS NEG, bread containing wheat flour added with 30% (*w*/*w*) sprouted barley–sprouted lentil (30:70% *w*/*w*) sourdough fermented by *Leuconostoc pseudomesenteroides* DSM 20193; SB-SL SWB EPS POS, bread containing wheat flour added to 30% (*w*/*w*) dextran-containing sprouted barley–sprouted lentil (30:70% *w*/*w*) sourdough fermented by *L. pseudomesenteroides* DSM 20193. fibers. The initial cell density of the strain was ca. 7 log cfu/g. The sourdough was fermented at 20 °C for 24 h. TDF, total dietary fibers; IDF, insoluble dietary fibers; SDF, soluble dietary. For each bread, bars with different superscript letters differ significantly ($p < 0.05$).

The predicted glycemic index (pGI) of breads containing sourdough from sprouted grains was lower than the reference bread. It is well known that sprouting can increase the release of reducing sugars and, consequently, HI and pGI thanks to starch hydrolysis by alpha-amylase and beta-amylase enzymes [70,71]. However, the germination conditions used in this work (16 °C and 3 days), such as those reported previously [32], may be useful to produce sprouted flours with a low glycemic index. Moreover, sourdough fermentation, per se, decreased the HI through the synthesis of organic acids [72]. The pGI value of dextran-containing SB-SL SWB bread (52%) was lower ($p < 0.05$) than the corresponding EPS NEG (56.5%) and the CWB (66.9%) (Figure 3). This could be attributed to the effects of β-glucans soluble fibers from barley on reducing GI [39] which the content here could be 0.05–0.27% of dough weight (based on 2–10% beta-glucan content of barley) [73] and high concentration of fibers of the legume flours [47]. β-glucans are non-starch polysaccharides composed of glucose molecules in long linear polymers with mixed β-(1→4) and β-(1→3) links (from 30% to 70%). Their *MW* ranges from 50 to 2000 kDa. The mixed linkages are important for their solubility and viscosity properties, and their viscosity is a function of the content of dissolved β-glucans, and of their *MW* [74], and further depends on differences on raw materials and processing [39]. Although the β-glucan content, solubility, viscosity, and *MW* were not considered in this study, some considerations should be made. The fermentation and bread-making process partially affects β-glucan level and *MW* [75,76]. A significant degradation of β-glucan occurring during fermentation may reduce the nutritional functionality of the residual β-glucan since it is supposed to be dependent on *MW*. High *MW* and insoluble β-glucan are considered positive for various physiological functions [38,40,74], comprising the reduction of glycemic response by increasing the viscosity of food. Foods with increased viscosity have been demonstrated to increase gastric transition time and slow absorption [38,77].

3.6. Volatile Organic Compounds

HS-SPME-GC–MS analysis was applied to characterize bread VOCs. Sixty-nine VOCs were found. Significantly different VOCs were grouped into ten different chemical classes (Table 2).

The addition of SB-SL sourdough to the bread caused significant ($p < 0.05$) changes in bread VOCs compared to CWB with positive repercussions on the global aroma profile of bread, influencing consumer acceptance [78]. As previously reported, composite sourdough wheat breads exhibited more complex aroma volatile profiles compared to bakers' yeast breads (Table 2) [79,80]. SB-SL SWB breads presented a higher concentration of alcohols and organic acids compared to CWB. Among the seven alcohol compounds, ethanol, 1-hexanol alcohol, and benzylalcohol were the most abundant. Acetic acid and hexanoic acid were the most representative carboxylic acids, especially in SB-SL SWB breads. Furfural was the most abundant aldehyde, particularly in SB-SL SWB EPS POS bread. 2-methylpropanal, 2-methylbutanal, 3-methylbutanal, benzaldehyde and 2-nonenal were the most abundant among the 11 aldehydes found in the SB-SL SWB breads. In general, the addition of sourdoughs did not affect the ethylacetate ester content; on the contrary, the concentration of acetoin (ketone) was affected and it appeared more abundant in CWB. A total of 12 heterocyclic compounds significantly differed among breads, 2-pentyl-furan and 2-methylpyrazine being the most representative. Moreover, in SB-SL SWB was observed a higher concentration of maltol. Dextran-containing SB-SL bread and its negative counterpart showed similar VOCs content. Therefore, these two breads should have, theoretically, received comparable scores for legume and bitterness flavor intensity. However, the SB-SL SWB EPS POS bread was perceived as more sweet, likely for the effect of the residual reducing sugars which contribute to the sweet taste and promote during baking the complex of Maillard reactions and the caramelization [28]. Wang et al. [28] reported a flavor-masking effect for dextran-enriched sorghum sourdough bread (0.56% bread weight) which showed decreased perception of bitterness, sourness, and aftertaste, compared to EPS NEG sorghum sourdough bread. The concentration of odorants in the final product can be driven by the ingredients and baking process [80] and sprouting and sourdough fermentation can significantly modify flavor and texture of raw materials [81]. Moreover, other volatile compounds with different origins are generated during baking [82].

Table 2. Quantification of total volatile organic compounds (VOCs) divided by chemical classes in control and sourdough breads (DY 162). Normalized data with the internal standard are reported, calculated as ratio peak area/total peak area percent. CWB, control wheat bread started with baker's yeast; SB-SL SWB EPS NEG, bread containing wheat flour added with 30% (*w/w*) sprouted barley–sprouted lentil (30:70%) sourdough fermented by *Leuconostoc pseudomesenteroides* DSM 20193; SB-SL SWB EPS POS, bread containing wheat flour added to 30% (*w/w*) dextran-containing sprouted barley–sprouted lentil (30:70%, *w/w*) sourdough fermented by *L. pseudomesenteroides* DSM 20193. The initial cell density of the strain was ca. 7 log cfu/g. The sourdough was fermented at 20 °C for 24 h.

Compounds	Odor	CWB	SL-SB SWB EPS NEG	SL-SB SWB EPS POS
Alcohols				
Ethanol	Alcoholic	2.03 ± 0.096 b	3.65 ± 0.074 a	2.7 ± 0.109 b
3-methylbutanol	Balsamic, alcoholic, malty	0.65 ± 0.014 a	0.52 ± 0.025 ab	0.36 ± 0.004 b
1-hexanol	Green grass, woody, sweet, flowery, mild	0.18 ± 0.014 b	1.00 ± 0.163 a	1.18 ± 0.031 a
1-heptanol	Green	n.d. [1]	0.01 ± 0.001 b	0.04 ± 0.015 a
2-ethylhexanol	Green, vegetable	0.27 ± 0.026 a	n.d.	n.d.
Benzylalcohol	Pleasant, aromatic	0.34 ± 0.135 b	0.39 ± 0.041 b	0.90 ± 0.016 a
1-nonanol	Citrus	0.01 ± 0.003 b	0.19 ± 0.089 a	0.27 ± 0.005 a
Total		3.56 ± 0.13	5.8 ± 0.63	5.45 ± 0.19
Organic Acids				
Acetic acid	Sour, acid, pungent	0.05 ± 0.004 b	0.50 ± 0.201 a	0.56 ± 0.013 a
Butanoic acid	Sweaty, rancid	n.d.	0.02 ± 0.001	0.03 ± 0
Hexanoic acid	Sweaty, cheesy, fatty, goat-like	0.04 ± 0.001 b	0.26 ± 0.073 a	0.28 ± 0.012 a
Total		0.10 ± 0.02 b	0.78 ± 0.03 a	0.87 ± 0.43 a
Aldehydes				
2-methylpropanal	Malty	0.1 ± 0.026 b	0.19 ± 0.003 a	0.19 ± 0.009 a
2-methylbutanal	Almond, malty	0.01 ± 0.001 b	0.24 ± 0.078 a	0.13 ± 0.015 ab
3-methylbutanal	Malty, roasty, cucumber-like	0.01 ± 0.003 b	0.37 ± 0.122 a	0.23 ± 0.01 ab
Hexanal	Green, grassy	0.52 ± 0.017 a	0.39 ± 0.069 b	0.33 ± 0.015 b
Octanal	Citrus, flowery	0.02 ± 0.029 b	0.18 ± 0.161 ab	0.34 ± 0.032 a
Nonanal	Citrus, soapy	0.37 ± 0.036a	0.26 ± 0.061b	0.44 ± 0.016a
Furfural	Almond, soil	0.26 ± 0.020b	1.11 ± 0.430a	1.90 ± 0.039a
Benzaldehyde	Almond, caramel	0.07 ± 0.009b	0.32 ± 0.08a	0.38 ± 0.002a
2-nonenal	Fatty, tallowy, green	0.02 ± 0.006b	0.17 ± 0.076ab	0.28 ± 0.002a
Acetaldehyde	Fruity	0.05 ± 0.003b	0.23 ± 0.074a	0.24 ± 0.003a
Benzeneacetaldehyde	flowery, honey-like	n.d.	0.02 ± 0.008a	0.03 ± 0.002a
Total		1.43 ± 0.03b	3.48 ± 0.72ab	4.49 ± 0.10a
Esters				
Ethylacetate	Sweet, fruity, pineapple	0.09 ± 0.003ab	0.11 ± 0.05a	0.05 ± 0b
Total		0.09 ± 0.00 ab	0.11 ± 0.00 a	0.05 ± 0.05 b
Ketones				
Acetoin	Butter, butterscotch, cream, yogurt	0.12 ± 0.005 a	0.06 ± 0.015 b	0.06 ± 0.009 b
Total		0.12 ± 0.005 a	0.06 ± 0.015 b	0.06 ± 0.009 b
Hydrocarbons				
d-limonene	Citrus	0.04 ± 0.032 b	0.18 ± 0.011 a	0.19 ± 0.031 a
Styrene	Pungent	n.d.	0.06 ± 0.013 a	0.01 ± 0.021 b
Nonadecane	n.f. [2]	n.d.	0.02 ± 0.014 a	0.01 ± 0.017 a
Pentadecane	n.f.	n.d.	n.d.	0.02 ± 0 ab
4 h-pyran-4-one,2,3-dihydro-3,5	Caramelized	n.d.	0.01 ± 0.014 b	0.05 ± 0.008 a
Total		0.04 ± 0.032 b	0.27 ± 0.07 a	0.28± 0.03 a

Table 2. *Cont.*

Compounds	Odor	CWB	SL-SB SWB EPS NEG	SL-SB SWB EPS POS
Furans				
2-pentyl- furan	Butter, green bean, floral	0.08 ± 0.024 b	0.32 ± 0.278 ab	0.28 ± 0.063 a
2-furancarboxaldehyde,5-methyl	n.f.	0.11 ± 0.015 b	0.63 ± 0.29 a	0.81 ± 0.002 a
Total		0.21 ± 0.02 b	0.95 ± 0.03 a	1.09 ± 0.07 a
Pyrazines				
2-methylpyrazine	Roasted, burnt, sweet	0.19 ± 0.004 b	0.32 ± 0.065 a	0.30 ± 0.004 a
2,5-dimethylpyrazine	Crust-like, popcorn	0.09 ± 0.01 b	0.14 ± 0.041 ab	0.20 ± 0.007 a
2,6-dimethylpyrazine	Roasted	0.06 ± 0 c	0.12 ±0.051 b	0.21 ± 0.009 a
2-ethyl-6-methylpyrazine	Nutty	0.03 ± 0.007 b	0.15 ± 0.066 a	0.17 ± 0.004 a
2-ethyl-5-methylpyrazine	Baked	0.03 ± 0.017 b	0.07 ± 0.005 a	0.06 ± 0.009 a
2-ethyl-3-methylpyrazine	Nutty, roasted, sweety	0.08 ± 0.011 b	0.12 ± 0.016 a	0.11 ± 0.001 a
3-ethyl-2,5-dimethylpyrazine	Baked, earthy, potato-like	0.02 ± 0.001 b	0.08 ± 0.020 a	0.09 ± 0.001 a
Pyrazinamide	n.f.	0.02 ± 0.003 b	0.03 ± 0.014 ab	0.04 ± 0.001 a
Total		0.62 ± 0.01 b	1.17 ± 0.03 a	1.27 ± 0.07 a
Pyrrolines				
2-acetyl-1-pyrroline	Cracker-like	0.02 ± 0.010 b	0.09 ± 0.040 a	0.10± 0.013 a
3-hydroxy-2-methyl-4-pyrone (maltol)	Caramel, sweet	0.04 ± 0.014 b	0.22 ± 0.15 a	0.32 ± 0.047 a
Total		0.06 ± 0.01 b	0.31 ± 0.08 a	0.42 ± 0.03 a

[1] n.d., not detected. [2] n.f., not found. Data are mean values ± standard deviation. a–c, Values in the same row with different letters differ significantly ($p < 0.05$).

3.7. Sensory Profile

Overall, sourdough breads achieved higher scores for several attributes such as elasticity, color, acidic taste and acidic odor compared to the CWB (Figure 4).

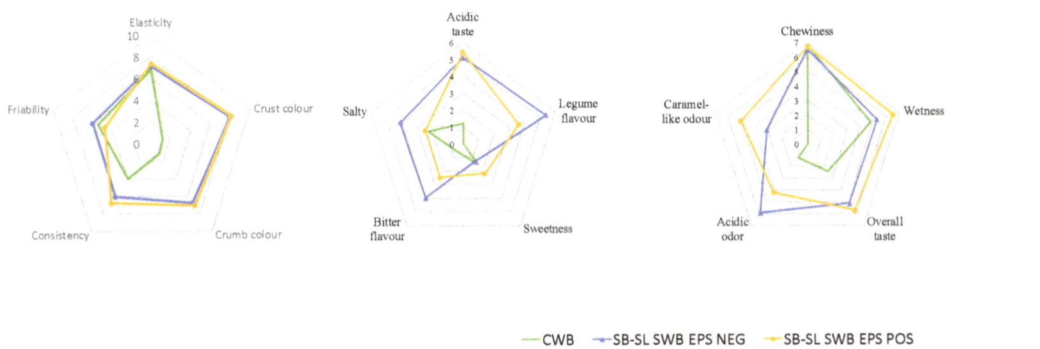

Figure 4. Spider web chart of the sensory analysis data for control and sourdough breads (DY 162). CWB, control wheat bread started with baker's yeast; SB-SL SWB EPS NEG, bread containing wheat flour added with 30% (*w*/*w*) sprouted barley–sprouted lentil (30:70% *w*/*w*) sourdough fermented by *Leuconostoc pseudomesenteroides* DSM 20193; SB-SL SWB EPS POS, bread containing wheat flour added to 30% (*w*/*w*) dextran-containing sprouted barley–sprouted lentil (30:70% *w*/*w*) sourdough fermented by *L. pseudomesenteroides* DSM 20193. The initial cell density of the strain was ca. 7 log cfu/g. The sourdough was fermented at 20 °C for 24 h.

Sourdough fermentation had an impact on flavor perception of final breads, which resulted in a bitter taste and legume flavor probably originating by liberation of small molecular weight polyphenol, bitter peptides and amino acids and the intense acidification due to endogenous and/or microbial enzymatic activities, and that could not be appreciated by the consumers [28,83]. Dextran-containing SB-SL bread was characterized by consistency, sweetness and caramel-like odor with significantly less perceived legume flavor and bitterness, while EPS NEG was judged as the most friable by the panelists, highlighting the fundamental role of dextran in improving bread texture and masking of unpleasant notes in composite bread [28]. As already reported, fiber-rich baked goods could help to reach the recommended dietary fiber intake, but they typically have a lower sensorial quality compared to baked goods produced from more refined ingredients [84]. In spite of this, composite SB-SL sourdough breads were more appreciated than CWB, as in previous studies [33,47].

3.8. Correlations between VOCs, Technological, Nutritional and Sensory Features of Breads

Relationship among measured chemical, technological, nutritional and sensory parameters from CWB and composite SWB breads, which could affect consumer acceptance were elaborated through Principal Component Analysis (PCA) (Figure 5).

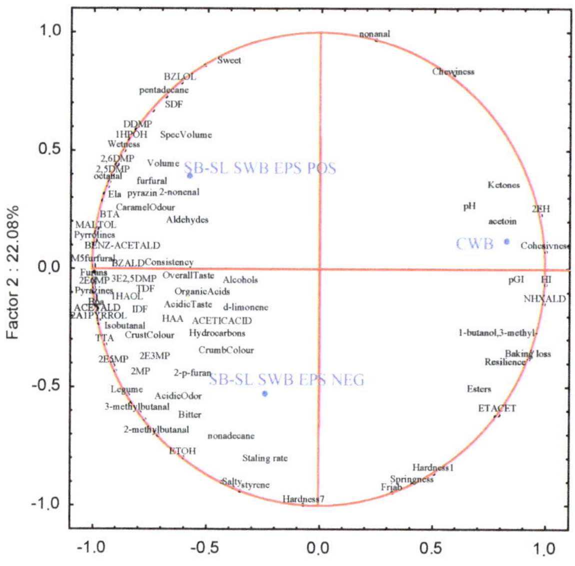

Figure 5. Correlations between physicochemical (pH, TTA, VOCs), technological, nutritional and sensory features of control and sourdough breads (DY 162). CWB, control wheat bread started with baker's yeast; SB-SL SWB EPS NEG, bread containing wheat flour added with 30% (*w*/*w*) sprouted barley-sprouted lentil (30:70% *w*/*w*) sourdough fermented by *Leuconostoc pseudomesenteroides* DSM 20193; SB-SL SWB EPS POS, bread containing wheat flour added to 30% (*w/w*) dextran-containing sprouted barley–sprouted lentil (30:70% *w*/*w*) sourdough fermented by *L. pseudomesenteroides* DSM 20193. The initial cell density of the strain was ca. 7 log cfu/g. The sourdough was fermented at 20 °C for 24 h. pH, pH bread crumb; TTA, TTA bread crumb; ETOH, ethanol; isopentanol, 1-butanol,3-methyl-; 1HAOL, 1-hexanol; 1HPOH, 1-heptanol; 2EH, 2-ethylhexanol; BZLOL, benzylalcohol; BTA, butanoic acid; HAA, hexanoic acid; NHXALD, hexanal; BZALD, benzaldehyde; BENZ-ACETALD, benzeneacetaldehyde; ETACET, ethylacetate; DDMP, 4 h-pyran-4-one,2,3-dihydro-3,5; 2-p-furan, 2-pentyl-furan; 2MP, 2-methylpyrazine; pyrazin, pyrazinamide; 2,5DMP, 2,5-dimethylpyrazine; 2,6DMP, 2,6-dimethylpyrazine; 2E6MP, 2-ethyl-6-methylpyrazine; 2E5MP, 2-ethyl-5-methylpyrazine; 2E3MP, 2-ethyl-3-methylpyrazine; 3E2,5DMP, 3-ethyl-2,5-dimethylpyrazine; 2A1PYRROL, 2-acetyl-1-pyrroline; MALTOL, 3-hydroxy-2-methyl-4-pyrone; Bpa, black pixel area; HI, hydrolysis index; pGI, predicted glycemic index; Ela, elasticity; Friab, friability; Legume, legume flavor; Sweet, sweetness; Bitter, bitter flavor.

The two PCs explained ca. 100% of the total variance of the data. Composite sourdough breads showed peculiar profiles and fell into different zones of the plane. Factor 1 clearly separated SB-SL SWB EPS NEG from dextran-containing bread. Factor 2 differentiated SWB and control breads. The chewiness and cohesiveness together with the highest values of some VOCs (e.g., total ketones, acetoin, 2-ethylhexanol) and nutritional (hydrolysis index and predicted GI) data, mainly characterized CWB. The dextran-containing SB-SL SWB not grouped together the EPS NEG because of the high volume, sweetness, wetness of the crumb, SDF, low hardness and springiness. The SB-SL SWB EPS NEG bread separated from the EPS POS for the higher hardness after 1 and 7 days, staling rate, legume and bitter flavor. Overall, the incorporation of 30% dextran-enriched SB-SL sourdough confirmed that the synergistic use of the sprouting process and LAB fermentation improves the nutritional and functional quality of cereal and legume grains with a positive effect on the undesirable beany flavor [33,35] compared to the EPS NEG counterpart.

4. Conclusions

In conclusion, the bioprocessing procedure developed successfully improved the quality of composite wheat bread, creating a good nutritional and sensory quality. The modulation of the fermentation parameters (native or sprouted cereal and legume flour type, DY, temperature) can stimulate metabolic activities of selected LAB strains, making them suitable for the production of sourdoughs enriched in EPS, which can be applied in the bread-making process as a "clean label" strategy.

The amount of dextran produced by *L. pseudomesenteroides* DSM20193 in a blend of sprouted barley and sprouted lentil flours effectively counteracted the quality deficiencies induced by wheat flour substitution in the composite sourdough bread. Dextran-enriched sprouted barley–sprouted lentil sourdough, resulting in the best overall system, could be used at a high level (30% of the dough weight) in wheat bread baking, resulting in bread with an enhanced nutritional quality (low HI and pGI), functionality (high soluble and total fibers content) and appreciable sensory attributes.

Supplementary Materials: The following are available online at https://www.mdpi.com/article/10.3390//foods10071489/s1, Table S1: Recipes for control and sourdough breads. CWB, control wheat bread started with baker's yeast; SB-SL SWB EPS NEG, bread containing wheat flour added with 30% (*w/w*) sprouted barley–sprouted lentil (30:70%) sourdough fermented by Leuconostoc pseudomesenteroides DSM 20193; SB-SL SWB EPS POS, bread containing wheat flour added to 30% (*w/w*) dextran-containing sprouted barley–sprouted lentil (30:70% *w/w*) sourdough fermented by L. pseudomesenteroides DSM 20193. The strain was inoculated at ca. 7 log cfu/g and sourdough fermented at 20 °C for 24 h. Doughs for bread making had DY 162. Table S2: Acidity (pH) and viscosity (Pa s^{-1}) values for doughs without sucrose addition (EPS NEG) and sucrose-supplemented (EPS POS) (16% of flour weight) doughs obtained from barley (B), sprouted barley (SB) and blends of sprouted barley with native lentil (SB-L) or sprouted lentil (SB-SL) flours at different ratios (60:40 and 70:30 ratios), before (T0 h) and after (T24 h) fermentation at 20 °C and 25 °C by *Lactobacillus plantarum* DPPMAB24W (B24W) as non EPS producing control or with the selected dextran-producing strains *Weissella confusa* SLA4 (SLA4), *Weissella paramesenteroides* SLA5 (SLA5), *Leuconostoc pseudomesenteroides* DSM 20193 (20193) and *Weissella confusa* DSM 20194 (20194). Doughs prepared without a starter and without the addition of sucrose were used as control (CT).

Author Contributions: Conceptualization, G.P., M.A., M.D.A. and M.C.; Formal analysis, G.P., G.C. and M.C.; Funding acquisition, C.G.R. and M.D.A.; Investigation, G.P. and M.C.; Methodology, C.G.R., M.A. and R.C.; Project administration, M.D.A. and M.C.; Resources, C.G.R., M.A., R.C. and M.C.; Supervision, M.A., R.C. and M.C.; Validation, G.P. and M.C.; Writing—original draft, G.P. and M.C.; Writing—review and editing, C.G.R., R.C., M.G. and M.D.A. All authors have read and agreed to the published version of the manuscript.

Funding: This work was supported by UE-FSE-FSER, PON RI 2014-2020 Action I.1—"Innovative doctoral of industrial interest"—XXXII cycle, PhD Program "Soil and Food Sciences— DOT1302942"—Grant n°1.

Conflicts of Interest: The authors declare no conflict of interest.

References

1. Nkhata, S.G.; Ayua, E.; Kamau, E.H.; Shingiro, J.B. Fermentation and germination improve nutritional value of cereals and legumes through activation of endogenous enzymes. *Food Sci. Nutr.* **2018**, *6*, 2446–2458. [CrossRef]
2. Mathur, H.; Beresford, T.P.; Cotter, P.D. Health benefits of Lactic Acid Bacteria (LAB) Fermentates. *Nutrients* **2020**, *12*, 1679. [CrossRef] [PubMed]
3. Röcken, W.; Voysey, P.A. Sourdough fermentation in bread making. *J. Appl. Microbiol.* **1995**, *79*, 38S–48S.
4. Katina, K.; Liukkonen, K.H.; Kaukovirta-Norja, A.; Adlercreutz, H.; Heinonen, S.M.; Lampi, A.M.; Pihlava, J.-M.; Poutanen, K. Fermentation-induced changes in the nutritional value of native or germinated rye. *J. Cereal Sci.* **2007**, *46*, 348–355. [CrossRef]
5. Leroy, F.; De Vuyst, L. Lactic acid bacteria as functional starter cultures for the food fermentation industry. *Trends Food Sci. Technol.* **2004**, *15*, 67–78. [CrossRef]
6. Gobbetti, M.; De Angelis, M.; Di Cagno, R.; Calasso, M.; Archetti, G.; Rizzello, C.G. Novel insights on the functional/nutritional features of the sourdough fermentation. *Int. J. Food Microbiol.* **2019**, *302*, 103–113. [CrossRef]
7. Marco, M.L.; Heeney, D.; Binda, S.; Cifelli, C.J.; Cotter, P.D.; Foligné, B.; Gänzle, M.; Kort, R.; Pasin, G.; Pihlanto, A.; et al. Health benefits of fermented foods: Microbiota and beyond. *Curr. Opin. Biotechnol.* **2017**, *44*, 94–102. [CrossRef] [PubMed]
8. Liukkonen, K.H.; Katina, K.; Wilhelmsson, A.; Myllymäki, O.; Lampi, A.M.; Kariluoto, S.; Piironen, V.; Heinonen, S.M.; Nurmi, T. Adlercreutz, H.; et al. Process-induced changes on bioactive compounds in whole grain rye. *Proc. Nutr. Soc.* **2003**, *62*, 117–122. [CrossRef]
9. Rizzello, C.G.; Portincasa, P.; Montemurro, M.; Di Palo, D.M.; Lorusso, M.P.; De Angelis, M.; Bonfrate, L.; Genot, B.; Gobbetti, M. Sourdough fermented breads are more digestible than those started with baker's yeast alone: An in vivo challenge dissecting distinct gastrointestinal responses. *Nutrients* **2019**, *11*, 2954. [CrossRef] [PubMed]
10. Montemurro, M.; Coda, R.; Rizzello, C.G. Recent advances in the use of sourdough biotechnology in pasta making. *Foods* **2019**, *8*, 129. [CrossRef] [PubMed]
11. De Vuyst, L.; Harth, H.; Van Kerrebroeck, S.; Leroy, F. Yeast diversity of sourdoughs and associated metabolic properties and functionalities. *Int. J. Food Microbiol.* **2016**, *239*, 26–34. [CrossRef]
12. Gobbetti, M.; Rizzello, C.G.; Di Cagno, R.; De Angelis, M. How the sourdough may affect the functional features of leavened baked goods. *Food Microbiol.* **2014**, *37*, 30–40. [CrossRef]
13. Bockwoldt, J.A.; Fellermeier, J.; Steffens, E.; Vogel, R.F.; Ehrmann, M.A. β-Glucan production by *Levilactobacillus brevis* and *Pediococcus claussenii* for in situ enriched rye and wheat sourdough breads. *Foods* **2021**, *10*, 547. [CrossRef]
14. Galle, S.; Arendt, E.K. Exopolysaccharides from sourdough lactic acid bacteria. *Crit. Rev. Food Sci. Nutr.* **2014**, *54*, 891–901. [CrossRef]
15. Monsan, P.; Bozonnet, S.; Ecile Albenne, C.; Joucla, G. Homopolysaccharides from lactic acid bacteria. *Int. Dairy J.* **2001**, *11*, 675–685. [CrossRef]
16. Korakli, M.; Gänzle, M.G.; Vogel, R.F. Metabolism by bifidobacteria and lactic acid bacteria of polysaccharides from wheat and rye, and exopolysaccharides produced by *Lactobacillus sanfranciscensis*. *J. Appl. Microbiol.* **2002**, *92*, 958–965. [CrossRef]
17. Tieking, M.; Gänzle, M.G. Exopolysaccharides from cereal-associated lactobacilli. *Trends Food Sci. Technol.* **2005**, *16*, 79–84. [CrossRef]
18. Lynch, K.M.; Zannini, E.; Coffey, A.; Arendt, E.K. Lactic acid bacteria exopolysaccharides in foods and beverages: Isolation, properties, characterization, and health benefits. *Annu. Rev. Food Sci. Technol.* **2018**, *9*, 155–176. [CrossRef] [PubMed]
19. Lynch, K.M.; Coffey, A.; Arendt, E.K. Exopolysaccharide producing lactic acid bacteria: Their techno-functional role and potential application in gluten-free bread products. *Food Res. Int.* **2018**, *110*, 52–61. [CrossRef]
20. Salazar, N.; Gueimonde, M.; De Los Reyes-Gavilán, C.G.; Ruas-Madiedo, P. Exopolysaccharides produced by lactic acid bacteria and bifidobacteria as fermentable substrates by the intestinal microbiota. *Crit. Rev. Food Sci. Nutr.* **2016**, *56*, 1440–1453. [CrossRef]
21. Saadat, Y.R.; Khosroushahi, A.Y.; Gargari, B.P. A comprehensive review of anticancer, immunomodulatory and health beneficial effects of the lactic acid bacteria exopolysaccharides. *Carbohydr. Polym.* **2019**, *217*, 79–89. [CrossRef]
22. González-Garcinuño, Á.; Tabernero, A.; Domínguez, Á.; Galán, M.A.; Martin del Valle, E.M. Levan and levansucrases: Polymer, enzyme, micro-organisms and biomedical applications. *Biocatal. Biotransform.* **2018**, *36*, 233–244. [CrossRef]
23. Hundschell, C.S.; Wagemans, A.M. Rheology of common uncharged exopolysaccharides for food applications. *Curr. Opin. Food Sci.* **2019**, *27*, 1–7. [CrossRef]
24. Wolter, A.; Hager, A.S.; Zannini, E.; Galle, S.; Gänzle, M.G.; Waters, D.M.; Arendt, E.K. Evaluation of exopolysaccharide producing *Weissella cibaria* MG1 strain for the production of sourdough from various flours. *Food Microbiol.* **2014**, *37*, 44–50. [CrossRef]
25. Valerio, F.; Bavaro, A.R.; Di Biase, M.; Lonigro, S.L.; Logrieco, A.F.; Lavermicocca, P. Effect of amaranth and quinoa flours on exopolysaccharide production and protein profile of liquid sourdough fermented by *Weissella cibaria* and *Lactobacillus plantarum*. *Front. Microbiol.* **2020**, *11*, 967. [CrossRef]
26. Wang, Y.; Sorvali, P.; Laitila, A.; Maina, N.H.; Coda, R.; Katina, K. Dextran produced *in situ* as a tool to improve the quality of wheat-faba bean composite bread. *Food Hydrocoll.* **2018**, *84*, 396–405. [CrossRef]

27. Wang, Y.; Compaoré-Sérémé, D.; Sawadogo-Lingani, H.; Coda, R.; Katina, K.; Maina, N.H. Influence of dextran synthesized *in situ* on the rheological, technological and nutritional properties of whole grain pearl millet bread. *Food Chem.* **2019**, *285*, 221–230. [CrossRef]
28. Wang, Y.; Trani, A.; Knaapila, A.; Hietala, S.; Coda, R.; Katina, K.; Maina, N.H. The effect of *in situ* produced dextran on flavour and texture perception of wholegrain sorghum bread. *Food Hydrocoll.* **2020**, *106*, 105913. [CrossRef]
29. Falcinelli, B.; Calzuola, I.; Gigliarelli, L.; Torricelli, R.; Polegri, L.; Vizioli, V.; Benincasa, P.; Marsili, V. Phenolic content and antioxidant activity of wholegrain breads from modern and old wheat (*Triticum aestivum* L.) cultivars and ancestors enriched with wheat sprout powder. *Ital. J. Agron.* **2018**, *13*, 297–302. [CrossRef]
30. Marti, A.; Cardone, G.; Pagani, M.A.; Casiraghi, M.C. Flour from sprouted wheat as a new ingredient in bread-making. *LWT* **2018**, *89*, 237–243. [CrossRef]
31. Benincasa, P.; Falcinelli, B.; Lutts, S.; Stagnari, F.; Galieni, A. Sprouted grains: A comprehensive review. *Nutrients* **2019**, *11*, 421. [CrossRef] [PubMed]
32. Rico, D.; Peñas, E.; García, M.D.C.; Martínez-Villaluenga, C.; Rai, D.K.; Birsan, R.I.; Frias, J.; Martín-Diana, A.B. Sprouted barley flour as a nutritious and functional ingredient. *Foods* **2020**, *9*, 296. [CrossRef]
33. Montemurro, M.; Pontonio, E.; Gobbetti, M.; Rizzello, C.G. Investigation of the nutritional, functional and technological effects of the sourdough fermentation of sprouted flours. *Int. J. Food Microbiol.* **2019**, *302*, 47–58. [CrossRef] [PubMed]
34. Hernandez-Aguilar, C.; Dominguez-Pacheco, A.; Tenango, M.P.; Valderrama-Bravo, C.; Hernández, M.S.; Cruz-Orea, A.; Ordonez-Miranda, J. Lentil sprouts: A nutraceutical alternative for the elaboration of bread. *J. Food Sci. Technol.* **2019**, *57*, 1817–1829. [CrossRef]
35. Perri, G.; Coda, R.; Rizzello, C.G.; Celano, G.; Ampollini, M.; Gobbetti, M.; De Angelis, M.; Calasso, M. Sourdough fermentation of whole and sprouted lentil flours: In situ formation of dextran and effects on the nutritional, texture and sensory characteristics of white bread. *Food Chem.* **2021**, *355*, 129638. [CrossRef] [PubMed]
36. Galle, S.; Schwab, C.; Dal Bello, F.; Coffey, A.; Gänzle, M.G.; Arendt, E.K. Influence of in-situ synthesized exopolysaccharides on the quality of gluten-free sorghum sourdough bread. *Int. J. Food Microbiol.* **2012**, *155*, 105–112. [CrossRef]
37. Katina, K.; Maina, N.H.; Juvonen, R.; Flander, L.; Johansson, L.; Virkki, L.; Tenkanen, M.; Laitila, A. In situ production and analysis of *Weissella confusa* dextran in wheat sourdough. *Food Microbiol.* **2009**, *26*, 734–743. [CrossRef] [PubMed]
38. Sullivan, P.; Arendt, E.; Gallagher, E. The increasing use of barley and barley by-products in the production of healthier baked goods. *Trends Food Sci. Technol.* **2013**, *29*, 124–134. [CrossRef]
39. European Food Safety Authority (EFSA) 2011a. Scientific opinion on the substantiation of health claims related to beta-glucans from oats and barley and maintenance of normal blood LDL-cholesterol concentrations (ID1236, 1299); increase in satiety leading to a reduction in energy intake (ID 851, 852), reduction of post-prandial glycemic responses (ID 821 824), and "digestive function" (ID 850) pursuant to Article 13(1) of Regulation (EC) No 1924/2006. *EFSA J.* **2011**, *9*, 2207.
40. European Food Safety Authority (EFSA). Scientific opinion on the substantiation of a health claim related to barley beta-glucans and lowering of blood cholesterol and reduced risk of (coronary) heart disease pursuant to Article 14 of Regulation (EC) No 1924/2006. *EFSA J.* **2011**, *9*, 2471. [CrossRef]
41. Food and Drug Administration, Department of Health and Human Services. *Federal Register—71 FR 29248: Health Claims;* Soluble Dietary Fiber from Certain Foods and Coronary Heart Disease Final Rule; FDA: Silver Spring, MD, USA, 2006; Volume 71, pp. 29248–29250.
42. Faris, M.E.A.I.E.; Takruri, H.R.; Issa, A.Y. Role of lentils (*Lens culinaris* L.) in human health and nutrition: A review. *Med. J. Nutr. Metab.* **2013**, *6*, 3–16. [CrossRef]
43. Santos, S.C.; Silva, B.; Valente, L.M.P.; Gruber, S.; Vasconcelos, M.W. The effect of sprouting in lentil (*Lens culinaris*) nutritional and microbiological profile. *Foods* **2020**, *9*, 400. [CrossRef] [PubMed]
44. Xu, Y.; Wang, Y.; Coda, R.; Säde, E.; Tuomainen, P.; Tenkanen, M.; Katina, K. In situ synthesis of exopolysaccharides by *Leuconostoc* spp. and *Weissella* spp. and their rheological impacts in fava bean flour. *Int. J. Food Microbiol.* **2017**, *248*, 63–71. [CrossRef] [PubMed]
45. Perri, G.; Calabrese, F.M.; Rizzello, C.G.; De Angelis, M.; Gobbetti, M.; Calasso, M. Sprouting process affects the lactic acid bacteria and yeasts of cereal, pseudocereal and legume flours. *LWT* **2020**, *126*, 109314. [CrossRef]
46. Xu, Y.; Coda, R.; Holopainen-Mantila, U.; Laitila, A.; Katina, K.; Tenkanen, M. Impact of in situ produced exopolysaccharides on rheology and texture of fava bean protein concentrate. *Food Res. Int.* **2019**, *115*, 191–199. [CrossRef]
47. Rizzello, C.G.; Calasso, M.; Campanella, D.; De Angelis, M.; Gobbetti, M. Use of sourdough fermentation and mixture of wheat, chickpea, lentil and bean flours for enhancing the nutritional, texture and sensory characteristics of white bread. *Int. J. Food Microbiol.* **2014**, *180*, 78–87. [CrossRef] [PubMed]
48. Regulation (EC) No 1924/2006 of the European Parliament and of the Council of 20 December 2006 on Nutrition and Health Claims Made on Foods. Official Journal L 404, 30 December 2006, p. 9–25. Available online: https://eur-lex.europa.eu/eli/reg/2006/1924/corrigendum/2007-01-18/oj (accessed on 26 June 2021).
49. AACC International. *Approved Methods of the American Association of Cereal Chemists*, 11th ed.; AACC: St. Paul, MN, USA, 2000.
50. Katina, K.; Heiniö, R.L.; Autio, K.; Poutanen, K. Optimization of sourdough process for improved sensory profile and texture of wheat bread. *LWT* **2006**, *39*, 1189–1202. [CrossRef]

51. Curiel, J.A.; Coda, R.; Centomani, I.; Summo, C.; Gobbetti, M.; Rizzello, C.G. Exploitation of the nutritional and functional characteristics of traditional Italian legumes: The potential of sourdough fermentation. *Int. J. Food Microbiol.* **2015**, *196*, 51–61. [CrossRef]
52. Crowley, P.; Grau, H.; Arendt, E.K. Influence of additives and mixing time on crumb grain characteristics of wheat bread. *Cereal Chem.* **2000**, *77*, 370–375. [CrossRef]
53. Liljeberg, H.; Åkerberg, A.; Björck, I. Resistant starch formation in bread as influenced by choice of ingredients or baking conditions. *Food Chem.* **1996**, *56*, 389–394. [CrossRef]
54. Goñi, I.; Garcia-Alonso, A.; Saura-Calixto, F. A starch hydrolysis procedure to estimate glycemic index. *Nutr. Res.* **1997**, *17*, 427–437. [CrossRef]
55. Tobaruela, E.D.C.; Santos, A.D.O.; de Almeida-Muradian, L.B.; Araujo, E.D.S.; Lajolo, F.M.; Menezes, E.W. Application of dietary fiber method AOAC 2011.25 in fruit and comparison with AOAC 991.43 method. *Food Chem.* **2018**, *238*, 87–93. [CrossRef]
56. Pico, J.; Khomenko, I.; Capozzi, V.; Navarini, L.; Bernal, J.; Gómez, M.; Biasioli, F. Analysis of volatile organic compounds in crumb and crust of different baked and toasted gluten-free breads by direct PTR-ToF-MS and fast-GC-PTR-ToF-MS. *J. Mass Spectrom.* **2018**, *53*, 893–902. [CrossRef] [PubMed]
57. Cavallo, N.; De Angelis, M.; Calasso, M.; Quinto, M.; Mentana, A.; Minervini, F.; Cappelle, S.; Gobbetti, M. Microbial cell-free extracts affect the biochemical characteristics and sensorial quality of sourdough bread. *Food Chem.* **2017**, *237*, 159–168. [CrossRef]
58. Korakli, M.; Vogel, R.F. Structure/function relationship of homopolysaccharide producing glycansucrases and therapeutic potential of their synthesised glycans. *Appl. Microbiol. Biotechnol.* **2006**, *71*, 790–803. [CrossRef] [PubMed]
59. Galle, S.; Schwab, C.; Arendt, E.; Gänzle, M. Exopolysaccharide-forming *Weissella* strains as starter cultures for sorghum and wheat sourdoughs. *J. Agric. Food Chem.* **2010**, *58*, 5834–5841. [CrossRef]
60. Wisselink, H.W.; Weusthuis, R.A.; Eggink, G.; Hugenholtz, J.; Grobben, G.J. Mannitol production by lactic acid bacteria: A review. *Int. Dairy J.* **2002**, *12*, 151–161. [CrossRef]
61. Zannini, E.; Mauch, A.; Galle, S.; Gänzle, M.; Coffey, A.; Arendt, E.K.; Taylor, G.P.; Waters, D.M. Barley malt wort fermentation by exopolysaccharide-forming *Weissella cibaria* MG1 for the production of a novel beverage. *J. Appl. Microbiol.* **2013**, *115*, 1379–1387. [CrossRef]
62. Wang, T.L.; Domoney, C.; Hedley, C.L.; Casey, R.; Grusak, M.A. Can we improve the nutritional quality of legume seeds? *Plant Physiol.* **2003**, *131*, 886–891. [CrossRef]
63. Duszkiewicz-Reinhard, W.A.N.D.A.; Gujska, E.; Khan, K. Reduction of stachyose in legume flours by lactic acid bacteria. *J. Food Sci.* **1994**, *59*, 115–117. [CrossRef]
64. Yoon, M.Y.; Hwang, H.J. Reduction of soybean oligosaccharides and properties of α-D-galactosidase from *Lactobacillus curvatus* R08 and *Leuconostoc mesenteriodes* JK55. *Food Microbiol.* **2008**, *25*, 815–823. [CrossRef] [PubMed]
65. Waters, D.M.; Mauch, A.; Coffey, A.; Arendt, E.K.; Zannini, E. Lactic acid bacteria as a cell factory for the delivery of functional biomolecules and ingredients in cereal-based beverages: A review. *Crit. Rev. Food Sci. Nutr.* **2015**, *55*, 503–520. [CrossRef] [PubMed]
66. Ferrero, C. Hydrocolloids in wheat breadmaking: A concise review. *Food Hydrocoll.* **2017**, *68*, 15–22. [CrossRef]
67. Kohajdová, Z.; Karovičová, J.; Magala, M. Effect of lentil and bean flours on rheological and baking properties of wheat dough. *Chem. Pap.* **2013**, *67*, 398–407. [CrossRef]
68. Fadda, C.; Sanguinetti, A.M.; Del Caro, A.; Collar, C.; Piga, A. Bread staling: Updating the view. *Compr. Rev. Food Sci. Food Saf.* **2014**, *13*, 473–492. [CrossRef]
69. Gray, J.A.; Bemiller, J.N. Bread staling: Molecular basis and control. *Compr. Rev. Food Sci. Food Saf.* **2003**, *2*, 1–21. [CrossRef] [PubMed]
70. Farzaneh, V.; Ghodsvali, A.; Bakhshabadi, H.; Zare, Z.; Carvalho, I.S. The impact of germination time on some selected parameters through malting process. *Int. J. Biol. Macromol.* **2017**, *94*, 663–668. [CrossRef]
71. Charoenthaikij, P.; Jangchud, K.; Jangchud, A.; Prinyawiwatkul, W.; No, H.K. Composite wheat–germinated brown rice flours: Selected physicochemical properties and bread application. *Int. J. Food Sci. Technol.* **2012**, *47*, 75–82. [CrossRef]
72. De Angelis, M.; Damiano, N.; Rizzello, C.G.; Cassone, A.; Di Cagno, R.; Gobbetti, M. Sourdough fermentation as a tool for the manufacture of low-glycemic index white wheat bread enriched in dietary fibre. *Eur. Food Res. Technol.* **2009**, *229*, 593–601. [CrossRef]
73. Henry, R.J. Pentosan and (1→ 3), (1→ 4)-β-glucan concentrations in endosperm and wholegrain of wheat, barley, oats and rye. *J. Cereal Sci.* **1987**, *6*, 253–258. [CrossRef]
74. Wood, P.; Beer, M.U.; Butler, G. Evaluation of role of concentration and molecular weight of oat β-glucan in determining effect of viscosity on plasma glucose and insulin following an oral glucose load. *Br. J. Nutr.* **2000**, *84*, 19–23. [CrossRef]
75. Marklinder, I.; Johansson, L. Sourdough fermentation of barley flours with varied content of mixed-linked (1→ 3),(1→ 4) β-d-glucans. *Food Microbiol.* **1995**, *12*, 363–371. [CrossRef]
76. Mariotti, M.; Garofalo, C.; Aquilanti, L.; Osimani, A.; Fongaro, L.; Tavoletti, S.; Hager, A.S.; Clementi, F. Barley flour exploitation in sourdough bread-making: A technological, nutritional and sensory evaluation. *LWT* **2014**, *59*, 973–980. [CrossRef]
77. Wood, P.J. Cereal β-glucans in diet and health. *J. Cereal Sci.* **2007**, *46*, 230–238. [CrossRef]
78. Ruiz, J.A.; Quilez, J.; Mestres, M.; Guasch, J. Solid-phase microextraction method for headspace analysis of volatile compounds in bread crumb. *Cereal Chem.* **2003**, *80*, 255–259. [CrossRef]

9. Plessas, S.; Bekatorou, A.; Gallanagh, J.; Nigam, P.; Koutinas, A.A.; Psarianos, C. Evolution of aroma volatiles during storage of sourdough breads made by mixed cultures of *Kluyveromyces marxianus* and *Lactobacillus delbrueckii* ssp. *bulgaricus or Lactobacillus helveticus*. *Food Chem.* **2008**, *107*, 883–889. [CrossRef]
0. Hansen, A.; Schieberle, P. Generation of aroma compounds during sourdough fermentation: Applied and fundamental aspects. *Trends Food Sci. Technol.* **2005**, *16*, 85–94. [CrossRef]
31. Heiniö, R.L.; Katina, K.; Wilhelmson, A.; Myllymäki, O.; Rajamäki, T.; Latva-Kala, K.; Liukkonen, K.H.; Poutanen, K. Relationship between sensory perception and flavour-active volatile compounds of germinated, sourdough fermented and native rye following the extrusion process. *LWT* **2003**, *36*, 533–545. [CrossRef]
32. Pétel, C.; Courcoux, P.; Génovesi, N.; Rouillé, J.; Onno, B.; Prost, C. Free Sorting and association Task: A variant of the free sorting method applied to study the impact of dried sourdough as an ingredient on the related bread odor. *J. Food Sci.* **2017**, *82*, 985–992. [CrossRef]
33. Zhao, C.J.; Schieber, A.; Gänzle, M.G. Formation of taste-active amino acids, amino acid derivatives and peptides in food fermentations—A review. *Food Res. Int.* **2016**, *89*, 39–47. [CrossRef]
34. Poutanen, K.; Flander, L.; Katina, K. Sourdough and cereal fermentation in a nutritional perspective. *Food Microbiol.* **2009**, *26*, 693–699. [CrossRef] [PubMed]

foods

Article

Characterization of Dextran Produced by the Food-Related Strain *Weissella cibaria* C43-11 and of the Relevant *Dextransucrase* Gene

Palmira De Bellis [1,*,†], Massimo Ferrara [1,†], Anna Rita Bavaro [1], Vito Linsalata [1], Mariaelena Di Biase [1], Biagia Musio [2], Vito Gallo [2], Giuseppina Mulè [1] and Francesca Valerio [1]

1 Institute of Sciences of Food Production (ISPA), National Research Council (CNR), Via G. Amendola 122/O, 70126 Bari, Italy
2 Department of Civil, Environmental, Land, Building and Chemical Engineering, Dicatech, Polytechnic University of Bari, Via Edoardo Orabona 4, 70125 Bari, Italy
* Correspondence: mirella.debellis@ispa.cnr.it; Tel.: +39-080-5929459; Fax: +39-080-5929374
† These authors contributed equally to this work.

Abstract: A metabolic feature of lactic acid bacteria (LAB) is the production of exopolysaccharides (EPSs), which have technological and functional properties of interest to the food sector. The present study focused on the characterization of the *Weissella cibaria* strain C43-11, a high EPS producer in the presence of sucrose, in comparison with a low-producing strain (C2-32), and on possible genetic regulatory elements responsible for the modulation of *dextransucrase* (*dsr*) genes expression. NMR analysis of the polymeric material produced by the C43-11 strain indicated the presence of dextran consisting mainly of a linear scaffold formed by α-(1–6) glycosidic linkages and a smaller amounts of branches derived from α-(1–2), α-(1–3), and α-(1–4) linkages. Molecular analysis of the *dsr* genes and the putative transcriptional promoters of the two strains showed differences in their regulatory regions. Such variations may have a role in the modulation of *dsr* expression levels in the presence of sucrose. The strong upregulation of the *dsr* gene in the C43-11 strain resulted in a high accumulation of EPS. This is the first report showing differences in the regulatory elements of the *dsr* gene in *W. cibaria* and indicates a new perspective of investigation to identify the regulatory mechanism of EPS production.

Keywords: lactic acid bacteria; exopolysaccharides; homopolysaccharides; dextran; sucrose; gene expression; nuclear magnetic resonance; monosaccharide composition

Citation: De Bellis, P.; Ferrara, M.; Bavaro, A.R.; Linsalata, V.; Di Biase, M.; Musio, B.; Gallo, V.; Mulè, G.; Valerio, F. Characterization of Dextran Produced by the Food-Related Strain *Weissella cibaria* C43-11 and of the Relevant *Dextransucrase* Gene. *Foods* **2022**, *11*, 2819. https://doi.org/10.3390/foods11182819

Academic Editor: Diego Bonatto

Received: 28 July 2022
Accepted: 9 September 2022
Published: 13 September 2022

1. Introduction

Lactic acid bacteria (LAB) represent, together with yeasts, the microorganisms with the greatest number of applications in food preparation. The study of lactic populations has allowed the selection of strains with specific biotechnological characteristics that are useful in improving food quality. The metabolic activity of LAB activates a series of biochemical processes that decisively influence the rheological, nutritional, functional, sensory, and shelf life properties of the products [1–3]. LAB are frequently used in the fermentation of leavened doughs for the production of baked goods to improve the technological properties of these products, allowing the reduction/elimination of additives and improvers ("label free") [4,5], and are also used to produce functional products (reduced salt/fat content, enhanced bioactive compounds, "yeast-free", and increased protein content) [6–10].

One metabolic feature of LAB is the production of exopolysaccharides (EPSs) during their fermentation activity. These compounds can be homo- (HoPS) or hetero-polysaccharides (HePS) depending on the type of monosaccharides constituting the polymer chain, and can have several technological and functional properties. Generally, EPSs improve the technological and sensorial properties of food products, and can act as prebiotics since they are not digested by human enzymes and are available as a carbon source for probiotic bacteria in the intestine and/or can protect the probiotic bacteria during the gastrointestinal transit by acting as a barrier [11]. Moreover, high molecular weight (MW) EPSs are postbiotics with antiviral and immunomodulatory activities [12,13].

EPS include dextran, a homopolysaccharide with a linear structure containing at least 50% of D-glucopyranosyl units attached by α-(1–6) linkages with varying percentages of α-(1–2), α-(1–3) or α-(1–4) branches [14,15], which can be used in foods since it has been generally recognized as safe (GRAS) by the Food and Drug Administration. In recent years, dextrans have assumed technological importance because they can improve the rheology and texture of many fermented food products and baked goods. For example, the European Commission authorizes the use of a *Leuconostoc mesenteroides* preparation containing dextran as a food ingredient in bakery products [16] to improve the softness, crumb texture, and loaf volume. Dextrans can also replace gelling agents or fatty substances currently used as thickeners in bread because they influence the viscoelastic properties of the dough and positively affect bread consistency and shelf-life, thus satisfying consumer demands for reduced use of food additives [9,17–22]. In addition to improving food texture, dextrans also provide the product with functional properties, acting as prebiotics and having beneficial effects on human health (antioxidant activity, cholesterol reduction, possible immunomodulatory and antitumor activity) [23–25].

The LAB *Weissella* and *Leuconostoc* are known to produce significant amounts of EPS, and the *Weissella* genus produces particularly high levels of dextran [13,26,27].

Several studies have investigated the molecular processes involved in polysaccharide synthesis by LAB [28–31], and some authors have explored the genome of *W. cibaria* strains to examine some metabolic traits, in particular EPS production [32,33]. Genomic analysis showed the presence of a single extracellular dextransucrase (*dsr*) enzyme encoded by the *dsr* gene involved in the dextran synthesis. Dextransucrases are members of the GH70 family according to the CAZy classification (http://www.cazy.org (accessed on 22 February 2022); they synthesize dextran by hydrolysis of the glycosidic bond of sucrose, subsequently releasing energy to catalyze the transfer of D-glucopyranosyl residues to the growing chain of the polymer, and at the same time releasing fructose [34]. Several studies have characterized the Drs enzymes [14], but few investigate regulation of the expression of the *dsr* gene [27,35]. Yu et al. [35] examined the effect of sucrose on *W. cibaria* compared with *Lactobacillus* spp., while Hu and Gänzle [36] studied the effect of temperature on the expression of the *dsr* gene in *W. cibaria*.

Recently, *W. cibaria* strains isolated from wheat semolina have been characterized for their ability to produce EPS [37,38]. In particular, the C43-11 strain has been selected as the greatest EPS producer of those tested, and has been used to obtain liquid sourdoughs enriched in EPS suitable for use as a fat replacer in baked goods. These sourdoughs were used to make focaccia, allowing a 20% reduction in the amount of added fat and improving the technological characteristics of the final product [9].

Therefore, the present work (1) aimed to chemically characterize the EPS produced by C43-11 in comparison with the low-producing C2-32 strain; (2) to investigate the possible genetic factors responsible for EPS production by the two strains, and (3) to analyze the effect of sucrose on the expression of the *dsr* gene in the two strains.

2. Materials and Methods

2.1. Microorganisms and Growth Conditions

W. cibaria strains C43-11 and C2-32, isolated from wheat semolina [37], were cultured in MRS broth (Biolife Italiana S.r.l., Milan, Italy) and incubated at 30 °C. For long-term storage, stock cultures were prepared by mixing 8 mL of culture with 2 mL of Bacto glycerol (Difco, Becton Dickinson, Co., Sparks, MD, USA) and freezing 1 mL portions of this mixture at −80 °C. Cultures were subcultured twice and incubated at 30 °C for 24 h before use.

2.2. EPS Production in mMRS

To investigate the EPS production, the growth conditions used in Valerio et al. [38] were applied. Briefly, the modified MRS broth (mMRS) was obtained as reported in Minervini et al. [39] by adding fresh yeast extract (5% *w*/*v*) and maltose (1% *w*/*v*) to the commercial MRS medium (Biolife Italiana S.r.l., Milan, Italy), final pH 5.6. The commercial medium contained 20.0 g/L glucose, 5.0 g/L yeast extract, 2.0 g/L K_2HPO_4, 5.0 g sodium acetate, 0.2 g/L $MgSO_4{\cdot}7H_2O$, 0.05 g/L $MnSO_4{\cdot}H_2O$, 2.0 g/L di-ammonium citrate, 10.0 g/L beef extract, and 10.0 g/L peptone 1 g/L Tween® 80. The medium, supplemented (mMRS + S) or not (mMRS) with 10% (*w*/*v*) sucrose, was inoculated at 4% (*v*/*v*) with 24 h cultures of each strain and incubated for 24 h at 30 °C. The cultures were labelled as follows: C43-11 + S and C2-32 + S when strains were grown in mMRS + S, and C43-11 and C2-32 when strains were grown in mMRS with no addition of sucrose. For each culture, three replicates were prepared. After 6 h, 10 h, and 24 h incubation, aliquots were taken to determine strain growth, EPS production, and RNA extraction. Cell enumeration was carried out by plate count on MRS agar after 48 h incubation at 30 °C.

2.3. Purification and Quantification of EPS

EPS purification from the bacterial cultures was performed as reported in Valerio et al. [38]. Briefly, liquid cultures of strains in mMRS + S and mMRS were heat treated at 100 °C for 15 min to inactivate enzymes responsible for EPS degradation [40] and centrifuged (9000× *g*, 10 min, 4 °C). The resulting supernatants were treated with three volumes of chilled ethanol 96–99% (*v*/*v*) and the solutions were stored overnight at 4 °C. The precipitated EPS were collected by centrifugation (11,325× *g*, 20 min, 4 °C), dissolved in distilled water, dialyzed (12–14 kDa) against distilled water at 4 °C for 48 h, and lyophilized. Lyophilized samples were used for EPS quantification, monosaccharide composition, and NMR analysis.

The concentration of EPS (g/L) was determined on the lyophilized samples rehydrated with distilled water at the initial volume according to the phenol-sulfuric method [41] using glucose as a standard (limit of detection, LOD: 0.078 g/L).

2.4. EPS Monosaccharide Composition

EPS monosaccharide composition was determined by acid hydrolysis of the lyophilized and rehydrated samples using 5% (*v*/*v*) perchloric acid (70%) at 100 °C for 5 h [42]. After filtration, hydrolyzed samples were analyzed using a HPLC Dionex DX500 system equipped with a GP50 gradient pump, an ED40 Electrochemical Detector in Pulsed Amperometric Detection (HPAEC-PAD), and Dionex PeakNet 5.11 chromatographic software. The chromatographic separation of sugars was carried out using a Dionex CarboPac PA1 column heated at 30 °C and a Carbopac PA1 guard column in isocratic mode with an elution of 150 mm NaOH at a flow rate of 1.0 mL/min [43]. Identification and quantification of the main monosaccharides was performed by integrating calibration peaks obtained from the relevant standards solutions of arabinose, glucose, fructose, and mannose, all purchased from Merck KGaA (Darmstadt, Germany).

2.5. NMR Spectroscopy

2.5.1. Chemicals

3-(Trimethylsilyl)-2,2,3,3-tetradeutero-propionic acid sodium salt (TSP-d_4, CAS No. 24493-21-8, 99%D, Armar Chemicals, Döttingen, Switzerland) and deuterium oxide (D_2O, CAS. No. 7789-20-0, 99.86%D, Eurisotop, Saclay, France) were used for sample preparation. NMR tubes (Norell 509-UP 7) were provided by Norell, Landisville NJ, United States. The following compounds were used as a reference: dextran (analytical standard Set Mp 1000–400,000, CAS. No. 9004-54-0, Fluka, Sigma-Aldrich, Switzerland), fructooligosaccharides from chicory (FOS, Sigma Aldrich, St. Louis, MO, USA), inulin from chicory (CAS. No. 9005-80-5, Sigma Aldrich, St. Louis, MO, USA), L-(+)-arabinose (CAS No. 5328-37-0, ≥98%), D-(+)-galactose (CAS No. 59-23-4, ≥99%), D-(+)-glucose (CAS No. 50-99-7, ≥99.5%), D-(−)-

fructose (CAS No. 57-48-7, ≥99%), D-(+)-mannose (CAS No. 3458-28-4, ≥99%), sucrose (CAS No. 57-50-1, ≥99.5%), and D-(+)-maltose monohydrate (CAS No. 6363-53-7, ≥99%).

2.5.2. Sample Preparation

Twenty milligrams of dialyzed and lyophilized samples (C43-11 + S and C2-32 + S), prepared as described in paragraph 2.3, were added to 600 µL of TSP-d_4/D_2O solution [3-(trimethylsilyl) propionic-2,2,3,3-d_4 acid sodium salt in D_2O (0.20%$_w$)]. The resulting solution was vortexed (Advanced Vortex Mixer ZX3, VELP Scientifica Srl, Italy) for 5 min at 2500 rpm and then transferred to the NMR tube.

2.5.3. NMR Experiment

NMR spectra were recorded through a Bruker Avance 400 MHz spectrometer equipped with a 5 mm inverse probe. Pulse lengths were calibrated before each experiment, and the probe tuning and matching were adjusted for each sample. The temperature of the probe was set at 303 K.

The chemical shifts are reported in ppm and referenced to the TSP-d_4 signal. The following acquisition parameters were used to record 1D ^{1}H NOESY NMR: pulse program = noesygppr1d; size of fid (TD) = 64 K; spectral width (SW) = 20 ppm; transmitter offset = 4.7 ppm; power level for pre-saturation (pl9) calculated at 25 Hz based on 90° hard pulse; dummy scans (ds) = 4; number of scans (ns) = 128; acquisition time = 4.09 s; mixing time (d8) = 0.010 s; recycle delay (d1) = 5 s. The following acquisition parameters were used to record ^{1}H NMR: pulse program = zg; size of fid (TD) = 64 K; spectral width (SW) = 20 ppm; transmitter offset = 5.00 ppm; dummy scans (ds) = 0; number of scans (ns) = 16; acquisition time = 4.09 s; recycle delay (d1) = 10.0 s. The following acquisition parameters were used to record ^{13}C NMR: pulse program = zgpg; size of fid (TD) = 131 K; spectral width (SW) = 240 ppm; transmitter offset = 110 ppm; dummy scans (ds) = 4; number of scans (ns) = 512; acquisition time = 2.71 s; recycle delay (d1) = 10 s. The following acquisition parameters were used to record TOCSY spectra: pulse program = mlevphpr; size of fid (TD) = 4 K × 256; spectral width (SW) = 20 ppm; transmitter offset = 4.70 ppm; power level for pre-saturation (pl9) calculated at 25 Hz based on 90° hard pulse; dummy scans (ds) = 16; number of scans (ns) = 32; spin lock (d9) = 75.00 ms.

NMR raw data (Free Induction Decays, FIDs) were processed using the software MestReNova 11.0 (Mestrelab Research SL, Santiago de Compostela, Spain). The phase and the baseline were optimized manually to make the resonances in the Fourier transformed data positive for all of the resonances except for the residual signal of water, which was subjected to a pre-saturation step. The zero-order (PH0) and first-order (PH0) phase parameters were adjusted opportunely after having set the pivot parameter to the biggest peak in the spectrum. The baseline correction was applied to flatten the baseline on the Fourier transformed data. Multipoint baseline correction was adopted upon selecting the points corresponding to baseline regions (also known as control points) which were then used by the software to build a baseline model using the interpolation algorithm.

2.6. Analysis of the Dsr Gene

The genomic DNA of the two strains was extracted from 24 h MRS broth cultures using a Wizard Genomic DNA Purification Kit (Promega Corporation, Madison, WI, USA) according to the supplier's specifications. DNA quality and quantity were assessed using an ND 1000 spectrophotometer (NanoDrop Technologies, Inc., Wilmington, NC, USA) and by 1% (*w*/*v*) agarose gel electrophoresis in TAE buffer. The complete *dsr* gene (ca. 4700 bp), which included the promoter region, was amplified using the primers dsr-full_F (5′-AACACGAAAAGACGCTTGCG-3′) and dsr-full_R (5′-TGAGTAGGGCTGGGGTACTG-3′). The primer pair was designed with Oligo Perfect v1.0 software (Invitrogen, Life Technologies, Grand Island, NY, USA) based on the *dsr* gene from the *W. cibaria* MG1 genome available in GenBank (accession number JWHU00000000).

Each 20 µL reaction mixture contained 10 µL of 2X InvitrogenTM Platinum SupereFi II PCR master mix (Thermo Fisher Scientific, Waltham, MA, USA), 0.5 µM of each primer and 1 µL (50 ng) of genomic DNA. The reaction mixtures were first incubated for 30 s at 98 °C, and then cycled for 30 cycles according to the following temperature profiles: 15 s at 98 °C, 15 s at an annealing temperature of 60 °C, 2 min at 72 °C, followed by a final extension for 5 min at 72 °C. The PCR products were analyzed by agarose gel electrophoresis (1.5%) in a TAE buffer stained with GelRed (Biotium, Hayward, CA, USA). The size of the amplified DNA fragments was estimated by comparison with a GelPilot 1 kb Plus Ladder (Qiagen GmbH, Hilden, Germany).

In addition to primers for full length amplification of the *dsr* gene, WcibDex fw (5′-GCATCTTTCAATACTTGAGG-3′) and WcibDex rev (5′-CATGACTTGTTGGCATAGC-3′) [44], dsr1a (5′-GGCATGCAAGTAATGGCTGA-3′), dsr1b (5′-ATTGATCCGGCACAAA ATCAGCC-3′), dsr1c (5′-CTAGTCAAGACGTCGTGCTGGTC-3′), dsr1d (5′-TTCTGCCTGA ACTTGTGGA-3′), and dsr1e (5′-ACAGTGCAAGTGCGGTAGTTGAT-3′) internal primers for gene sequencing were used.

Moreover, in order to amplify the promoter region, a primer pair was designed using Oligo Perfect v1.0 software (Invitrogen, Life Technologies, Grand Island, NY, USA) and based on the *dsr* sequence of *W. cibaria* C43.11. The promoter region (ca. 500 bp) was amplified using the primers dsrPROM_F (5′-CCGGCAGCTACCCAAAACTT-3′) and dsrPROM_R (5′-AYGGGGTGTCTGAATTRGGC-3′) and the PCR conditions described above. PCR products were purified with a QIA quick PCR Purification Kit (QIAGEN, Inc., Toronto, ON, Canada), and quantified by an ND 1000 spectrophotometer (NanoDrop Technologies, Inc., Wilmington, NC, USA). Sequencing of the *dsr* gene and promoter sequence was performed by using a Big Dye Terminator v3.1 Cycle Sequencing Kit (Applied Biosystems, Thermo Fisher Scientific, Waltham, MA, USA) on an ABI Prism 3100 Genetic Analyzer (Applied Biosystems Waltham, MA, USA).

Nucleotide sequences were aligned using Clustal Omega (https://www.ebi.ac.uk/Tools/msa/clustalo/ (accessed on 10 March 2022)) and compared to the *dsr* sequences of MG1 and SP7 (NCBI acc.n. NZ_JWHU01000000.1 and NZ_SDGK00000000.1) strains. Nucleotide sequences of the *dsr* gene and promoter region of C43-11 and C2-32 strains are available in the NCBI nucleotide sequence database under accession numbers ON791804, ON791805, ON791806, and ON791807.

The putative regulatory elements and start site for transcription were predicted in silico within the promoter region of the *dsr* gene by using PBROM [45].

2.7. Transcription Analysis

2.7.1. RNA Extraction and cDNA Synthesis

At selected time points (T0, T6, T10, T24), total RNA was isolated from 2 mL aliquots of liquid cultures in mMRS + S and mMRS. At T0, RNA was extracted from 24 h broth cultures used for inoculation. Cells were harvested by centrifugation (4000× *g* for 5 min at 18 °C) and snap-frozen in liquid nitrogen. RNA was extracted using the CTAB-based method [46] and modified as follows. Cell pellets were re-suspended in 800 µL CTAB buffer and transferred into 2 mL tubes containing a loop full of acid-washed 425–600 µm glass beads (Sigma-Aldrich, St. Louis, MO, USA) and two iron balls (0.5 cm diameter Ø). The mechanical disruption of cells was performed in a Retsch MM301 mixer mill (Retsch Gmbh, Haan, Germany) at 15/s frequency for 30 s and cell lysates were centrifuged at 16,060× *g* for 10 min. RNA was purified from the supernatant using chloroform-isoamyl alcohol (24:1), as described by Wang and Stegemann [46].

Possible contamination by genomic DNA was removed by DNase I (Thermo Fisher Scientific, Waltham, MA, USA) digestion, according to the manufacturer's instructions. RNA concentration and quality were determined using a NanoDrop ND-1000 spectrophotometer and RNA integrity was checked by gel electrophoresis on 1.5% (*w/v*) agarose gel stained with GelRed. The RNA was reverse transcribed to cDNA using the SuperScript IV First-

Strand Synthesis System (Invitrogen, Waltham, MA, USA) following the manufacturer's instructions. The cDNA was stored at −20 °C.

2.7.2. Real-Time (RT) qPCR

RT-PCR was performed using a 7500 Fast Real-Time PCR System (Applied Biosystems Waltham, MA, USA) using Power Up SYBR Green Master Mix (Applied Biosystem, Foster City, CA, USA) according to the manufacturer's instructions. The primers used for the quantification of *dsr* gene expression were GLUC_Rq_F (5′-AATGTTGACACGAGCGACCT-3′) and GLUC_Rq_R (5′-CCGGAAGCACCACTAGTTGT-3′). qPCR primers were designed using Oligo Perfect v1.0 software based on the *dsr* gene sequence of *W. cibaria* C43-11; they amplified a PCR product of 194 bp. Relative gene expression profiles of *dsr* at time points T0, T6, T10, and T24 were determined by comparison to the expression of the *recA* gene used as a reference gene. The primers used for the quantification of gene expression of the *recA* were Wcib-recA-F (5′-GTAACCCAGAAACGACGCCT-3′) and Wcib-recA-R (5′-AGATTTCAACTTCCACTTCACGG-3′). These primers, reported by Koirala et al. [30], were modified after comparison with the *recA* gene sequences of *W. cibaria* (NCBI acc.n. JWHU01000016.1) The PCR product of 195 bp was amplified. qPCR was performed in triplicate in a volume of 10 µL containing 5 µL of Power Up SYBR Green Master Mix, 200 nm of each primer, and 1 µL of cDNA. The conditions for amplification were as follows: 50 °C for 2 min, 95 °C for 2 min followed by 40 cycles of 95 °C for 15 s, and 60 °C for 40 s. A dissociation step was performed after the RT-PCR run. Relative expression (RQ) of the *dsr* gene at all-time points was calculated using the $2^{-\Delta\Delta CT}$ method [47]. RQ values were calculated by comparison with *dsr* gene expression levels in mMRS.

2.8. Statistical Analysis

Data are presented as mean values ± standard deviations. Statistical analysis of the data was performed using STATISTICA 12.0 software (StatSoft, Inc., Tulsa, OK, USA). Data concerning strain loads, EPS production, and monosaccharides concentration were compared by applying a one-way ANOVA ($p < 0.05$), substituting data below the detection limit (LOD) with half of the LOD [48]. Significant differences ($p < 0.05$) among groups were determined by using a post hoc Tukey test.

3. Results

3.1. EPS Quantification and Characterization

3.1.1. EPS Production

The EPS production by *W. cibaria* C43-11 and C2-32 was evaluated in mMRS with or without sucrose at 10% (*w*/*v*). Both strains grew in the presence of the two substrates (Table 1), reaching ca. 9 log cfu/mL after 24 h ($p > 0.05$), although strain C43-11 showed a higher cell number ($p < 0.05$) after 6 h in the presence of sucrose. EPS production for C43-11 was mainly observed in the presence of sucrose, reaching ca. 12 g/L after 24 h, while low amounts close to the LOD were produced by C43-11 with no added sucrose, and by strain C2-32 in both media (Table 1).

Table 1. Cell count (log cfu/mL) and EPS production by *W. cibaria* C43-11 and C2-32 in mMRS with (+S) or without sucrose after 0 h, 6 h, 10 h, and 24 h incubation at 30 °C.

Culture	T0	T6	T10	T24
		Cell Count (log cfu/mL)		
C2-32	7.08 ±0.01 [b]	7.20 ± 0.08 [b]	7.57 ± 0.03 [b]	8.52 ± 0.62 [a]
C2-32 + S	7.07 ± 0.10 [b]	7.40 ± 0.02 [b]	7.50 ± 0.09 [b]	8.59 ± 0.78 [a]
C43-11	7.32 ± 0.03 [a]	7.39 ± 0.01 [b]	7.68 ± 0.03 [b]	8.77 ± 0.78 [a]
C43-11 + S	7.320 ± 0.06 [a]	7.64 ± 0.06 [a]	7.89 ± 0.02 [a]	9.27 ± 0.29 [a]
		EPS (g/L)		
C2-32	<LOD [b,*]	<LOD [c,*]	<LOD [b,*]	<LOD [b,*]
C2-32 + S	<LOD [b,*]	0.12 ± 0.004 [b]	0.14 ± 0.00 [b]	0.13 ± 0.00 [b]
C43-11	<LOD [b,*]	<LOD [c,*]	0.12 ± 0.00 [b]	<LOD [b,*]
C43-11 + S	0.23 ± 0.00 [a]	3.55 ± 0.02 [a]	5.02 ± 0.31 [a]	11.74 ± 3.25 [a]

Data are represented as mean ± standard deviation. [a–c] Values with different superscript letters within each column are significantly different ($p < 0.05$). cfu = colony forming unit. * LOD = 0.078 g/L.

3.1.2. Monosaccharide Composition

The HPAEC-PAD results (Table 2) showed that the polymeric material produced by C43-11 in the presence of sucrose (C43-11 + S) was composed of approximately 94% glucose, 1% mannose, and 5% fructose. This finding is consistent with the production of dextran as the major polysaccharide product. Differently, the strain C2-32 (C2-32 + S) produced a low amount of material consisting of approximately 54% glucose, 43% mannose, and 3% fructose. When no sucrose was added to the medium, the strain C43-11 produced a material consisting of 32% glucose, 63% mannose, and 5% fructose, while the strain C2-32 produced a polysaccharide with a composition of 47% glucose, 46% mannose, and 7% fructose (C2-32). Finally, no arabinose was detected in any of the cases.

Table 2. Glucose, mannose, and fructose (mg/L) produced by strains C43-11 and C2-32 in mMRS + S and mMRS bacterial cultures after quantified analysis by HPAEC-PAD.

Culture	Glucose	Mannose	Fructose
		mg/L	
C43-11 + S	16299.9 ± 2114.1 [a]	203.4 ± 55.4 [ab]	779.9 ± 2.7 [a]
C2-32 + S	259.7 ± 75.9 [b]	207.0 ± 19.0 [ab]	13.8 ± 4.7 [b]
C43-11	138.5 ± 23.3 [b]	280.5 ± 23.3 [a]	21.0 ± 8.5 [b]
C2-32	125.8 ± 22.7 [b]	124.9± 10.6 [b]	20.0 ± 4.0 [b]

Data are represented as mean ± standard deviation. [a,b] Values with different superscript letters within each column are significantly different ($p < 0.05$).

3.1.3. EPS Characterization (NMR)

Based on the findings described in the previous paragraph, the NMR study focused on the characterization of the polymeric material produced by C43-11 in the presence of sucrose (C43-11 + S, Table 2). The sample's macromolecular profile was confirmed by the poor sharpness of the signals in the 1D ^{1}H NOESY NMR spectra [49]. The absence of free monosaccharides (arabinose, galactose, glucose, fructose, and mannose) and disaccharides (sucrose and maltose) was confirmed by comparison with spectra of reference compounds (see Figure S1 in Supplementary Materials for further details). Dextran, inulin, and fructooligosaccharide (FOS) were considered among the possible EPSs produced by LAB species. Dextrans with different molecular weights (MW: 25, 80, and 270 kDa) were taken into consideration to disclose possible variations in the chemical shifts of the signals due to the size of the polysaccharide, but no significant differences in spectral features were recognized.

In Figure 1, the 1D ^{1}H NOESY NMR spectra of FOS, inulin, dextran (MW 270 kDa), and purified samples from C43-11 + S and C2-32 + S are reported.

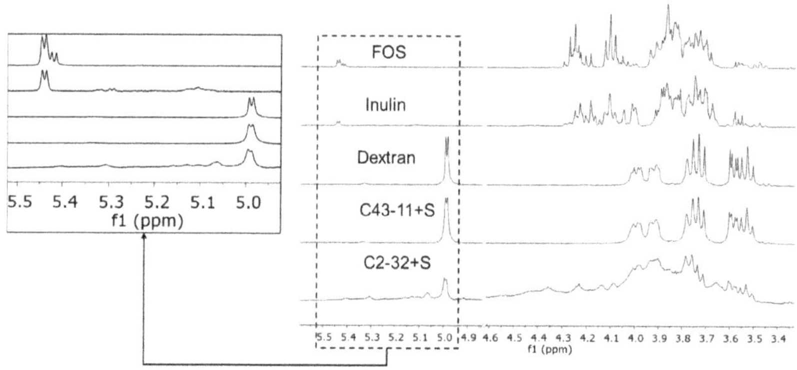

Figure 1. Typical 1D ^{1}H NOESY NMR spectra (D_2O, 400 MHz, 303 K) of fructooligosaccharide (FOS), inulin, dextran (MW 270 kDa) and purified samples from C43-11 + S and C2-32 + S.

Comparison of the spectra indicated that spectra labeled as dextran and C43-11 + S were almost superimposable. Thus, the signal pattern in the C43-11 + S spectrum could be mainly attributed to the glucose units constituting the polymeric chain of the dextran structure. Characteristic signals of anomeric hydrogen of dextran were identified in the region between 4.99 and 5.40 ppm, the most intense of which is attributable to α-(1–6) glycosidic linkage at 4.99 ppm. Weak signals at 5.40, 5.33, and 5.23 ppm were assigned to the anomeric hydrogen atoms involved in the α-(1–4), α-(1–3), and α-(1–2) linkages, respectively. These assignments were consistent with the correlations found in the ^{1}H TOCSY experiment (Figure 2) which, in agreement with the data reported in the literature, supports the hypothesis that these hydrogens belong to the same polymeric scaffold [50]. Signal intensities of the anomeric hydrogens were consistent with a dextran structure predominantly consisting of α-(1–6) glycosidic linkage and a smaller amount of α-(1–4), α-(1–3) and α-(1–2) glycosidic linkages, in a ratio that was assessed at approximately 97 [α-(1–6)]: 3 [α-(1–4), α-(1–3), α-(1–2)].

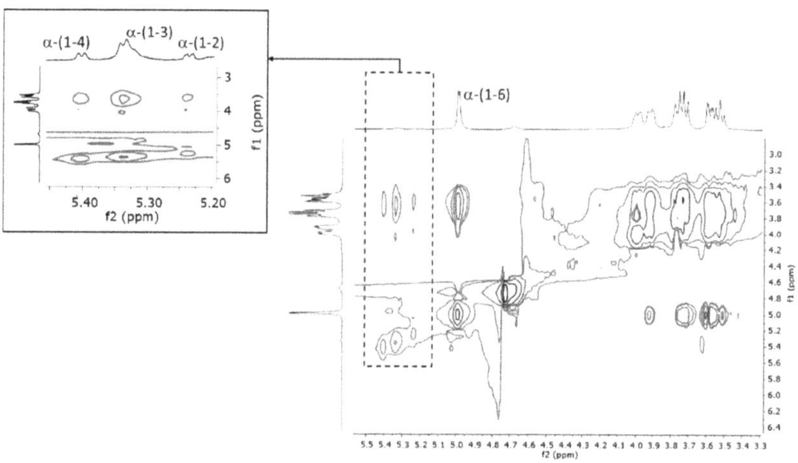

Figure 2. The TOCSY spectrum of EPS extracted from C43–11 + S recorded at 400 MHz in D_2O at 303 K. Anomeric protons are labeled as α-(1–6), α-(1–2), α-(1–3), and α-(1–4).

The low fructose content revealed via analysis of the monosaccharide composition of sample C43–11 + S (5%, Table 2) was detected in the 1D ^{1}H NOESY NMR spectrum in the form of very weak signals at 4.25 and 4.35 ppm (see Figure S2 in Supplementary Materials for further details). While no conclusive statements can be made based on their low intensities, the chemical shift values of these two signals may suggest the presence of heteropolymers similar to inulin or/and FOS.

The predominance of dextran in the EPS produced by C43-11 in the presence of sucrose was further confirmed by $^{13}C\{^{1}H\}$ NMR analysis. As represented in Figure 3, characteristic signals related to dextran were found at 97.2 (C-1), 72.89 (C-3), 70.89 (C-2), 69.70 (C-4), 69.07 (C-5), and 65.13 (C-6).

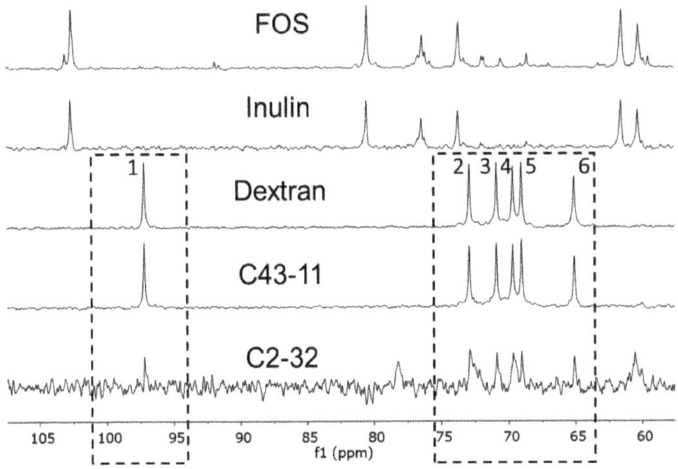

Figure 3. Comparison of typical $^{13}C\{^{1}H\}$ NMR spectra (D_2O, 400 MHz, 303 K) of fructooligosaccharide (FOS), inulin, dextran (MW 270 kDa) and purified samples from C43-11 + S and C2-32 + S. The signals ascribed to the dextran structure are indicated in dashed squares.

The ^{1}H NMR signals in the spectrum of the material produced by C2-32 in the presence of sucrose (Figure 1) cannot be ascribed exclusively to dextran or to materials containing fructose (inulin and FOS). The presence of dextran can be supposed (Figure 3), but the low yield of the C2-32 + S sample did not allow a more detailed investigation of the structural features. Further investigations and appropriate growth conditions are needed to obtain information on the nature of the polysaccharides produced by the C2-32 strain.

3.2. Analysis of the Dsr Gene

Amplification of the *dsr* gene with primers *dsr*-full_F/*dsr*-full_R produced an amplicon of the expected length (4702 bp) that included the promoter region and the complete *dsr* gene of *W. cibaria* strains (4377 bp).

Pairwise sequence comparison of *dsr* genes of *W. cibaria* C43-11 and C2-32 strains revealed a 96% identity. The *dsr* genes of C43-11 and MG1, known as a high EPS producer, [33], showed 99% identity, while *dsr* genes of C2-32 and SP7 showed 98% identity (data not shown).

The region upstream the *dsr* gene in both strains was determined by DNA sequencing and compared to the DNA sequence of those of *W. cibaria* MG1 and SP7 strains. The pairwise sequence comparison of the *dsr* gene promoter region of *W. cibaria* C43-11 and C2-32 strains showed a 94% identity (Figure 4).

DNA sequencing showed several nucleotide differences, in particular an AT insertion at position 55-56 of the C2-32 promoter sequence. A Blast N analysis against *W. cibaria* genomes available at the NCBI Genebank revealed the presence of this insertion also in other *W. cibaria* strains CSM1, CSM2, CMS3, CMU, and CXO-1 (data not shown). The sequence analysis for σ70-dependent DNA motif revealed a different pattern for −10 and −35 motifs. In detail, for strain C43-11 the transcriptional start site was predicted at an adenine (A) and the −10 (5′-TTGTGAAAT-3′) and −35 (5′-GTGCTA-3′) motifs were identified. Differently, for strain C2-32, the transcriptional start site was predicted at a thymine (T) and the −10 (5′-ATCAATAAT-3′) and −35 (5′-ATAACG-3′) motifs were identified.

```
C2-32       CATAATATCTTTATGGTCGATGATTTGACAGTTAGTAGCTGTTAATTCATCGGCATTTTT    60
SP7         CATAATATCTTTATGGTCGATGATTTGACAGTTAGTAGCTGTTAATTCATCGGCATTTTT    60
MG1         CATAATATCTTTATGGTCGATGATTTGACAGTTAGTAGCGATTAATTCATCGGC--TTTT    58
C43-11      CATAATATCTTTATGGTCGATGATTTGACAGTTAGTAGCGATTAATTCATCGGC--TTTT    58
            **************************************  *************   ****

C2-32       TTATTTACTTGTGCTACCAATAACGGTGGCTTGTGAAATATCAATAATTAATTTTTGTAT    120
SP7         TTATTTACTTGTGCTACCAATAACGGTGGCTTGTGAAATATCAATAATTAATTTTTGTAT    120
MG1         TTATTTACTTGTGCTACCAATAGCGGTGGCTTGTGAAATATCAATAATTAATTTTTGTAT    118
C43-11      TTATTTACTTGTGCTACCAATAGCGGTGGCTTGTGAAATATCAATAATTAATTTTTGTAT    118
            ********************* **************************************

C2-32       GTTTTTTATATTTTATATATATTAGATTTACAGTTGTAAACTAAAGGGTACCCGTGGAAC    180
SP7         GTTTTTTATATTTTATATATATTAGATTTACAGTTGTAAACTAAAGGGTACCCGTGGAAC    180
MG1         GTTTTTAATATTTTATCTATATTAGATTTACAGTTGTAAACTAAAGGGTACCCGTGGAAC    178
C43-11      GTTTTTAATATTTTATCTATATTAGATTTACAGTTGTAAACTAAAGGGTACCCGTGGAAC    178
            ****** ********* *******************************************

C2-32       TAATGTTCTGAGTAAGGATGGGGTACTGTGACTTAAGGCAATTTGGATAAAAAGCACTAA    240
SP7         TAATGTTCTGAGTAAGGATGGGGTACTGTGACTTAAGGCAATTTGGATAAAAAGCACTAA    240
MG1         TAATATTTTGAGTAGGGCTGGGGTACTGTGACTTAATGCAATTTGGATAAAAAGGACTAA    238
C43-11      TAATATTTTGAGTAGGGCTGGGGTACTGTGACTTAATGCAATTTGGATAAAAAGGACTAA    238
            **** ** ****** ** ***************** ***************** *****

C2-32       TTTAACTAACGAGTCTCATTATGCTTGCATATGAATAAAATGGAACCGGTTCTAAAATAA    300
SP7         TTTAACTAACGAGTCTCATTATGCTTGCATATGAATAAAATGGAACCGGTTCTAAAATAA    300
MG1         TTTGACTAACAAGTTTCATTATGCCTGCATATGAATAAAATGGAACCGGTTCTAAAATGA    298
C43-11      TTTGACTAACGAGTTTCATTATGCCTGTATATGAATAAAATGGAACCGGTTCTAAAATGA    298
            *** ****** *** ********* ** ***************************** *

C2-32       CATTGTGAAAAAATTGACAGAGAGATTATGTCCGCGTTAAGGTGAACCCTACACCAAGGC    360
SP7         CATTGTGAAAAAATTGACAGAGAGATTATGTCCGCGTTAAGGTGAACCCTACACCAAGGC    360
MG1         CATTGTGAAAAAGTTGACAGAAAGATTATGCCCGCGTTAAGGTGAACCCTGTACCAAGGC    358
C43-11      CATTGTGAAAAAGTTGACAGAAAGATTATGCCCGCGTTAAGGTGAACCCTGCACCAAGGC    358
            ************ ******** ******** *******************  ********

C2-32       CTTGGGGATATGAATAGAATACGGGGATATAATTT    395
SP7         CTTGGGGATATGAATAGAATACGGGGATATAATTT    395
MG1         CTTGGGGATATGAATAGAATACGGGGATATAATTT    393
C43-11      CTTGGGGATATGAATAGAATACGGGGATATAATTT    393
            ***********************************
```

Figure 4. Alignment of the promoter regions of C2-32, C43-11, SP7, and MG1 strains with the in silico predicted −35 and −10 motifs (**underlined text**), transcriptional start sites (**bold letters**), and lexA (**light gray**) DNA-binding motif indicated.

Analysis of known transcription factor binding sites revealed the presence of a different pattern for C2-32 compared to C43-11. In particular, the C2-32 promoter region hosts additional lexA DNA-binding motif sites (Figures 4 and 5) that are not present in the promoter region of the C43-11 *dsr* gene.

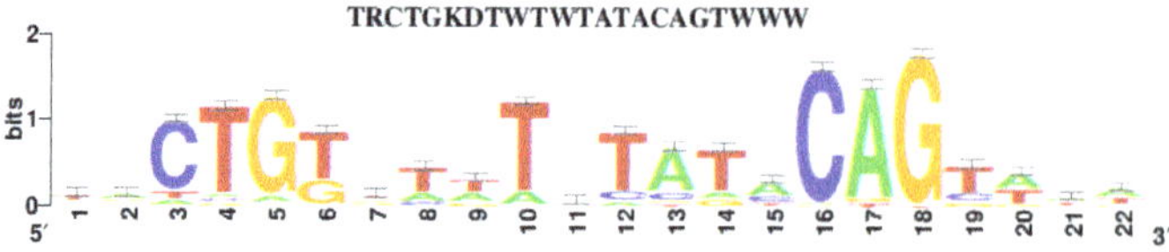

Figure 5. Position weight matrix (PWM) of LexA transcription factor binding (from RegulonDB—http://regulondb.ccg.unam.mx/ (accessed on 10 March 2022)).

3.3. Transcription Analysis

To verify whether the differing EPS production between the two strains also resulted in a different expression level of dextransucrase, *dsr* gene expression was quantified at the mRNA abundance level. The relative expression of *dsr* was monitored in the presence and absence of sucrose during a 24 h incubation period and is shown in Figure 6. RT-qPCR demonstrated that *dsr* expression is strongly induced by the presence of sucrose only in C43-11, with the highest level occurring after 6 h (Fold Change > 10); subsequently, expression decreased up until 24 h. Conversely, for strain C2-32 the modulation of *dsr* expression did not respond to the supplementation of sucrose, showing only a slight upregulation after 10 h in the presence of sucrose (FC < 2, Figure 6).

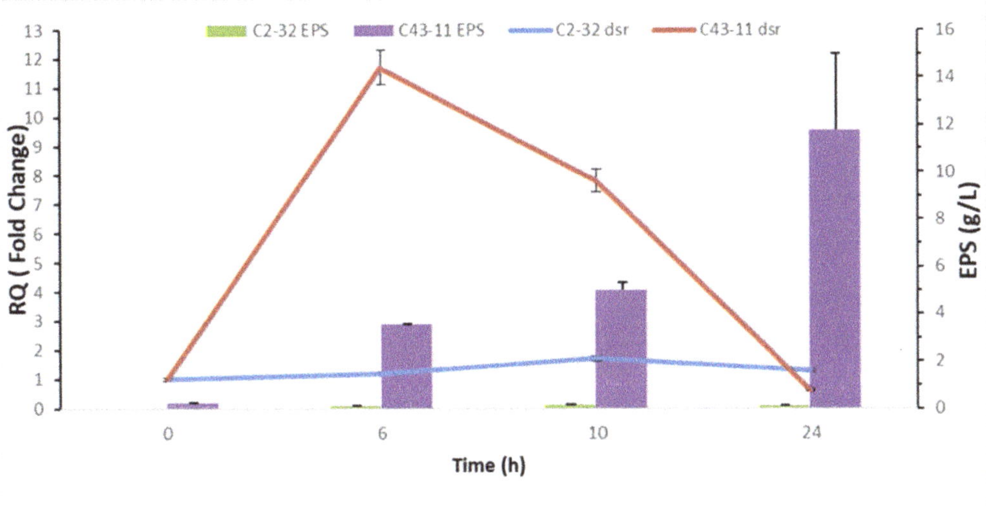

Figure 6. Relative Quantification (RQ, Fold Change = $2^{-\Delta\Delta Ct}$) of *dsr* gene expression in *W. cibaria* C43-11 and C2-32 grown in mMRS with sucrose (10% *w*/*w*). Reported RQ (lines) refer to *dsr* gene expression in mMRS with sucrose (mMRS + S) as compared to mMRS. RQ values are correlated with EPS production (g/L) for each strain in mMRS + S (bars).

4. Discussion

In recent years, microbial communities have been studied to select microorganisms with useful characteristics for the nutritional/functional improvement of fermented foods [31,51,52]. The use of LAB is considered an excellent tool for improving the nutritional/functional quality of foods via the production of bioactive compounds production, the bioavailability of dietary fibers and phytochemicals, and their technological and organoleptic characteristics, as well as the reduction of antinutritional factors [4,5]. Several studies have shown the positive impact of the EPS produced by LAB strains on the nutritional, technological, functional, and rheological features of foods [18,26,28,53]. In particular, *W. confusa* and *W. cibaria* have been studied for their ability to produce dextran [26].

In previous studies, the *W. cibaria* strain C43-11 was selected for its ability to produce high quantities of EPS [38], and was then used to produce a liquid sourdough applied in the production of focaccia bread as a fat replacer [9]. The current study has characterized the EPS produced by *W. cibaria* C43-11 and the *dsr* gene involved in dextran synthesis has been characterized. For this purpose, the *W. cibaria* C43-11 was compared with the strain C2-32, which belongs to the same species and has same origin but produces extremely low amounts of EPS [38]. Results showed that while both strains reached a cell load of about 9 log cfu/mL in mMRS containing sucrose or not, markedly higher EPS production was achieved only by C43-11 in the presence of sucrose. According to the results of Yu et al. [35], the difference in EPS production did not depend on the cell density but could be due to a higher gene expression.

The polymeric material was characterized by HPAEC-PAD and NMR analyses. HPAEC PAD revealed a high glucose content in the hydrolysed samples of strain C43-11 grown in the presence of sucrose, supporting the hypothesis of dextran production. At the same time, the presence of fructose and mannose in addition to glucose after EPS hydrolysis was observed, suggesting the presence of oligosaccharides or heteropolysaccharides in low amounts. The presence of mannose could originate from the baker's yeast used in the mMRS formulation, since it contains approximately 16% of this sugar [54], and could have been used as an acceptor in the synthesis of polymers other than dextran. Additionally, the release of cell wall components could be an effect of partial cell lysis caused by the heat treatment of the bacterial cultures during EPS purification. In particular, the monomers were detected in both *W. cibaria* culture filtrate samples, with or without sucrose and at similar concentrations, except for strain C43-11. In that case, glucose and fructose increased by approximately 100-fold and 30-fold, respectively, in the presence of sucrose. EPS production for strain C2-32 was scarce, and the amount and ratio of monomers was unchanged in the presence of sucrose, indicating that it did not influence the production of oligo- or heteropolysaccharides. Similarly, Yu et al. [35] studied the production of EPS by a *W. cibaria* strain isolated from kimchi in the presence of sucrose: NMR analysis showed the polymer to be α-1,6-dextran, and the ion chromatography revealed the presence of glucose, fructose and mannose at 62.1%, 34.2% and 3.8% in the MRS culture, and of 97% of glucose when the sucrose (2%, *w*/*w*) was added to the medium. Other authors also found low amounts of galactose, mannose, and glucose after acidic hydrolysis of the polymeric material obtained by a strain of dextran-producer *Leuconostoc mesenteroides* [27]. HePS can be produced by *Weissella* genus spp. and generally consist of mannose, glucose, fructose, arabinose, rhamnose, xylose, galactose, and in some cases containing *N*-acetyl-D-glucosamine and *N*-acetyl-D-galactosamine [26,55]. The production of polymers other than dextran by the *W. cibaria* strains could be further ascertained in a future study.

In agreement with the HPAEC-PAD results, NMR spectroscopy indicated that strain C43-11 produced dextran in the presence of sucrose. Dextran production was almost exclusive and was characterized by linear chains of glucose units bound predominantly by α-(1–6) glycosidic linkage. The signal assignment was in agreement with the data reported in the literature for dextran produced by wild and mutant strains of *Leuconostoc mesenteroides* [56]. It is known that the amount of α-(1–6) linkages in a dextran can vary from 50% to 97% of the total number of glycosidic linkages and the structural differences may depend on the strains and production conditions [50]. Moreover, Bounaix et al. [15,55] and Ahmed et al. [57] described the structure of dextran produced by *W. confusa* and *W. cibaria* strains containing 2.4–3.4% of α-(1–3) branch linkages. In the present study, traces of α-(1–4), α-(1–3) and α-(1–2) branched dextrans were also detected: approximately 1–3% compared to the α-(1–6) linkage. The formation of gluco-oligomers/polymers containing α-(1–4) and α-(1–2) has already been reported in the literature for LAB, like *Streptococcus*, *Lactobacillus*, and *Weissella* species incubated with sucrose and maltose [50,58].

To date, numerous studies have investigated the ability to produce EPS from the different bacterial isolates. The information gathered showed wide heterogeneity in EPS production capacity among strains. Based on the available information, the efficiency of the Dsr enzymes is influenced by different factors, including culture medium pH [42], the carbon source present in the growth medium [14], sucrose concentration [35], and temperature [36]. In particular, Bounaix et al. [15] and Besrour-Aouam et al. [14] indicated the optimal pH value of ca. 5.4 for Dsr activity. They also observed the influence of the carbon source (sucrose or glucose) on the synthesis of Dsr in different LAB species and *Weissella* strains and hypothesized that the regulation of *dsr* expression could depend on the species or strain. In fact, while *dsr* induction by sucrose is common in *Leuconostoc* spp. [15,59], Drs synthesis is reported as constitutive in other LAB, including *Lactobacillus reuteri* and *Streptococcus* spp. [60,61] and some *Weissella* strains [15]. In the present study, significant upregulation of the *dsr* gene was inducted in the presence of sucrose only in *W. cibaria* C43-11 strain, in agreement with the results obtained for *W. confusa* strain by Koirala et al. [30]. Moreover, Yu et al. [35] reported that EPS production was enhanced by increasing sucrose concentration in a *W. cibaria* strain; while Hu and Gänzle [36] and Besrour-Aouam et al. [27] reported the induction of the *dsr* gene by cold stress as well.

Although several studies have explored the factors affecting their differing ability to produce EPS, few have focused on the genetic characteristics of the regulatory elements of the *dsr* gene in *W. cibaria*. The present study focused on the expression of the *dsr* gene in two strains of *W. cibaria* to determine the regulatory elements that might be responsible for the difference in production capacity. In particular, the in silico analysis of the promoter region of the *dsr* gene in the two strains highlighted substantial differences concerning the predicted regulatory regions. Specifically, the AT insertion at position 55–56 of the C2-32 promoter sequence determined a shift in the transcriptional start site and −10 and −35 motifs. This evidence needs to be confirmed experimentally. The insertion generated a new DNA-binding motif, LexA, shared only with strain SP7 (100% match of the promoter region with C2-32 strains), and not with C43-11 or MG1. *W. cibaria* MG1 was the first and best studied EPS producer, also due to its promising potential for bakery and dairy applications [33,44,62,63], while the SP7 strain was reported as unable to produce EPS on modified MRS agar plates containing different carbohydrates [64]. LexA is known to repress the transcription of several genes involved in the cellular response to DNA damage or inhibition of DNA replication [65,66], as well as its synthesis [67]. This regulation is known as the SOS response [66]. To repress transcription, LexA blocks the access of RNA polymerase to target promoters [68]. These findings were consistent with expression pattern analysis of the *dsr* gene in the C2-32 strain. In the presence of sucrose, *dsr* gene expression was strongly upregulated in the C43-11 strain, while the expression of the gene was not modulated in C2-32. This shared behavior could be related to the presence in both strains of this additional DNA-binding motif not present in the C43-11 and MG1 strains. To the best of our knowledge, this is the first study reporting a difference in regulatory elements of *dsr* genes in *W. cibaria*. Our findings will require further investigation to verify whether dextran production in the C2-32 strain could be restored by stressor factors such as cold stress, anaerobic growth conditions, and oxidative stress.

5. Conclusions

This study investigated the EPS production ability of *W. cibaria* strain C43-11 in comparison with the low producing strain C2-32, highlighting the induction of EPS production in the presence of a high sucrose level only by the C43-11 strain. NMR analysis demonstrated the dextran production consisting of a linear chain of glucose units bound predominantly by α-(1–6) glycosidic linkage and in smaller amounts (1–3%) by α-(1–2), α-(1–3), and α-(1–4) glycosidic linkages. Molecular analysis revealed a differential regulation of the *dsr* gene involved in dextran biosynthesis. The in silico analysis of the *dsr* promoter region highlighted the presence of different stress-responsive regulatory elements in the low producer C2-32 strain that could be responsible for its low production rate in the tested conditions.

Therefore, this study provides new insights into the investigation of the molecular bases of differences in EPS production ability, and indicates a new perspective of investigation for the identification of the regulatory mechanism of EPS production. Furthermore, dextran produced by the C43-11 strain may be further investigated for its possible functional and technological properties in the food sector.

Supplementary Materials: The following supporting information can be downloaded at: https://www.mdpi.com/article/10.3390//foods11182819/s1. Figure S1. Typical 1D 1H NOESY NMR spectra (D2O, 400 MHz, 303 K) of free mono- (arabinose, galactose, glucose, fructose, and mannose) and disaccharides (sucrose and maltose), and purified samples from C43-11 + S and C2-32 + S Figure S2. Typical 1D 1H NOESY NMR spectra (D2O, 400 MHz, 303 K) of free mono- (arabinose, galactose, glucose, fructose, and mannose) and disaccharides (sucrose and maltose), and purified samples from C43-11 + S and C2-32 + S.

Author Contributions: Conceptualization, P.D.B., M.F. and F.V.; Data curation, P.D.B. and M.F.; Formal analysis, P.D.B., M.F., A.R.B., B.M. and V.G.; Investigation, P.D.B., M.F., A.R.B., V.L. and B.M.; Supervision, P.D.B. and F.V.; Visualization, M.F., M.D.B., V.G. and F.V.; Writing—original draft, P.D.B., M.F., M.D.B., B.M. and V.G.; Writing—review & editing, P.D.B., M.F., G.M. and F.V. All authors have read and agreed to the published version of the manuscript.

Funding: The work was carried out with the financial support of Project PON ARS01_00783 ALIFUN—"Development of functional foods for the traditional Italian food products innovation".

Institutional Review Board Statement: Not applicable.

Data Availability Statement: The nucleotide sequences of the *dsr* gene and promoter region of C43-11 and C2-32 strains are available in the NCBI nucleotide sequence database under accession numbers ON791804; ON791805; ON791806; ON791807.

Acknowledgments: The authors thank Lisa Lonigro for her technical assistance, Simonetta Martena and Marinella Cavallo for their technical support in the management of the ALIFUN Project, and Sarah Jane Christopher for the English editing of the manuscript.

Conflicts of Interest: The authors declare that they have no conflicts of interest.

References

1. Corsetti, A.; Settanni, L. Lactobacilli in sourdough fermentation. *Food Res. Int.* **2007**, *40*, 539–558. [CrossRef]
2. Leroy, F.; De Vuyst, L. Lactic acid bacteria as functional starter cultures for the food fermentation industry. *Trends Food Sci. Tech.* **2004**, *15*, 67–78. [CrossRef]
3. Kleerebezem, M.; Kuipers, O.P.; Smid, E.J. Lactic acid bacteria—A continuing journey in science and application. *FEMS Microbiol. Rev.* **2017**, *41* (Suppl. 1), S1–S2. [CrossRef] [PubMed]
4. Gobbetti, M.; Rizzello, C.G.; Di Cagno, R.; De Angelis, M. How the sourdough may affect the functional features of leavened baked goods. *Food Microbiol.* **2014**, *37*, 30–40. [CrossRef]
5. Gobbetti, M.; De Angelis, M.; Di Cagno, R.; Calasso, M.; Archetti, G.; and Rizzello, C.G. Novel insights on the functional/nutritional features of the sourdough fermentation. *Int. J. Food Microbiol.* **2019**, *302*, 103–113. [CrossRef]
6. Di Biase, M.; Bavaro, A.R.; Lonigro, S.L.; Pontonio, E.; Conte, A.; Padalino, L.; Minisci, A.; Lavermicocca, P.; Valerio, F. *Lactobacillus plantarum* ITM21B fermentation product and chickpea flour enhance the nutritional profile of salt reduced bakery products. *Int. J. Food Sci. Nutr.* **2019**, *70*, 701–713. [CrossRef]
7. De Bellis, P.; Rizzello, C.G.; Sisto, A.; Valerio, F.; Lonigro, S.L.; Conte, A.; Lorusso, V.; Lavermicocca, P. Use of a selected *Leuconostoc citreum* strain as a starter for making a "yeast-free" bread. *Foods* **2019**, *8*, 70. [CrossRef]
8. De Bellis, P.; Montemurro, M.; D'Imperio, M.; Rizzello, C.G.; Sisto, A.; Lavermicocca, P. Production of a yeast-free focaccia with reduced salt content using a selected *Leuconostoc citreum* strain and seawater. *LWT* **2020**, *134*, 109918. [CrossRef]
9. Bavaro, A.R.; Di Biase, M.; Conte, A.; Lonigro, S.L.; Caputo, L.; Cedola, A.; Del Nobile, M.A.; Logrieco, A.F.; Lavermicocca, P.; Valerio, F. *Weissella cibaria* short-fermented liquid sourdoughs based on quinoa or amaranth flours as fat replacer in focaccia bread formulation. *Int. J. Food Sci. Technol.* **2021**, *56*, 3197–3208. [CrossRef]
10. Valerio, F.; Di Biase, M.; Lattanzio, V.M.; Lavermicocca, P. Improvement of the antifungal activity of lactic acid bacteria by addition to the growth medium of phenylpyruvic acid, a precursor of phenyllactic acid. *Int. J. Food Microbiol.* **2016**, *222*, 1–7. [CrossRef]
11. Pan, L.; Han, Y.; Zhou, Z. In vitro prebiotic activities of exopolysaccharide from *Leuconostoc pseudomesenteroides* XG5 and its effect on the gut microbiota of mice. *J. Funct. Food* **2020**, *67*, 103853. [CrossRef]
12. Nácher-Vázquez, M.; Ballesteros, N.; Canales, Á.; Saint-Jean, S.R.; Pérez-Prieto, S.I.; Prieto, A.; Aznar, R.; López, P. Dextrans produced by lactic acid bacteria exhibit antiviral and immunomodulatory activity against salmonid viruses. *Carbohyd. Polym.* **2015**, *124*, 292–301. [CrossRef] [PubMed]
13. Zarour, K.; Llamas, M.G.; Prieto, A.; Ruas-Madiedo, P.; Dueñas, M.T.; de Palencia, P.F.; Aznar, R.; Kihal, M.; Lopez, P. Rheology and bioactivity of high molecular weight dextrans synthesised by lactic acid bacteria. *Carbohyd. Polym.* **2017**, *174*, 646–657. [CrossRef]

14. Besrour-Aouam, N.; Fhoula, I.; Hernández-Alcántara, A.M.; Mohedano, M.L.; Najjari, A.; Prieto, A.; Ruas Madiedo, P.; López, P.; Ouzari, H.I. The role of dextran production in the metabolic context of *Leuconostoc* and *Weissella* Tunisian strains. *Carbohyd. Polym.* **2021**, *253*, 117254. [CrossRef]
15. Bounaix, M.S.; Robert, H.; Gabriel, V.; Morel, S.; Remaud-Simeon, M.; Gabriel, B.; Fontagne-Faucher, C. Characterization of dextran-producing *Weissella* strains isolated from sourdoughs and evidence of constitutive dextransucrase expression. *FEMS Microbiol. Lett.* **2010**, *311*, 18–26. [CrossRef]
16. Byrne, D. *Commission Decision of 30 January 2001 on Authorising the Placing on the Market of a Dextran Preparation Produce by Leuconostoc Mesenteroides as a Novel Food Ingredients in Bakery Products under Regulation (EC) No. 258/97 of the European Parliament and of the Council Official Journal European Commission L44*; European Commission: Brussels, Belgium, 2001.
17. Xu, Y.; Wang, Y.; Coda, R.; Säde, E.; Tuomainen, P.; Tenkanen, M.; Katina, K. In situ synthesis of exopolysaccharides by *Leuconostoc* spp. and *Weissella* spp. and their rheological impacts in fava bean flour. *Int. J. Food Microbiol.* **2017**, *248*, 63–71. [CrossRef]
18. Wang, Y.; Compaoré-Sérémé, D.; Sawadogo-Lingani, H.; Coda, R.; Katina, K.; Maina, N.H. Influence of dextran synthesized in situ on the rheological, technological and nutritional properties of whole grain pearl millet bread. *Food Chem.* **2019**, *285*, 221–230. [CrossRef] [PubMed]
19. Buksa, K.; Kowalczyk, M.; Boreczek, J. Extraction, purification and characterisation of exopolysaccharides produced by newly isolated lactic acid bacteria strains and the examination of their influence on resistant starch formation. *Food Chem.* **2021**, *362*, 130221. [CrossRef] [PubMed]
20. Tieking, M.; Gänzle, M.G. Exopolysaccharides from cereal-associated lactobacilli. *Trends Food Sci. Technol.* **2005**, *16*, 79–84. [CrossRef]
21. Galle, S.; Schwab, C.; Dal Bello, F.; Coffey, A.; Gänzle, M.G.; Arendt, E.K. Influence of in-situ synthesized exopolysaccharides on the quality of gluten-free sorghum sourdough bread. *Int. J. Food Microbiol.* **2012**, *155*, 105–112. [CrossRef]
22. Galle, S.; Arendt, E.K. Exopolysaccharides from sourdough lactic acid bacteria. *Cri. Rev. Food Sci.* **2014**, *54*, 891–901. [CrossRef] [PubMed]
23. Korcz, E.; Kerényi, Z.; Varga, L. Dietary fibers, prebiotics, and exopolysaccharides produced by lactic acid bacteria: Potential health benefits with special regard to cholesterol-lowering effects. *Food Funct.* **2018**, *9*, 3057–3068. [CrossRef] [PubMed]
24. Nwodo, U.U.; Green, E.; Okoh, A.I. Bacterial exopolysaccharides: Functionality and prospects. *Int. J. Mol. Sci.* **2012**, *13*, 14002–14015. [CrossRef] [PubMed]
25. Zhao, D.; Jiang, J.; Liu, L.; Wang, S.; Ping, W.; Ge, J. Characterization of exopolysaccharides produced by *Weissella confusa* XG-3 and their potential biotechnological applications. *Int. J. Biol. Macromol.* **2021**, *178*, 306–315. [CrossRef]
26. Kavitake, D.; Devi, P.B.; Shetty, P.H. Overview of exopolysaccharides produced by Weissella genus—A review. *Int. J. Biol. Macromol.* **2020**, *164*, 2964–2973. [CrossRef]
27. Besrour-Aouam, N.; Mohedano, M.L.; Fhoula, I.; Zarour, K.; Najjari, A.; Aznar, R.; Prieto, A.; Ouzari, H.-I.; López, P. Different modes of regulation of the expression of dextransucrase in *Leuconostoc lactis* AV1n and *Lactobacillus sakei* MN1. *Front. Microbiol.* **2019**, *10*, 959. [CrossRef]
28. Zeidan, A.A.; Poulsen, V.K.; Janzen, T.; Buldo, P.; Derkx, P.M.; Øregaard, G.; Neves, A.R. Polysaccharide production by lactic acid bacteria: From genes to industrial applications. *FEMS Microbiol. Rev.* **2017**, *41* (Suppl. 1), S168–S200. [CrossRef]
29. Coda, R.; Xu, Y.; Moreno, D.S.; Mojzita, D.; Nionelli, L.; Rizzello, C.G.; Katina, K. Performance of *Leuconostoc citreum* FDR241 during wheat flour sourdough type I propagation and transcriptional analysis of exopolysaccharides biosynthesis genes. *Food Microbiol.* **2018**, *76*, 164–172. [CrossRef]
30. Koirala, P.; Maina, N.H.; Nihtilä, H.; Katina, K.; Coda, R. Brewers' spent grain as substrate for dextran biosynthesis by *Leuconostoc pseudomesenteroides* DSM20193 and *Weissella confusa* A16. *Microb. Cell Fact.* **2021**, *20*, 1–13. [CrossRef]
31. Milanović, V.; Osimani, A.; Garofalo, C.; Belleggia, L.; Maoloni, A.; Cardinali, F.; Mozzon, M.; Foligni, R.; Aquilanti, L.; Clementi, F. Selection of cereal-sourced lactic acid bacteria as candidate starters for the baking industry. *PLoS ONE* **2020**, *15*, e0236190. [CrossRef]
32. Kumari, M.; Kumar, R.; Singh, D.; Bhatt, S.; Gupta, M. Physiological and genomic characterization of an exopolysaccharide-producing *Weissella cibaria* CH_2 from cheese of the western Himalayas. *Food Biosci.* **2020**, *35*, 100570. [CrossRef]
33. Lynch, K.M.; Lucid, A.; Arendt, E.K.; Sleator, R.D.; Lucey, B.; Coffey, A. Genomics of *Weissella cibaria* with an examination of its metabolic traits. *Microbiology* **2015**, *161*, 914–930. [CrossRef] [PubMed]
34. Werning, M.L.; Notararigo, S.; Nácher, M.; Fernández de Palencia, P.; Aznar, R.; López, P. Biosynthesis, purification and biotechnological use of exopolysaccharides produced by lactic acid bacteria. In *Food Additives*; El-Samragy, Y., Ed.; Intech: Rijeka, Croatia, 2012; pp. 83–114.
35. Yu, Y.J.; Chen, Z.; Chen, P.T.; Ng, I.S. Production, characterization and antibacterial activity of exopolysaccharide from a newly isolated *Weissella cibaria* under sucrose effect. *J. Biosci. Bioeng.* **2018**, *126*, 769–777. [CrossRef] [PubMed]
36. Hu, Y.; Gänzle, M.G. Effect of temperature on production of oligosaccharides and dextran by *Weissella cibaria* 10 M. *Int. J. Food Microbiol.* **2018**, *280*, 27–34. [CrossRef] [PubMed]
37. Valerio, F.; Favilla, M.; De Bellis, P.; Sisto, A.; de Candia, S.; Lavermicocca, P. Antifungal activity of strains of lactic acid bacteria isolated from a semolina ecosystem against *Penicillium roqueforti*, *Aspergillus niger* and *Endomyces fibuliger* contami-nating bakery products. *Syst. Appl. Microbiol.* **2009**, *32*, 438–448. [CrossRef]

38. Valerio, F.; Bavaro, A.R.; Di Biase, M.; Lonigro, S.L.; Logrieco, A.F.; Lavermicocca, P. Effect of amaranth and quinoa flours on exopolysaccharide production and protein profile of liquid sourdough fermented by *Weissella cibaria* and *Lactobacillus plantarum*. *Front. Microbiol.* **2020**, *11*, 967. [CrossRef]
39. Minervini, F.; De Angelis, M.; Surico, R.F.; Di Cagno, R.; Gänzle, M.; Gobbetti, M. Highly efficient synthesis of exopolysaccharides by *Lactobacillus curvatus* DPPMA10 during growth in hydrolyzed wheat flour agar. *Int. J. Food Microbiol.* **2010**, *141*, 130–135. [CrossRef]
40. Cerning, J.C.M.C.; Renard, C.M.G.C.; Thibault, J.F.; Bouillanne, C.; Landon, M.; Desmazeaud, M.; Topisirovic, L. Carbon source requirements for exopolysaccharide production by *Lactobacillus casei* CG11 and partial structure analysis of the polymer. *Appl. Environ. Microb.* **1994**, *60*, 3914–3919. [CrossRef]
41. Dubois, M.; Gilles, K.A.; Hamilton, J.K.; Rebers, P.T.; Smith, F. Colorimetric method for determination of sugars and related substances. *Anal. Chem.* **1956**, *28*, 350–356. [CrossRef]
42. Rühmkorf, C.; Jungkunz, S.; Wagner, M.; Vogel, R.F. Optimization of homoexopolysaccharide formation by lactobacilli in gluten-free sourdoughs. *Food Microbiol.* **2012**, *32*, 286–294. [CrossRef]
43. Garbetta, A.; D'Antuono, I.; Melilli, M.G.; Sillitti, C.; Linsalata, V.; Scandurra, S.; Cardinali, A. Inulin enriched durum wheat spaghetti: Effect of polymerization degree on technological and nutritional characteristics. *J. Funct. Foods* **2020**, *71*, 104004. [CrossRef]
44. Galle, S.; Schwab, C.; Arendt, E.; Gänzle, M. Exopolysaccharide-forming *Weissella* strains as starter cultures for sorghum and wheat sourdoughs. *J. Agric. Food Chem.* **2010**, *58*, 5834–5841. [CrossRef] [PubMed]
45. Solovyev, V.; Salamov, A. Automatic Annotation of Microbial Genomes and Metagenomic Sequences. In *Metagenomics and Its Applications in Agriculture, Biomedicine and Environmental Studies*; Li, R.W., Ed.; Nova Science Publishers: Hauppauge, NY, USA, 2011; pp. 61–78.
46. Wang, L.; Stegemann, J.P. Extraction of high quality RNA from polysaccharide matrices using cetlytrimethylammonium bromide. *Biomaterials* **2010**, *31*, 1612–1618. [CrossRef] [PubMed]
47. Livak, K.J.; Schmittgen, T.D. Analysis of relative gene expression data using real-time quantitative PCR and the 2− ΔΔCT method. *Methods* **2001**, *25*, 402–408. [CrossRef] [PubMed]
48. Ogden, T.L. Handling results below the level of detection. *Ann. Occup. Hyg.* **2010**, *54*, 255–256.
49. Keeler, J. *Understanding NMR Spectroscopy*; John Wiley & Sons: Hoboken, NJ, USA, 2011; ISBN 978-0-470-74608-0.
50. Maina, N.H.; Tenkanen, M.; Maaheimo, H.; Juvonen, R.; Virkki, L. NMR spectroscopic analysis of exopolysaccharides produced by *Leuconostoc citreum* and *Weissella Confusa*. *Carbohyd. Res.* **2008**, *343*, 1446–1455. [CrossRef]
51. Verni, M.; Wang, C.; Montemurro, M.; De Angelis, M.; Katina, K.; Rizzello, C.G.; Coda, R. Exploring the microbiota of faba bean: Functional characterization of lactic acid bacteria. *Front. Microbiol.* **2017**, *8*, 2461. [CrossRef]
52. Sáez, G.D.; Sabater, C.; Fara, A.; Zárate, G. Fermentation of chickpea flour with selected lactic acid bacteria for improving its nutritional and functional properties. *J. Appl. Microbiol.* **2021**, *133*, 181–199. [CrossRef]
53. Baruah, R.; Maina, N.H.; Katina, K.; Juvonen, R.; Goyal, A. Functional food applications of dextran from *Weissella cibaria* RBA12 from pummelo (*Citrus maxima*). *Int. J. Food Microbiol.* **2017**, *242*, 124–131. [CrossRef]
54. Herman, R.H. Mannose metabolism. II. *Am. J. Clin. Nutr.* **1971**, *24*, 556–561. [CrossRef]
55. Bounaix, M.S.; Gabriel, V.; Morel, S.; Robert, H.; Rabier, P.; Remaud-Simeon, M.; Gabriel, B.; Fontagne-Faucher, C. Biodiversity of exopolysaccharides produced from sucrose by sourdough lactic acid bacteria. *J. Agr. Food Chem.* **2009**, *57*, 10889–10897. [CrossRef] [PubMed]
56. Siddiqui, N.N.; Aman, A.; Silipo, A.; Qader, S.A.U.; Molinaro, A. Structural analysis and characterization of dextran produced by wild and mutant strains of *Leuconostoc mesenteroides*. *Carbohyd. Polym.* **2014**, *99*, 331–338. [CrossRef] [PubMed]
57. Ahmed, R.Z.; Siddiqui, K.; Arman, M.; Ahmed, N. Characterization of high molecular weight dextran produced by *Weissella cibaria* CMGDEX3. *Carbohyd. Polym.* **2012**, *90*, 441–446. [CrossRef] [PubMed]
58. Dobruchowska, J.M.; Gerwig, G.J.; Kralj, S.; Grijpstra, P.; Leemhuis, H.; Dijkhuizen, L.; Kamerling, J.P. Structural characterization of linear isomalto-/malto-oligomer products synthesized by the novel GTFB 4,6-α-glucanotransferase enzyme from *Lactobacillus reuteri* 121. *Glycobiology* **2012**, *22*, 517–528. [CrossRef]
59. Quirasco, M.; Lopez-Munguia, A.; Remaud-Simeon, M.; Monsan, P.; Farres, A. Induction and transcription studies of the dextransucrase gene in *Leuconostoc mesenteroides* NRRL B-512F. *Appl. Environ. Microbiol.* **1999**, *65*, 5504–5509. [CrossRef]
60. Arskold, E.; Svensson, M.; Grage, H.; Roos, S.; Radstrom, P.; van Niel, E.W. Environmental influences on exopolysaccharide formation in *Lactobacillus reuteri* ATCC 55730. *Int. J. Food Microbiol.* **2007**, *116*, 159–167. [CrossRef]
61. Monsan, P.; Bozonnet, S.; Albenne, C.; Joucla, G.; Willemot, R.M.; Remaud-Siméon, M. Homopolysaccharides from lactic acid bacteria. *Int. Dairy J.* **2001**, *11*, 675–685. [CrossRef]
62. Wolter, A.; Hager, A.S.; Zannini, E.; Galle, S.; Gänzle, M.G.; Waters, D.M.; Arendt, E.K. Evaluation of exopolysaccharide producing *Weissella cibaria* MG1 strain for the production of sourdough from various flours. *Food Microbiol.* **2014**, *37*, 44–50. [CrossRef]

63. Zannini, E.; Mauch, A.; Galle, S.; Gänzle, M.; Coffey, A.; Arendt, E.K.; Taylor, J.P.; Waters, D.M. Barley malt wort fermentation by exopolysaccharide-forming *Weissella cibaria* MG1 for the production of a novel beverage. *J. Appl. Microbiol.* **2013**, *115*, 1379–1387. [CrossRef]
64. Patrone, V.; Al-Surrayai, T.; Romaniello, F.; Fontana, A.; Milani, G.; Sagheddu, V.; Puglisi, E.; Callegari, M.L.; Al-Mansour, H.; Mohamed Waheed Kishk, M.W.; et al. Integrated Phenotypic-Genotypic Analysis of Candidate Probiotic *Weissella Cibaria* Strains Isolated from Dairy Cows in Kuwait. *Probiotics Antimicro.* **2021**, *13*, 809–823. [CrossRef]
65. Fernandez De Henestrosa, A.R.; Ogi, T.; Aoyagi, S.; Chafin, D.; Hayes, J.J.; Ohmori, H.; Woodgate, R. Identification of additional genes belonging to the LexA regulon in *Escherichia coli*. *Mol. Microbiol.* **2000**, *35*, 1560–1572. [CrossRef] [PubMed]
66. D'Ari, R. The SOS system. *Biochimie* **1985**, *67*, 343–347. [CrossRef]
67. Brent, R.; Ptashne, M. The lexA gene product represses its own promoter. *Proc. Natl. Acad. Sci. USA* **1980**, *77*, 1932–1936. [CrossRef]
68. Little, J.W.; Mount, D.W.; Yanisch-Perron, C.R. Purified lexA protein is a repressor of the recA and lexA genes. *Proc. Natl. Acad. Sci. USA* **1981**, *78*, 4199–4203. [CrossRef] [PubMed]

foods

MDPI

Article

Improvement of Sourdough and Bread Qualities by Fermented Water of Asian Pears and Assam Tea Leaves with Co-Cultures of *Lactiplantibacillus plantarum* and *Saccharomyces cerevisiae*

Ramita Supasil, Uthaiwan Suttisansanee, Chalat Santivarangkna, Nattapol Tangsuphoom, Chanakan Khemthong, Chaowanee Chupeerach and Nattira On-nom *

Food and Nutrition Academic and Research Cluster, Institute of Nutrition, Mahidol University, Salaya, Phuttamonthon, Nakhon Pathom 73170, Thailand; ramita.sup1@gmail.com (R.S.); uthaiwan.sut@mahidol.ac.th (U.S.); chalat.san@mahidol.ac.th (C.S.); nattapol.tng@mahidol.ac.th (N.T.); chanakan.khe@mahidol.ac.th (C.K.); chaowanee.chu@mahidol.ac.th (C.C.)

* Correspondence: nattira.onn@mahidol.ac.th; Tel.: +66-(0)-2800-2380 (ext. 426)

Abstract: Qualities of sourdough and sourdough bread using fermented water from Asian pears and Assam tea leaves with *Lactiplantibacillus plantarum* 299v and *Saccharomyces cerevisiae* TISTR 5059 as starter cultures were evaluated. Changes in the growth of lactic acid bacteria and yeast, pH, sourdough height, total phenolic contents (TPCs) and antioxidant activities detected by ORAC, FRAP and DPPH radical scavenging assays were monitored during sourdough production. Mature sourdough was achieved within 4 h after 18 h retard fermentation and used for bread production. The bread was then analyzed to determine chemical and physical properties, nutritional compositions, TPCs, antioxidant activities and sensory properties as well as shelf-life stability. Results showed that fermented water significantly promoted the growth of yeast and increased TPCs and antioxidant activities of sourdough. Compared to common sourdough bread, fermented water sourdough bread resulted in 10% lower sugar and 12% higher dietary fiber with improved consumer acceptability; TPCs and antioxidant activities also increased by 2–3 times. The fermented water sourdough bread maintained microbial quality within the standard range, with adequate TPCs after storage at room temperature for 7 days. Fermented water from Asian pears and Assam tea leaves with *L. plantarum* 299v and *S. cerevisiae* TISTR 5059 as starter cultures improved dough fermentation and bread quality.

Keywords: *Pyrus pyrifolia*; *Camellia sinensis* var. *assamica*; fermentation; nutritive composition; physical properties; chemical properties; antioxidant activities; sensory properties; shelf-life

Citation: Supasil, R.; Suttisansanee, U.; Santivarangkna, C.; Tangsuphoom, N.; Khemthong, C.; Chupeerach, C.; On-nom, N. Improvement of Sourdough and Bread Qualities by Fermented Water of Asian Pears and Assam Tea Leaves with Co-Cultures of *Lactiplantibacillus plantarum* and *Saccharomyces cerevisiae*. *Foods* **2022**, *11*, 2071. https://doi.org/10.3390/foods11142071

Academic Editors: Carlo Giuseppe Rizzello and Palmira De Bellis

Received: 19 June 2022
Accepted: 11 July 2022
Published: 12 July 2022

Publisher's Note: MDPI stays neutral with regard to jurisdictional claims in published maps and institutional affiliations.

1. Introduction

Nowadays, consumers prefer to eat healthy food [1]. Sourdough bread offers health benefits with enhanced absorbable nutrients and vitamins [2,3] and reduced glycemic index value compared to common white bread [4]. Cereal flour and water as the main ingredients of traditional sourdough are spontaneously fermented, enhancing numbers of wild yeast and lactic acid bacteria [5]. The combination of culture in the fermentation provides complex growth patterns that can improve the functional and organoleptic properties of food [6]. *Saccharomyces cerevisiae* is a robust yeast that has a high fermentation efficiency [7]. It acts as a leavening agent in bread making [5]. *Lactiplantibacillus plantarum* has been recognized for its probiotic characteristics. It has been used widely in food fermentation processes of different types of foods [8]. In bread making, lactic acid bacteria contribute mainly to acidification and the production of flavor and other metabolic compounds of bread [5]. The study of Hu et al., (2022) reported that the bread fermented by a combination strains of *S. cerevisiae* and *L. plantarum* had 15.2% higher specific volume, softer bread crumb, more vigorous taste than the bread fermented by *S. cerevisiae* only [9]. In addition, functional ingredients may also be introduced in the early fermentation steps to further improve the functional properties of the final product [10]. The several studies showed that sourdough was developed using pacific white shrimp (*Litopenaeus vannamei*) protein hydrolysates and (−)-epigallocatechin gallate [11], pear and naval orange [12]. Sources of these functional ingredients as nutrients or bioactive compound-rich fruits and vegetables support bacterial growth and provide particular taste/texture to increase the health benefits of sourdough bread.

The abundant nutrients in Asian pears and Assam tea leaves can be used to support the growth of microorganisms [13,14]. Asian pears (*Pyrus pyrifolia*) are juicy fruits with favorable aroma and sweetness that enhance the sensory acceptability of developed products [15]. Pears are an excellent source of dietary fiber and a good source of ascorbic acid [16]. Moreover, it contains a high number of bioactive compounds including total phenolic compounds and total flavonoid compounds which significantly correlated with antioxidant activities [17]. Assam tea (*Camellia sinensis* var. *assamica*) leaves also possess a unique aroma and taste. It contains a variety of biologically active compounds, such as phenolics, minerals, vitamins and dietary fiber, and have several health benefits including antimicrobial and antioxidant activities [13,18,19]. As far as we know, a study of fermented water from Asian pears and Assam tea leaves using *Lactiplantibacillus plantarum* 299v and *Saccharomyces cerevisiae* TISTR 5059 as starter on sourdough fermentation and bread quality has not been reported. Recent investigations by our group indicated that fermented water from Asian pears and Assam tea leaves using *Lactiplantibacillus plantarum* 299v and *Saccharomyces cerevisiae* TISTR 5059 as starter cultures significantly promoted the growth of yeast and lactic acid bacteria, optimized acidity and improved total phenolic contents (TPCs) and antioxidant activities [20]. Previous findings suggested the potential of fermented water from Asian pears and Assam tea leaves as a functional ingredient to prepare sourdough and improve bread quality.

Hence, the aim of this work was to develop sourdough and bread with improved qualities using fermented water from Asian pears and Assam tea leaves using *L. plantarum* 299v and *S. cerevisiae* TISTR 5059 as starter cultures. Changes in microbiological quantification, chemical properties, TPCs and antioxidant activities were monitored during sourdough fermentation. Mature sourdough was used for breadmaking. Chemical and physical properties, nutritional compositions, TPCs, antioxidant activities and sensory evaluation of sourdough bread were investigated. Shelf-life stability was also determined.

2. Materials and Methods

2.1. Preparation of Fermented Water

Asian pears were purchased from Salaya Market, Phutthamonthon District, Nakhon Pathom Province, Thailand and steamed Assam tea leaves were received from Wiang Pa Pao District, Chiang Rai Province, Thailand in December 2020. For quality control of raw material, nutritional analysis was performed according to the protocols indicated in Section 2.7. The chemical composition of Asian pears was 84.73% moisture, 0.45% protein, 0.11% total fat and 14.11% total carbohydrates (comprising 5.18% glucose, 5.11% fructose, 0.75% sucrose and 2.80% dietary fiber), while steamed Assam tea leaves were composed of 77.32% moisture, 5.32% protein, 0.48% total fat and 15.44% total carbohydrates (comprising 0.75% sucrose and 8.03% dietary fiber). Fermented water was prepared following the method of Supasil et al., (2021) [20]. Briefly, Asian pears (50 g) and steamed Assam tea leaves (50 g) with suspensions of *L. plantarum* 299v (6 mL, Bio-Life Sciences Corp., Mississauga, Canada) and *S. cerevisiae* TISTR 5059 (6 mL, TISTR, Bangkok, Thailand) were mixed with distilled water (488 mL) and fermented in a sterile glass jar covered with a lid at 30 °C for 24 h. Then, solid remnants of Asian pears and Assam tea leaves were filtered using a sterile colander. The solution part of the fermented water was used in fermented water sourdough and fermented water sourdough bread development. In addition, a suspension of *L. plantarum* 299v and *S. cerevisiae* TISTR 5059 in sterile deionized water was prepared as control water and used in common sourdough and common sourdough bread development. Initial viable cell count of *L. plantarum* at 8.56 ± 0.16 Log CFU/mL and *S. cerevisiae* at 7.23 ± 0.05 Log CFU/mL in the suspension of selective cultures were maintained in both control and fermented water ($p \geq 0.05$).

2.2. Preparation of Sourdough

Fermented water sourdough was prepared by mixing unbleached wheat flour (200 g, Bread Flour, Thai Flour Mill Industry Co., Ltd., Samut Prakan, Thailand) with fermented water from Section 2.1 (200 g) in a sterile glass jar (5.2 inches diameter and 6.5 inches height) covered by a cheesecloth. The mixture was fermented at 30 °C for 1 h, followed by retard fermentation in a refrigerator (7 ± 2 °C) for 18 h. The final fermentation was performed at 30 °C to produce mature sourdough with the following characteristics: pH 3.5–4.3, lactic acid bacteria at least 8 Log Colony Forming Unit (CFU)/g and yeast at least 6 Log CFU/g according to Hammes et al., (2005) [21] and Mohd Roby et al., (2020) [22]. Development of bubbles, sweet smelling odor and double volume was observed. Common sourdough was prepared using a similar procedure to the fermented water sourdough but fermented water was replaced by control water from Section 2.1. The sourdough samples were observed for their height, collected for evaluation of microbiological and chemical properties at initiation and then hourly after retard fermentation until gaining the mature sourdough. Moreover, TPCs and antioxidant activities at the start and end of the fermentation were also analyzed.

2.3. Preparation of Bread

The fermented water sourdough bread and common sourdough bread were prepared using the mature fermented water sourdough and mature common sourdough from Section 2.2 as a leavening agent. The recipe of sourdough bread, modified from Mohd Roby et al., (2020) [22], consisted of unbleached wheat flour (100%, Bread Flour, Thai Flour Mill Industry Co., Ltd., Samut Prakan, Thailand), distilled water (44%), sugar (15%, Lin Caster Sugar, Thai Roong Ruang Sugar Group, Bangkok, Thailand), salt (2.5%, Prung Thip Iodized Table Salt, Thai Refined Salt Co., Ltd., Bangkok, Thailand), butter (5%, Allowrie Pure Creamy Unsalted Butter, KCG Corporation Co., Ltd., Bangkok, Thailand) and mature sourdough (44%). All ingredients, with the exception of butter, were mixed and kneaded for 5 min by a dough kneading machine (model Bear HMJ-A50E2, Bear Electric Appliance Co., Ltd., Shunde, Fosahn, China). The butter was then added to the mixture, kneaded for another 25 min, and allowed to proof for 4 h at 30 °C. The dough was then portioned (approximated 175 g), rested at room temperature (28 ± 2 °C) and 75 ± 5% relative humidity (RH) for 20 min, rolled for degassing and shaped into a roll. Three rolls were placed into a loaf pan (8.5 × 14.5 × 8.5 cm) before overnight leavening in a refrigerator (7 ± 2 °C) and final proofing at 30 °C for 6 h. The doughs were then baked at 180 °C for 30 min using an electric oven (model Tecno+, The Signature Brand Co., Ltd., Bangkok, Thailand) and cooled at room temperature (28 ± 2 °C and 75 ± 5% RH) for 1 h before packing in polypropylene plastic bags (Aro Commercial Co., Ltd., Bangkok, Thailand) with a proper heat seal for further analysis. The bread samples were analyzed for chemical and physical properties, nutritional compositions, TPCs, antioxidant activities and sensory evaluation.

2.4. Determination of Microbiological Quality

Sourdough samples (10 g fresh weight (FW)) were suspended in 90 mL of 0.85% (*w/v*) sodium chloride solution and enumerated for *L. plantarum* and *S. cerevisiae* by spread plate technique [23]. *L. plantarum* grew on De Man Rogosa Sharpe (MRS) (Difco™ & BBL™, BD Diagnostic, Sparks, MD, USA) agar (Hardy Diagnostics, Santa Maria, CA, USA) supplemented with 4 mg/L ciprofloxacin (Siam Pharmaceutical Co., Ltd., Bangkok, Thailand), incubated at 37 °C for 48 h [24]. *S. cerevisiae* grew on Yeast Peptone Dextrose (YPD) agar supplemented with 100 mg/L chloramphenicol (T.P. Drug Laboratories (1969) Co., Ltd., Bangkok, Thailand), incubated at 30 °C for 48 h [25]. Results were calculated and reported as Log CFU/g.

Microbiological analysis of the bread samples was performed according to the standard protocols of the Bacteriological Analytical Manual (BAM) [26]. Bread samples were blended using a blender (model HR2115, Philips (Thailand) Ltd., Bangkok, Thailand). Ten grams of blended sample were suspended in 90 mL of 0.1% (*v/v*) peptone water, plated on Dichloran-Glycerol 18 (DG18) agar and incubated at 25 °C for 5 days in the dark for yeast and mold evaluation. Total aerobic bacteria were evaluated following a spread plate technique [23] and BAM [26]. Briefly, the blended samples were suspended in 0.85% (*w/v*) sodium chloride solution, poured onto plate count agar and incubated at 37 °C for 48 h. Results were calculated and reported as Log CFU/g.

2.5. Determination of Chemical and Physical Quality

For pH measurement, the sourdough or blended bread samples (10 g) were mixed with 90 mL of 0.85% (*w/v*) sodium chloride solution and measured for pH using a calibrated pH meter (Ohaus Corporation, Morris Country, NJ, USA) [25]. Specific volume was calculated using loaf volume divided by loaf weight. Loaf weight was measured by an electronic weighing balance (Mettler Toledo, Toronto, Canada), while loaf volume was determined using the black sesame seeds replacement method [11,27]. The blended bread samples were measured for water activity by a water activity measurement instrument (model ms1-1 M, Novasina, Lachen, Switzerland). Texture of the bread samples including hardness, chewiness, springiness and cohesiveness was measured as described by Mohd Roby et al., (2020) [22] and the American Association of Cereal Chemists (AACC) (2001) [28] with some modifications as follows. Three slices of bread including the middle slice and one on either side were used for texture profile analysis (TPA) using a Texture Analyzer (TA.XT.plus®, Stable Micro System, Surrey, UK) equipped with an aluminum 36-mm cylindrical probe with the following parameters: 1.0 mm/s pre-test speed and 0.5 mm/s test speed, 10.0 mm/s post-test speed, 5 mm distance, 5.0 s time and 5.0 g trigger force.

2.6. Determination of Total Phenolic Contents and Antioxidant Activities

The sourdough samples were collected to evaluate TPCs and antioxidant activities at the start and end of the fermentation, while the bread samples were collected every day for 7 days. The collected samples were freeze-dried using a −50 °C and 0.086 mbar freeze dryer (model Lyovac GT2, GEA® Lyophil GmbH, Nordrhein-Westfalen, Germany) for 72 h and ground using a grinder (model MR-1268, MARA, Nonthaburi, Thailand) into fine powder. Extraction of all samples followed the method of Sripum et al., (2017) [29] with slight modifications as follows. The powdered sample (1 g) was dissolved in 40% (*v/v*) aqueous ethanol (20 mL) and incubated at 50 °C using a WNE45 water bath shaker (Memmert GmBh, Eagle, WI, USA) for 2 h. The supernatant was collected by centrifugation at 3800× *g* using a Hettich® ROTINA 38R centrifuge (Andreas Hettich GmbH, Tuttlingen, Germany) for 10 min and filtered through a 0.45 μM polyether sulfone membrane syringe filter. The filtrate was kept at −20 °C until analysis.

TPCs and antioxidant activities were assessed following Sripum et al., (2017) [29] with no modifications. Briefly, Folin-Ciocalteu phenol was used as the reagent, while gallic acid (0 to 200 μg/mL) was used as a standard for the determination of TPCs. A Synergy™ HT 96-well UV/visible microplate reader with Gen 5 data analysis software (BioTek Instruments, Inc., Winooski, VT, USA) was used to detect TPCs at 765 nm, and the results were reported as mg gallic acid equivalent (GAE)/g dry weight (DW). Antioxidant activities were analyzed for oxygen radical absorbance capacity (ORAC), ferric ion reducing antioxidant power (FRAP) and 2,2-diphenyl-1-picrylhydrazyl (DPPH) radical scavenging assays. Fluorescein, FRAP reagent and DPPH in 95% (*v/v*) aqueous ethanol were employed as reagents for the ORAC, FRAP and DPPH radical scavenging assays, respectively. Antioxidant activities were monitored at an excitation wavelength (λ_{ex}) of 485 nm and an emission wavelength (λ_{em}) of 528 nm for ORAC assay, 600 nm for FRAP assay and 520 nm for DPPH radical scavenging assay. Trolox was used as a standard in all antioxidant assays and results were reported as μmol Trolox equivalent (TE)/g DW. All chemicals and reagents were sourced from Sigma-Aldrich (St. Louis, MO, USA).

2.7. Determination of Nutritional Quality

Determination of nutritional compositions (moisture, fat, protein, carbohydrate, energy, dietary fiber, sugar and ash) was conducted at the Institute of Nutrition, Mahidol University (Nakhon Pathom, Thailand) using the international standard for laboratory quality systems with ISO/IEC 17025:2005 and the standard protocols of the Association of Official Analytical Chemists (AOAC) [30]. Moisture content was evaluated by drying the fresh samples in a hot-air oven (Memmert model UNE 500, Eagle, WI, USA) at 100 °C until constant weight (AOAC 930.04, 934.01). Total fat content was determined by acidic digestion and extracted with petroleum ether using a Soxtec System (Tecator model 1043, Hoganas, Sweden) (AOAC 948.15, 945.16). Protein content was analyzed by the Kjeldahl method utilizing digestion and distillation units (Buchi model K-435 and B-324, Flawil, Switzerland, respectively), and then calculated using a conversion factor of 6.25 (AOAC 992.23). Ash content was analyzed by incineration in a muffle furnace (Carbolite model CWF 1100, Hope, UK) at 550 °C (AOAC 930.30, 945.46). Total carbohydrate was calculated by the subtraction of moisture, fat, protein and ash contents from 100. Energy value was attained from the integration of total energy from carbohydrate, protein and fat as 4, 4 and 9 kcal/g samples, respectively. Total dietary fiber was evaluated by the enzyme gravimetric method (AOAC 991.43). Total sugar was determined using a protocol previously reported by Wannasaksri et al., (2021) [31] as a liquid chromatographic method utilizing ultra-fast liquid chromatography (UFLC from Shimadzu Corporation, Kyoto, Japan) with a detector (Alltech 800 evaporative light scattering detector from BÜCHI Corporation, New Castle, DE, USA) and column (5 μm, 250 mm × 4.6 mm Shodex Asahi Pak NH2P-504E from Shodex Group, Kanagawa, Japan).

2.8. Sensory Evaluation

Fermented water sourdough bread and common sourdough bread were evaluated for sensory properties by 10 trained panelists (4 male and 6 females, range: 18 to 60 years old, nonsmokers), who were familiar with sourdough bread. A 9-point hedonic scale rating 1 for dislike extremely, 5 for neither like nor dislike and 9 for like extremely was utilized to evaluate consumer attributes including appearance, color, taste, aroma, texture, sourness and overall acceptability [32]. The panelists were allowed to drink water for mouth cleansing between sample testing.

2.9. Shelf-Life Stability

The bread samples were packed in clear polypropylene plastic bags with a proper heat seal and stored at room temperature (28 ± 2 °C and 75 ± 5% RH). Samples were randomly selected to evaluate microbiological, TPCs and antioxidant activities at 0, 1, 3, 5 and 7 days. For microbiological properties, total aerobic bacteria, as well as yeast and mold, were enumerated by the pour plate technique, as described in Section 2.4, while TPCs and antioxidant activities were analyzed using the protocols listed in Section 2.6.

2.10. Statistical Analysis

All experiments were performed in triplicate using three independent batches of sourdough and bread, with data presented as mean ± standard deviation (SD) of each quality value and subjected to univariate data analysis using IBM SPSS Statistics for Windows version 26.0, IBM Corp., Armonk, New York, USA software. Mean differences of $p < 0.05$ were determined by one-way analysis of variance (ANOVA), followed by Duncan's multiple comparison test for more than two data sets or Student's unpaired t-test for two data sets.

3. Results and Discussion

3.1. *Effect of Fermented Water and Fermentation Time on Sourdough Quality*

3.1.1. Growth of *L. plantarum* and *S. cerevisiae*

The initial viable cell count of *L. plantarum* and *S. cerevisiae* in the fermented water sourdough was 7.27 ± 0.10 and 5.72 ± 0.04 Log CFU/g, respectively, which were not significantly different ($p \geq 0.05$) from the common sourdough (Figure 1 and Supplementary Table S1). It seemed that the viable cell counts of *L. plantarum* and *S. cerevisiae* increased rapidly in the fermented water sourdough than the common sourdough in the beginning stage (1 h) of fermentation. This could be due to the available nutrients in the fermented water or unbleached wheat flour as carbon or nitrogen sources that enhanced bacterial growth [20]. In the later stage (2–3 h) of fermentation, less rapid growth was observed. This might be because of pH which is an imporatant controlling factor for the survival and growth of the microorganisms [33]. The growth rate of *L. plantarum* and *S. cerevisiae* increased again in the fourth h of fermentation and reached a maximum number of viable cell count. Both common sourdough and fermented water sourdough achieved the criteria of mature sourdough pH between 3.5 to 4.3, lactic acid bacteria at least 8 Log CFU/g and yeast at least 6 Log CFU/g [21,22] within 4 h after 18 h retard fermentation. Viable cell count of *L. plantarum* reached a maximum of 7.94 ± 0.27 Log CFU/g in common sourdough and 8.14 ± 0.07 Log CFU/g in fermented water sourdough. A similar growth trend with lower viable cell count of *S. cerevisiae* was observed. Growth rate gradually increased and reached a maximum at 6.00 ± 0.04 Log CFU/g in common sourdough and 6.36 ± 0.03 Log CFU/g in fermented water sourdough. Similar results were reported by Minervini et al., (2016). They found that dough containing macerated pears exhibited 8 Log CFU/g of lactic acid bacteria and 6 Log CFU/g of yeast after 8 h fermentation at 30 °C [34], while Unban et al., (2019) reported that lactic acid bacteria detected in fermented Assam tea leaves (Cha-miang) ranged of 6–8 Log CFU/g [35], with yeast and mold ranging 6–10 Log CFU/g [36]. During fermentation, the fermented water sourdough exhibited significantly higher viable cell count of *L. plantarum* and *S. cerevisiae* than common sourdough. Catechin, a predominant polyphenol in Assam tea leaves, promoted the growth of *L. plantarum* [37]. López de Felipe et al., (2010) explained that absorption of catechin through the lactic acid bacterial membrane altered the function of proteins associated with glucose transport, resulting in increased glucose consumption and higher growth of *L. plantarum* [38]. Nutrients in the fermented water including glucose and fructose enhanced *S. cerevisiae* growth [39]. Another supportive study indicated that *L. plantarum* and *S. cerevisiae* stimulated growth of each other only in the presence of fructose, glucose and lactose as carbon sources but not with galactose, maltose, sucrose and starch [34].

3.1.2. pH and Height

The initial pH of fermented water sourdough was 5.51 ± 0.02, and significantly ($p < 0.05$) lower than common sourdough (5.98 ± 0.01) (Figure 2A and Supplementary Table S2). Throughout fermentation, the pH of both common sourdough and fermented water sourdough was significantly ($p < 0.05$) reduced to 4.45 ± 0.02 and 4.14 ± 0.03, respectively, within 4 h due to acid production of *L. plantarum* and *S. cerevisiae*. This result was supported by Jin et al., (2019), who observed that lactic acid was the main organic acid produced (up to 6.12 g/L) in a mixed 24 h fermentation of mango slurry using *L. plantarum* and *S. cerevisiae* DV10 [40]. Similar results were reported by Duan et al., (2011). They revealed that the increased growth rate of *L. plantarum* was influenced by shrimp waste peptides that elevated the rate of lactic acid production, resulting in lower pH [41], while persistent acids from fermented water including lactic acid, butyric acid, caffeic acid and sinapic acid were mainly found in fermented Assam tea leaves [42], with chlorogenic acid, ascorbic acid and sinapic acid found in Asian pears [16,43].

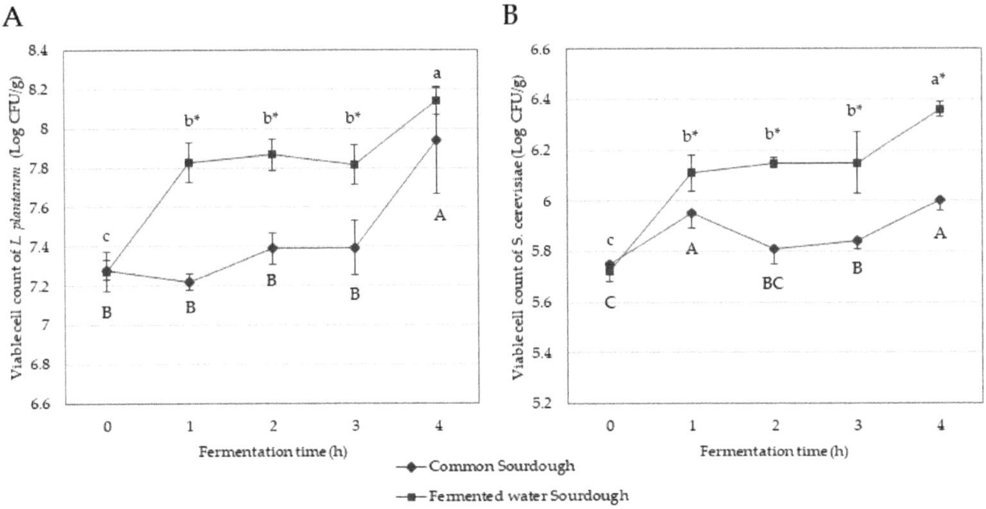

Figure 1. Growth of (**A**) *L. plantarum* and (**B**) *S. cerevisiae* in common sourdough and fermented water sourdough throughout fermentation. Results were expressed as mean values of triplicate determinations ($n = 3$). Different uppercase and lowercase letters denote significantly different viable cell count of microorganisms in common sourdough and fermented water sourdough, respectively, at different fermentation times using one-way ANOVA, followed by Duncan's multiple comparison test at $p < 0.05$, while * denotes significantly different viable cell count of microorganisms between common sourdough and fermented water sourdough detected at the same fermentation time using Student's unpaired *t*-test at $p < 0.05$.

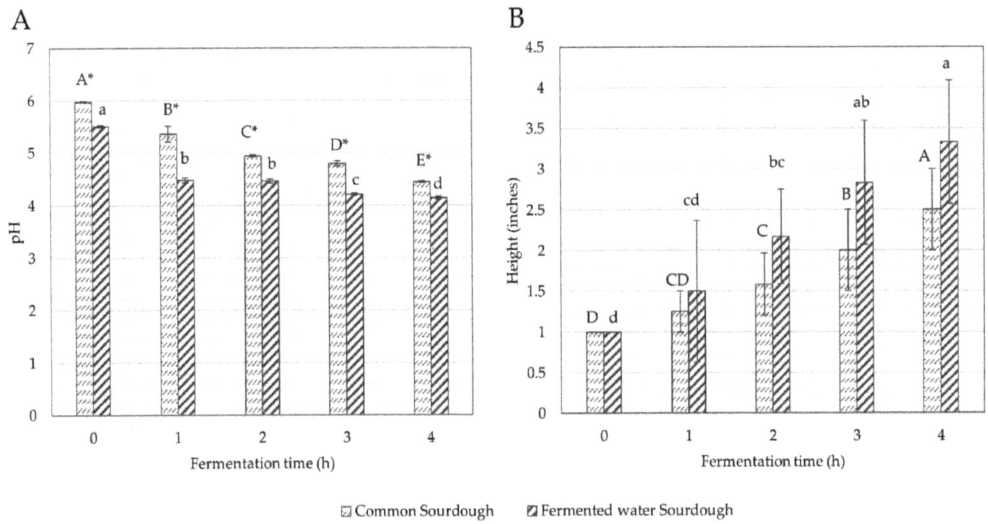

Figure 2. A trend in (**A**) pH and (**B**) height of common sourdough and fermented water sourdough over fermentation time periods. Results were expressed as mean values of triplicate determinations ($n = 3$). Different uppercase and lowercase letters denote significantly different values (pH or height) at $p < 0.05$ in common sourdough and fermented water sourdough, respectively, fermented at different time periods using one-way ANOVA, followed by Duncan's multiple comparison test, while * denotes significantly different values (pH or height) at $p < 0.05$ between common sourdough and fermented water sourdough fermented at the same time period using Student's unpaired *t*-test.

Likewise, initial heights of common sourdough and fermented water sourdough were insignificantly different (Figure 2B). Similar to microorganism growth, sourdough height increased during fermentation, reaching a maximum within 4 h after the 18 h retard fermentation (2.50 ± 0.50 inches in common sourdough and 3.33 ± 0.76 inches in fermented water sourdough). Bubbles and a 2–3 times increment in volume were observed in mature sourdough. The gas produced by *L. plantarum* and *S. cerevisiae* during fermentation resulted in doubling the volume of dough. The results were explained by Winters et al., (2019). They found that the combination of *L. plantarum* and *S. cerevisiae* had an increase in gas produced compared to the yeast alone [44].

3.1.3. Total Phenolic Contents and Antioxidant Activities

At the start of the fermentation, fermented water sourdough had significantly higher ($p < 0.05$) TPCs and antioxidant activities detected by ORAC, FRAP and DPPH radical scavenging assays than common sourdough (Table 1), and also exhibited significantly ($p < 0.05$) higher TPCs and antioxidant activities than common sourdough throughout the fermentation. At the end of the fermentation (4 h after the 18 h retard fermentation), TPCs and antioxidant activities of fermented water sourdough were 2–7 times higher than common sourdough due to the remaining TPCs and antioxidant activities in the fermented water [20]. Moreover, a significantly lower pH in fermented water sourdough than in common sourdough supported bioactive compound stabilization. Złotek et al., (2019) reported that polyphenols were auto-oxidized with increased pH [45].

Table 1. Total phenolic contents (TPCs) and antioxidant activities of common sourdough and fermented water sourdough fermented at different time periods (0 and 4 h).

		Common Sourdough		
		Antioxidant Activities		
Time (h)	**TPCs (mg GAE/g DW)**	**ORAC Assay (μmol TE/g DW)**	**FRAP Assay (μmol TE/g DW)**	**DPPH Radical Scavenging Assay (μmol TE/100 g DW)**
0	0.24 ± 0.02^{b}	9.88 ± 1.16^{b}	2.51 ± 0.07^{b}	0.16 ± 0.01^{b}
4	0.50 ± 0.05^{a}	17.64 ± 1.22^{a}	4.20 ± 0.29^{a}	0.50 ± 0.04^{a}
		Fermented Water Sourdough		
		Antioxidant Activities		
Time (h)	**TPCs (mg GAE/g DW)**	**ORAC assay (μmol TE/g DW)**	**FRAP assay (μmol TE/g DW)**	**DPPH radical scavenging assay (μmol TE/100 g DW)**
0	$1.76 \pm 0.03^{b,*}$	$91.36 \pm 4.31^{b,*}$	$19.40 \pm 0.27^{a,*}$	$0.81 \pm 0.08^{b,*}$
4	$2.52 \pm 0.24^{a,*}$	$121.55 \pm 6.93^{a,*}$	$17.74 \pm 0.74^{b,*}$	$1.19 \pm 0.08^{a,*}$

All data are shown as the mean ± standard deviation (SD) of triplicate determinations ($n = 3$). Different lowercase letters denote significantly different TPCs or antioxidant activities at $p < 0.05$ of the same type of sourdough fermented at different time periods, while * denotes significant differences at $p < 0.05$ between common sourdough and fermented water sourdough fermented at the same time period using Student's unpaired *t*-test. ORAC: oxygen radical absorbance capacity; FRAP: ferric ion reducing antioxidant power; DPPH: 2,2-diphenyl-1-picrylhydrazyl; GAE: gallic acid equivalent; TE: Trolox equivalent; DW: dry weight.

Moreover, TPCs and antioxidant activities of both sourdough types tended to increase throughout the fermentation (Table 1). Fermentation broke down the structure of unbleached wheat flour or fermented water [46,47], related to significant ($p < 0.05$) growth of *L. plantarum* in both types of sourdough during fermentation, resulting in the activation of complex polyphenol hydrolyzing enzymes that produced simpler and active polyphenols [48]. Katina et al., (2007) reported that amylase, proteases and xylanases derived from microbes and grains during sourdough fermentation released phenolics [49], while Złotek et al., (2019) reported that *L. plantarum* 299v enrichment significantly ($p < 0.05$) improved TPCs in legume sprout preparation through de novo synthesis induction [45]. The results showed that fermentation significantly increases TPCs in both fermented sourdough and common sourdough. Corresponded to increased TPCs, the antioxidant activities of fermented sourdough and common sourdough measured with ORAC and DPPH radical scavenging assay were greatly increased after fermentation. Since ORAC assay is a method to measure antioxidants with hydrogen atom transfer (HAT) mechanism and DPPH radical scavenging assay was for antioxidants with both HAT and single electron transfer (SET) mechanisms, these results suggested that increased phenolics after fermentation might act as antioxidants with HAT mechanism rather than SET mechanism. However, the FRAP activities were slightly reduce in fermented water sourdough after fermentation. Since FRAP assay is for antioxidants following SET mechanism, this result suggested the potential degradation of antioxidants with SET mechanism. A similar result was report by Kinga et al., (2021). They found that after fermentation of natural fruit meads the antioxidant activity measured by FRAP assay (SET mechanism) was reduced by 18%, while those by $ABTS^+$ assay increased by 14% (both HAT and SET mechanisms) [50].

3.2. *Effect of Fermented Water on Bread Quality*

3.2.1. Chemical and Physical Quality

Water activity, pH, specific volume and texture of bread are shown in Table 2. No significant ($p \geq 0.05$) differences in water activity in all bread samples (0.82 ± 0.00) were observed but fermented water sourdough bread showed a greater pH decrease than the control bread obtained from common sourdough ($p < 0.05$). A similar result was reported by Mohd Roby et al., (2020). They found that encapsulated kombucha sourdough bread had a significantly ($p < 0.05$) lower pH than liquid traditional sourdough bread [22], while Karimi et al., (2020) observed a significantly ($p < 0.05$) lower pH in sourdough bread containing (−)-epigallocatechin gallate and pacific white shrimp (*Litopenaeus vannamei*) protein hydrolysates than common sourdough bread (without added ingredients) [11]. A significantly ($p < 0.05$) lower pH of fermented water sourdough bread than common sourdough bread, possibly related to acids produced during sourdough development. Yu et al., (2019) reported that organic acid produced by lactic acid bacteria decreased pH values in bread [27], while Bartkiene et al., (2017) found a strong negative correlation between pH and amylolytic enzyme activity, with activities reported at a high level in *L. plantarum* sourdough [51].

Table 2. pH, specific volume and texture profile of common sourdough bread and fermented water sourdough bread.

Properties	Common Sourdough Bread	Fermented Water Sourdough Bread	Significance
pH	4.12 ± 0.01 *	4.02 ± 0.01	t = 22.627; sig. = 0.000
Specific volume (mL/g)	5.22 ± 0.05	5.28 ± 0.11	t = −0.873; sig. = 0.432
Water activity	0.82 ± 0.00	0.82 ± 0.00	t = 0.000; sig. = 1.000
Texture profile			
Hardness (g)	268.06 ± 17.81	248.54 ± 30.49	t = 0.958; sig. = 0.392
Chewiness	205.69 ± 1.78	192.62 ± 17.87	t = 1.261; sig. = 0.332
Springiness (cm)	0.92 ± 0.03	0.92 ± 0.02	t = −0.085; sig. = 0.936
Cohesiveness (g)	0.84 ± 0.02	0.85 ± 0.02	t = −0.387; sig. = 0.719

All data are shown as the mean ± standard deviation (SD) of triplicate determinations ($n = 3$). The * denotes significantly different values at $p < 0.05$ of the same property in common sourdough bread and fermented water sourdough bread using Student's unpaired *t*-test.

Specific volume and texture profiles of chewiness, hardness, springiness and cohesiveness showed insignificant ($p \geq 0.05$) differences between common sourdough bread and fermented water sourdough bread. Mohd Roby et al., (2020) reported a similar result, indicating that encapsulated kombucha sourdough bread exhibited comparably specific loaf volume as liquid traditional sourdough bread [22]. The specific volume of our fermented water sourdough bread was 1.2 times higher than encapsulated kombucha sourdough bread [22], suggesting that the fermented water sourdough bread acidified dough had an improved gas-holding gluten capacity [12,52].

3.2.2. Nutritional Quality

Nutritional compositions of common sourdough bread and fermented water sourdough bread are shown in Table 3. Energy, moisture content, protein, total fat, carbohydrate and ash contents insignificantly differed between common sourdough bread and fermented water sourdough bread. However, fermented water sourdough bread showed significantly ($p < 0.05$) lower total sugar content than common sourdough bread, related to higher numbers of *L. plantarum* and *S. cerevisiae* in fermented water sourdough (Figure 1). Most starches are hydrolyzed by lactic acid bacteria during sourdough fermentation, resulting in digestible sugar for yeast consumption, especially glucose, as the primary carbon source for yeast survival and, thus, lower total sugar content [53,54]. Yoon et al., (2003) also suggested that fermentation by *S. cerevisiae* completely removed disaccharides such as maltose, sucrose and turanose, while cellobiose, lactose and melibiose levels were maintained [55]. Similarly, Jin et al., (2019) indicated that total soluble solids and reducing sugar content of mango pulp fermented with *S. cerevisiae* DV10 (single and co-culture) significantly ($p < 0.05$) decreased after 24 h fermentation, while these values remained unchanged in mango pulp fermented by only *L. plantarum* [40]. Total dietary fiber in fermented water sourdough bread was significantly ($p < 0.05$) higher than in common sourdough bread (Table 3) because the insoluble dietary fiber in the fermented water could not be digested by microorganisms [52].

Table 3. Nutritional values of common sourdough bread and fermented water sourdough bread.

Nutrients (per 100 g FW)	Common Sourdough Bread	Fermented Water Sourdough Bread	Significance
Energy (kcal)	271.24 ± 0.66	273.75 ± 1.07	t = −2.831; sig. = 0.105
Moisture (%)	33.76 ± 0.28	33.16 ± 0.45	t = 1.620; sig. = 0.247
Protein (N × 6.25) (g)	9.73 ± 0.07	9.79 ± 0.10	t = −0.697; sig. = 0.558
Total fat (g)	2.80 ± 0.11	2.83 ± 0.15	t = −0.232; sig. = 0.838
Total carbohydrate (g)	51.78 ± 0.33	52.28 ± 0.70	t = -0.912; sig. = 0.458
Total dietary fiber (g)	1.54 ± 0.00	1.73 ± 0.04 *	t = −6.333; sig. = 0.024
Total sugar (g)	8.94 ± 0.07 *	8.09 ± 0.01	t = 16.866; sig. = 0.003
Ash (g)	1.93 ± 0.02	1.94 ± 0.01	t = −0.632; sig. = 0.592

All data are shown as the mean ± standard deviation (SD) of triplicate determinations ($n = 3$). The * denotes significantly different contents at $p < 0.05$ of the same nutritional composition in common sourdough bread and fermented water sourdough bread using Student's unpaired *t*-test.

3.2.3. Sensory Evaluation

Consumer acceptability is an important factor when developing food products because this influences consumer purchase willingness. A hedonic scale sensory analysis was performed by trained panelists ($n = 10$), with results depicted in Figure 3. Both common and fermented water sourdough bread samples were well accepted by the panelists, with all attribute scores higher than 6 [56–58]. The fermented water sourdough bread obtained higher scores for aroma, taste, texture and overall liking than common sourdough bread. This can be due to the fruity and unique aroma of pear and Assam tea, respectively. The advantage of using aromatic raw materials in sourdough bread was reported also in a study by Karimi et al., (2020). It was suggested that sourdough bread containing pacific white shrimp (*Litopenaeus vannamei*) protein hydrolysates and (−)-epigallocatechin gallate had significantly ($p < 0.05$) higher preference scores of taste, flavor, softness, chewiness and overall acceptability than the control bread [11]. Insignificantly different sourness scores between the bread samples were observed although the pH values of fermented water sourdough bread were lower than common sourdough bread (Table 2). It indicated that flavor and acidic taste from fermentation with Asian pears and Assam tea leaves in fermented water sourdough bread did not adversely affect sensory perceptions. Sensory scores on appearance and color of common sourdough bread were higher than fermented sourdough bread due to the mild-yellow color of fermented sourdough bread from the color of fermented water. The overall liking score obtained for fermented sourdough bread was significantly higher than common sourdough bread. The fermented sourdough bread recorded a very satisfactory score of 7.3, considering that it was a new product.

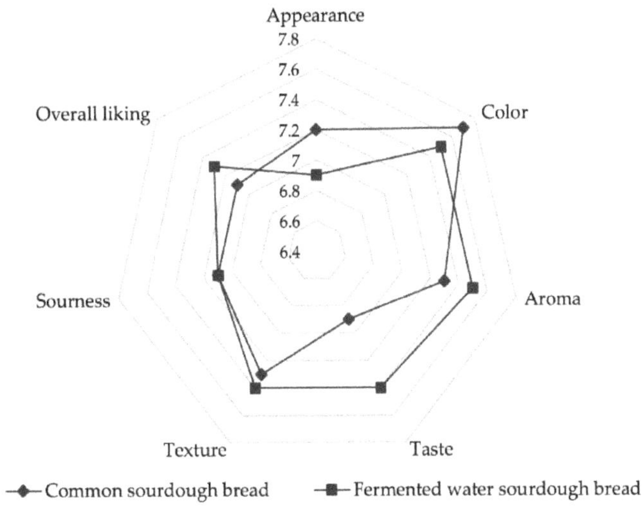

Figure 3. Sensory properties of common sourdough bread and fermented water sourdough bread.

3.3. Shelf-Life Stability

3.3.1. Microbiological Quality

Microbial analysis results of common sourdough bread and fermented water sourdough bread under different storage times are shown in Table 4. The initial viable cell counts of total aerobic bacteria, yeast and mold in both common sourdough bread and fermented water sourdough bread on a baking day passed the criteria of the Thai community product standard. This states that total aerobic bacteria must be less than 4 Log CFU/g and less than 2 Log CFU/g for yeast and mold [59]. During storage, aerobic bacteria grew faster than yeast and mold. At day 7 of storage time, total aerobic bacteria were detected at 4.26 Log CFU/g in common sourdough bread, while only 3.38 Log CFU/g was detected in fermented water sourdough bread. This was due to bacteriocin which is produced from *L. plantarum* during fermentation. Behera, Ray and Zdolec (2018) stated that *L. plantarum* produced bacteriocin of high activity and a wide range of antimicrobial activity and their properties could increase shelf life of products [60]. Epigallocatechin gallate (EGCG), a predominant phenolic in Assam tea, also possessed antimicrobial activities [61,62].

Yeast and mold were found at less than 1 Log CFU/g in both common sourdough bread and fermented water sourdough bread over the whole storage period (Table 4). The presence of *L. plantarum* in both mature fermented water sourdough and common sourdough prevented the growth of fungi by producing antifungal substances such as cyclic dipeptides, hydroxy fatty acids or phenyl and substituted phenyl derivates (3-phenyllactic acid, 4-hydroxyphenyl acetic acid and benzoic acid) [63]. Our results showed that fermented water sourdough increased bread shelf-life by 2 days at room temperature (7 days) compared to common sourdough bread (5 days).

Table 4. Total aerobic bacteria and yeast and mold of common sourdough bread and fermented water sourdough bread over storage duration.

Storage Days	Common Sourdough Bread		Fermented Water Sourdough Bread	
	Total Aerobic Bacteria (Log CFU/g)	Yeast and Mold (Log CFU/g)	Total Aerobic Bacteria (Log CFU/g)	Yeast and Mold (Log CFU/g)
0	0.00 ± 0.00 d	Less than 1	2.57 ± 0.01 b,*	Less than 1
1	0.00 ± 0.00 d	Less than 1	2.37 ± 0.13 bc,*	Less than 1
3	2.98 ± 0.03 c,*	Less than 1	2.30 ± 0.14 c	Less than 1
5	3.23 ± 0.01 b	Less than 1	3.21 ± 0.02 a	Less than 1
7	4.26 ± 0.06 a,*	Less than 1	3.38 ± 0.02 a	Less than 1

All data are shown as the mean ± standard deviation (SD) of duplicate determinations. Different lowercase letters denote significantly different viable cell count of microorganisms at $p < 0.05$ of the same type of bread under different storage times using one-way ANOVA, followed by Duncan's multiple comparison test, while * denotes significantly different viable cell count of microorganisms at $p < 0.05$ between common sourdough bread and fermented water sourdough bread at the same storage time using Student's unpaired *t*-test.

3.3.2. Total Phenolic Contents and Antioxidant Activities

TPCs and antioxidant activities detected by ORAC, FRAP and DPPH radical scavenging assays of fermented water sourdough bread were significantly ($p < 0.05$) higher than the common sourdough bread throughout shelf-life storage times (Table 5). The lower pH of fermented water sourdough bread was beneficial for the presence of TPCs and antioxidant activities because polyphenols were auto-oxidized at higher pH [38]. However, TPCs and antioxidant activities of both fermented water sourdough bread and common sourdough bread significantly ($p < 0.05$) declined during storage. The degradation of TPCs and antioxidant activities was due to high temperature, change in pH or oxygen availability [64].

Table 5. Total phenolic contents (TPCs) and antioxidant activities detected by ORAC, FRAP, and DPPH radical scavenging assays of bread prepared by common sourdough and fermented water sourdough over storage duration.

Common Sourdough Bread				
		Antioxidant Activities		
Storage Days	TPCs (mg GAE/g DW)	ORAC Assay (µmol TE/g DW)	FRAP Assay (µmol TE/g DW)	DPPH Radical Scavenging Assay (µmol TE/100 g DW)
0	0.28 ± 0.02 a	6.04 ± 0.60 ab	0.98 ± 0.07 a	0.16 ± 0.01 a
1	0.27 ± 0.03 a	6.44 ± 0.60 a	0.83 ± 0.03 b	0.12 ± 0.01 c
3	0.24 ± 0.01 b	6.10 ± 0.57 ab	0.74 ± 0.04 c	0.13 ± 0.01 b
5	0.25 ± 0.01 b	5.88 ± 0.29 ab	0.70 ± 0.02 d	0.14 ± 0.01 b
7	0.23 ± 0.02 b	5.79 ± 0.33 b	0.75 ± 0.03 c	0.12 ± 0.01 c
Fermented Water Sourdough Bread				
		Antioxidant Activities		
Storage Days	TPCs (mg GAE/g DW)	ORAC assay (µmol TE/g DW)	FRAP assay (µmol TE/g DW)	DPPH radical scavenging assay (µmol TE/100 g DW)
0	0.68 ± 0.02 a*	14.89 ± 1.44 c,*	3.12 ± 0.22 a,*	0.33 ± 0.03 a,*
1	0.66 ± 0.03 ab,*	26.73 ± 0.65 a,*	3.08 ± 0.29 a,*	0.29 ± 0.01 b,*
3	0.64 ± 0.03 b*	26.54 ± 2.63 a,*	2.77 ± 0.18 bc,*	0.29 ± 0.03 b,*
5	0.61 ± 0.02 c,*	23.45 ± 0.93 b,*	2.66 ± 0.08 c,*	0.29 ± 0.02 b,*
7	0.57 ± 0.05 d,*	15.26 ± 1.47 c,*	2.90 ± 0.11 b,*	0.27 ± 0.02 c,*

All data are shown as the mean ± standard deviation (SD) of triplicate determinations ($n = 3$). Different lowercase letters denote significantly different TPCs or antioxidant activities at $p < 0.05$ of the same type of sourdough bread stored at different time periods, while * denotes significant differences at $p < 0.05$ between common sourdough bread and fermented water sourdough bread stored at the same time period using Student's unpaired *t*-test. ORAC: oxygen radical absorbance capacity; FRAP: ferric ion reducing antioxidant power; DPPH: 2,2-diphenyl-1-picrylhydrazyl; GAE: gallic acid equivalent; TE: Trolox equivalent; DW: dry weight.

4. Conclusions

This study showed that fermented water sourdough passed the criteria of mature sourdough after 4 h, following 18 h retard fermentation, with significantly higher growth of *S. cerevisiae*, optimal pH and improved antioxidant activity compared with common sourdough. The fermented water sourdough bread had 10% less sugar, 12% higher dietary fiber and 2 to 3 times higher total phenolic contents and antioxidant activities detected by ORAC, FRAP and DPPH radical scavenging assays, compared with common sourdough bread. Sensory evaluation determined that the fermented water sourdough bread was preferred in terms of aroma, taste, texture and overall liking when compared with common sourdough bread. The fermented water sourdough bread presented good stability during room temperature (28 ± 2 °C and $75 \pm 5\%$ RH) storage for 7 days. At the end of the storage period, adequate amounts of bioactive compounds and good microbial quality suggested that the bread could be safely consumed. The finding demonstrated that fermented water prepared from Asian pears and Assam tea leaves using *L. plantarum* 299v and *S. cerevisiae* TISTR 5059 as starter cultures was successfully applied to develop sourdough and bread with improved qualities. However, the significant improvement found in the current study needs to be validated at the industrial application level.

Supplementary Materials: The following supporting information can be downloaded at: https://www.mdpi.com/article/10.3390//foods11142071/s1, Supplementary Table S1. Growth of *L. plantarum* and *S. cerevisiae* in common sourdough and fermented water sourdough throughout fermentation; Supplementary Table S2. A trend in pH and height of common sourdough and fermented water sourdough over fermentation time periods.

Author Contributions: Conceptualization, R.S., U.S., C.S., N.T., C.C. and N.O.-n.; methodology, R.S., C.K. and N.O.-n.; validation, R.S. and N.O.-n.; formal analysis, R.S., U.S., C.K. and N.O.-n.; investigation, R.S., U.S. and N.O.-n.; resources, U.S., C.S., C.C. and N.O.-n.; data curation, R.S., U.S. and N.O.-n.; writing—original draft preparation, R.S. and N.O.-n.; writing—review and editing, U.S., C.S., N.T., C.C. and N.O.-n.; supervision, N.O.-n.; project administration, N.O.-n.; funding acquisition, N.O.-n. All authors have read and agreed to the published version of the manuscript.

Funding: This research was funded by the 2020 Plant Genetic Conservation Project Under the Royal Initiation of Her Royal Highness Princess Maha Chakri Sinrindhorn, Mahidol University, 2021–2022, Grant Number: RSPG03/2564.

Institutional Review Board Statement: The study was conducted in accordance with the Declaration of Helsinki and approved by the Center Institutional Review Board (MU-CIRB) of Mahidol University (protocol code MU-CIRB 2021/042.2101 and date of approval 21 Jan 2021).

Informed Consent Statement: Informed consent was obtained from all subjects involved in the study.

Data Availability Statement: Data are contained within this article.

Acknowledgments: The authors gratefully acknowledge the support and generosity of Institute of Nutrition, Mahidol University for facilities support.

Conflicts of Interest: The authors declare no conflict of interest.

References

1. Khan, R.S.; Grigor, J.; Winger, R.; Win, A. Functional food product development—Opportunities and challenges for food manufacturers. *Trends Food Sci. Technol.* **2013**, *30*, 27–37. [CrossRef]
2. Gobbetti, M.; Rizzello, C.G.; Di Cagno, R.; De Angelis, M. How the sourdough may affect the functional features of leavened baked goods. *Food Microbiol.* **2014**, *37*, 30–40. [CrossRef]
3. Katina, K.; Arendt, E.; Liukkonen, K.H.; Autio, K.; Flander, L.; Poutanen, K. Potential of sourdough for healthier cereal products. *Trends Food Sci. Technol.* **2005**, *16*, 104–112. [CrossRef]
4. Singh, S.; Sethi, S.; Gupta, S.; Kaur, C.; Wood, E. Fermentation of Multigrain Dough—An Approach to Reduce Glycemic Index for Healthy Bread. *EUREKA Life Sci.* **2019**, *5*, 19–31. [CrossRef]
5. Plessas, S.; Fisher, A.; Koureta, K.; Psarianos, C.; Nigam, P.; Koutinas, A.A. Application of Kluyveromyces marxianus, *Lactobacillus delbrueckii* ssp. bulgaricus and *L. helveticus* for sourdough bread making. *Food Chem.* **2008**, *106*, 985–990. [CrossRef]

6. do Amaral Santos, C.C.A.; da Silva Libeck, B.; Schwan, R.F. Co-culture fermentation of peanut-soy milk for the development of a novel functional beverage. *Int. J. Food Microbiol.* **2014**, *186*, 32–41. [CrossRef]
7. Reis, V.R.; Bassi, A.P.G.; da Silva, J.C.G.; Ceccato-Antonini, S.R. Characteristics of *Saccharomyces cerevisiae* yeasts exhibiting rough colonies and pseudohyphal morphology with respect to alcoholic fermentation. *Braz. J. Microbiol.* **2014**, *44*, 1121–1131. [CrossRef] [PubMed]
8. Dalcanton, F.; Carrasco, E.; Pérez-Rodríguez, F.; Posada-Izquierdo, G.D.; Falcão de Aragão, G.M.; García-Gimeno, R.M. Modeling the Combined Effects of Temperature, pH, and Sodium Chloride and Sodium Lactate Concentrations on the Growth Rate of *Lactobacillus plantarum* ATCC 8014. *J. Food Qual.* **2018**, *2018*, 1726761. [CrossRef]
9. Hu, Y.; Zhang, J.; Wang, S.; Liu, Y.; Li, L.; Gao, M. Lactic acid bacteria synergistic fermentation affects the flavor and texture of bread. *J. Food Sci.* **2022**, *87*, 1823–1836. [CrossRef] [PubMed]
10. Holzapfel, W.H. Appropriate starter culture technologies for small-scale fermentation in developing countries. *Int. J. Food Microbiol.* **2002**, *75*, 197–212. [CrossRef]
11. Karimi, N.; Nikoo, M.; Ahmadi Gavlighi, H.; Piri Gheshlaghi, S.; Regenstein, J.M.; Xu, X. Effect of pacific white shrimp (*Litopenaeus vannamei*) protein hydrolysates (SPH) and (−)-epigallocatechin gallate (EGCG) on sourdough and bread quality. *LWT* **2020**, *131*, 109800. [CrossRef]
12. Yu, Y.; Wang, L.; Qian, H.; Zhang, H.; Qi, X. Contribution of spontaneously-fermented sourdoughs with pear and navel orange for the bread-making. *LWT* **2018**, *89*, 336–343. [CrossRef]
13. Chupeerach, C.; Aursalung, A.; Watcharachaisoponsiri, T.; Whanmek, K.; Thiyajai, P.; Yosphan, K.; Sritalahareuthai, V.; Sahasakul, Y.; Santivarangkna, C.; Suttisansanee, U. The Effect of Steaming and Fermentation on Nutritive Values, Antioxidant Activities, and Inhibitory Properties of Tea Leaves. *Foods* **2021**, *10*, 117. [CrossRef] [PubMed]
14. Ungsongtham, P.; Puengpan, V.; Aksornchoo, P.; Supanuwat, J. *Nutritional Values in Fruits*, 1st ed.; Agricultural Cooperative Federation of Thailand: Nonthaburi, Thailand, 2010.
15. Jiang, G.; Ameer, K.; Eun, J.-B. Physicochemical, antioxidant, microstructural, and sensory properties of sesame bars sweetened with pear juice concentrate. *J. Food Sci. Technol.* **2020**, *57*, 4551–4561. [CrossRef]
16. Reiland, H.; Slavin, J. Systematic Review of Pears and Health. *Nutr. Today* **2015**, *50*, 301–305. [CrossRef]
17. Kumar, V. Assessment of Phytochemical and Antioxidant Potential of Underutilized Pear (*Pyrus pyrifolia*) and Plum (*Prunus domestica*) from Indigenous Himalayan Region of Himachal Pradesh. *Int. J. Pharm. Sci. Res.* **2017**, *8*, 1000–1006.
18. Astill, C.; Birch, M.R.; Dacombe, C.; Humphrey, P.G.; Martin, P.T. Factors affecting the caffeine and polyphenol contents of black and green tea infusions. *J. Agric. Food Chem.* **2001**, *49*, 5340–5347. [CrossRef]
19. Gramza-Michałowska, A.; Kobus-Cisowska, J.; Kmiecik, D.; Korczak, J.; Helak, B.; Dziedzic, K.; Górecka, D. Antioxidative potential, nutritional value and sensory profiles of confectionery fortified with green and yellow tea leaves (*Camellia sinensis*). *Food Chem.* **2016**, *211*, 448–454. [CrossRef]
20. Supasil, R.; Thangsiri, S.; Whanmek, K.; Khemthong, C.; Tangsuphoom, N.; Santivarangkna, C.; Suttisansanee, U.; On-nom, N. The improvement of total phenolic content and antioxidant activity in fermented water from Asian pears (Pyrus pyrifolia) and Assam tea (*Camellia sinensis* var. *assamica*) leaves with co-cultures between *Lactobacillus plantarum* and *Saccharomyces cerevisiae*. In Proceedings of the 23rd Food Innovation Asia Conference 2021 (FIAC 2021), Bangkok, Thailand, 17–19 June 2021.
21. Hammes, W.P.; Brandt, M.J.; Francis, K.L.; Rosenheim, J.; Seitter, M.F.H.; Vogelmann, S.A. Microbial ecology of cereal fermentations. *Trends Food Sci. Technol.* **2005**, *16*, 4–11. [CrossRef]
22. Mohd Roby, B.H.; Muhialdin, B.J.; Abadl, M.M.T.; Mat Nor, N.A.; Marzlan, A.A.; Lim, S.A.H.; Mustapha, N.A.; Meor Hussin, A.S. Physical properties, storage stability, and consumer acceptability for sourdough bread produced using encapsulated kombucha sourdough starter culture. *J. Food Sci.* **2020**, *85*, 2286–2295. [CrossRef]
23. Bender, D.; Fraberger, V.; Szepasvári, P.; D'Amico, S.; Tömösközi, S.; Cavazzi, G.; Jäger, H.; Domig, K.J.; Schoenlechner, R. Effects of selected lactobacilli on the functional properties and stability of gluten-free sourdough bread. *Eur. Food Res. Technol. Z. Lebensm.-Unters. Forschung. A* **2018**, *244*, 1037–1046. [CrossRef] [PubMed]
24. Bujalance, C.; Jiménez-Valera, M.; Moreno, E.; Ruiz-Bravo, A. A selective differential medium for *Lactobacillus plantarum*. *J. Microbiol. Methods* **2006**, *66*, 572–575. [CrossRef] [PubMed]
25. Xu, D.; Zhang, Y.; Tang, K.; Hu, Y.; Xu, X.; Gänzle, M.G. Effect of Mixed Cultures of Yeast and Lactobacilli on the Quality of Wheat Sourdough Bread. *Front. Microbiol.* **2019**, *10*, 2113. [CrossRef] [PubMed]
26. FDA. *Bacteriological Analytical Manual (BAM)*; US Food and Drug Administration: Silver Spring, MD, USA, 2001.
27. Yu, Y.; Wang, L.; Qian, H.; Zhang, H.; Li, Y.; Wu, G.; Qi, X.; Xu, M.; Rao, Z. Effect of selected strains on physical and organoleptic properties of breads. *Food Chem.* **2019**, *276*, 547–553. [CrossRef]
28. AACC. *Approved Methods of the American Association of Cereal Chemists*, 11th ed.; Cereals & Grains Association: St. Paul, MN, USA, 2001.
29. Sripum, C.; Kukreja, R.; Charoenkiatkul, S.; Kriengsinyos, W.; Suttisansanee, U. The effect of extraction conditions on antioxidant activities and total phenolic contents of different processed Thai Jasmine rice. *Int. Food Res. J.* **2017**, *24*, 1644–1650.
30. AOAC. *Official Method of Analysis of AOAC Internation*, 21st ed.; AOAC International: Arlington, VA, USA, 2019.
31. Wannasaksri, W.; Temviriyanukul, P.; Aursalung, A.; Sahasakul, Y.; Thangsiri, S.; Inthachat, W.; On-Nom, N.; Chupeerach, C.; Pruesapan, K.; Charoenkiatkul, S.; et al. Influence of Plant Origins and Seasonal Variations on Nutritive Values, Phenolics and Antioxidant Activities of *Adenia viridiflora* Craib., an Endangered Species from Thailand. *Foods* **2021**, *10*, 2799. [CrossRef]

32. Meilgaard, M.; Civille, G.V.; Carr, B.T. *Sensory Evaluation Techniques*, 4th ed.; CRC Press: Boca Raton, FL, USA, 2006.
33. Fleet, G.H.; Heard, G.M. *Yeast-Growth during Fermentation*; Harwood Academic Publishers: Chur, Switzerland, 1992.
34. Sieuwerts, S.; Bron, P.A.; Smid, E.J. Mutually stimulating interactions between lactic acid bacteria and *Saccharomyces cerevisiae* in sourdough fermentation. *LWT* **2018**, *90*, 201–206. [CrossRef]
35. Unban, K.; Khatthongngam, N.; Shetty, K.; Khanongnuch, C. Nutritional biotransformation in traditional fermented tea (Miang) from north Thailand and its impact on antioxidant and antimicrobial activities. *J. Food Sci. Technol.* **2019**, *56*, 2687–2699. [CrossRef]
36. Ketwal, S.; Chueamchaitrakun, P.; Theppakorn, T.; Wongsakul, S. Contents of total polyphenol, microorganisms and antioxidant capacities of pickled tea (miang) commercially available in Chiang Rai, Thailand. In Proceedings of the 16th Food Innovation Asia Conference (FIAC 2014), Bangkok, Thailand, 12–13 June 2014.
37. Somsong, P.; Santivarangkna, C.; Tiyayon, P.; Hsieh, C.-M.; Srichamnong, W. Assessing Polyphenol Components and Antioxidant Activity during Fermented Assam Tea Ball Processing. *Sustainability* **2020**, *12*, 5853. [CrossRef]
38. López de Felipe, F.; Curiel, J.A.; Muñoz, R. Improvement of the fermentation performance of *Lactobacillus plantarum* by the flavanol catechin is uncoupled from its degradation. *J. Appl. Microbiol.* **2010**, *109*, 687–697. [CrossRef]
39. Kayikci, Ö.; Nielsen, J. Glucose repression in *Saccharomyces cerevisiae*. *FEMS Yeast Res.* **2015**, *15*, fov068. [CrossRef]
40. Jin, X.; Chen, W.; Chen, H.; Chen, W.; Zhong, Q. Combination of *Lactobacillus plantarum* and *Saccharomyces cerevisiae* DV10 as Starter Culture to Produce Mango Slurry: Microbiological, Chemical Parameters and Antioxidant Activity. *Molecules* **2019**, *24*, 4349. [CrossRef] [PubMed]
41. Duan, S.; Zhang, Y.X.; Lu, T.T.; Cao, D.X.; Chen, J.D. Shrimp Waste Fermentation Using Symbiotic Lactic Acid Bacteria. *Adv. Mater. Res.* **2011**, *194–196*, 2156–2163. [CrossRef]
42. Unban, K.; Khatthongngam, N.; Pattananandecha, T.; Saenjum, C.; Shetty, K.; Khanongnuch, C. Microbial Community Dynamics During the Non-filamentous Fungi Growth-Based Fermentation Process of Miang, a Traditional Fermented Tea of North Thailand and Their Product Characterizations. *Front. Microbiol.* **2020**, *11*, 1515. [CrossRef] [PubMed]
43. Lee, S.-H.; Cho, J.-Y.; Jeong, H.Y.; Jeong, D.E.; Kim, D.; Cho, S.-Y.; Kim, W.-S.; Moon, J.-H. Comparison of bioactive compound contents and in vitro and ex vivo antioxidative activities between peel and flesh of pear (*Pyrus pyrifolia* Nakai). *Food Sci. Biotechnol.* **2015**, *24*, 207–216. [CrossRef]
44. Winters, M.; Panayotides, D.; Bayrak, M.; Rémont, G.; Viejo, C.G.; Liu, D.; Le, B.; Liu, Y.; Luo, J.; Zhang, P.; et al. Defined co-cultures of yeast and bacteria modify the aroma, crumb and sensory properties of bread. *J. Appl. Microbiol.* **2019**, *127*, 778–793. [CrossRef]
45. Złotek, U.; Świeca, M.; Reguła, J.; Jakubczyk, A.; Sikora, M.; Gawlik-Dziki, U.; Kapusta, I. Effects of probiotic *L. plantarum* 299v on consumer quality, accumulation of phenolics, antioxidant capacity and biochemical changes in legume sprouts. *Int. J. Food Sci.* **2019**, *54*, 2437–2446. [CrossRef]
46. Adebo, O.A.; Gabriela Medina-Meza, I. Impact of Fermentation on the Phenolic Compounds and Antioxidant Activity of Whole Cereal Grains: A Mini Review. *Molecules* **2020**, *25*, 927. [CrossRef]
47. Lee, L.S.; Kim, S.H.; Kim, Y.B.; Kim, Y.C. Quantitative analysis of major constituents in green tea with different plucking periods and their antioxidant activity. *Molecules* **2014**, *19*, 9173–9186. [CrossRef]
48. Dueñas, M.; Fernández, D.; Hernández, T.; Estrella, I.; Muñoz, R. Bioactive phenolic compounds of cowpeas (*Vigna sinensis* L.). Modifications by fermentation with natural microflora and with *Lactobacillus plantarum* ATCC 14917. *J. Sci. Food Agric.* **2005**, *85*, 297–304. [CrossRef]
49. Katina, K.; Laitila, A.; Juvonen, R.; Liukkonen, K.H.; Kariluoto, S.; Piironen, V.; Landberg, R.; Aman, P.; Poutanen, K. Bran fermentation as a means to enhance technological properties and bioactivity of rye. *Food Microbiol.* **2007**, *24*, 175–186. [CrossRef] [PubMed]
50. Adamenko, K.; Kawa-Rygielska, J.; Kucharska, A.; Głowacki, A.; Piórecki, N. Changes in the Antioxidative Activity and the Content of Phenolics and Iridoids during Fermentation and Aging of Natural Fruit Meads. *Biomolecules* **2021**, *11*, 1113. [CrossRef] [PubMed]
51. Bartkiene, E.; Bartkevics, V.; Pugajeva, I.; Krungleviciute, V.; Mayrhofer, S.; Domig, K. The contribution of *P. acidilactici*, *L. plantarum*, and *L. curvatus* starters and L-(+)-lactic acid to the acrylamide content and quality parameters of mixed rye—Wheat bread. *LWT* **2017**, *80*, 43–50. [CrossRef]
52. Katina, K.; Heiniö, R.L.; Autio, K.; Poutanen, K. Optimization of sourdough process for improved sensory profile and texture of wheat bread. *LWT—Food Sci. Technol.* **2006**, *39*, 1189–1202. [CrossRef]
53. Burton, P.; Lightowler, H.J. Influence of bread volume on glycaemic response and satiety. *Br. J. Nutr.* **2006**, *96*, 877–882. [CrossRef] [PubMed]
54. Kingsbury, J.M.; Goldstein, A.L.; McCusker, J.H. Role of nitrogen and carbon transport, regulation, and metabolism genes for *Saccharomyces cerevisiae* survival in vivo. *Eukaryot. Cell* **2006**, *5*, 816–824. [CrossRef] [PubMed]
55. Yoon, S.H.; Mukerjea, R.; Robyt, J.F. Specificity of yeast (*Saccharomyces cerevisiae*) in removing carbohydrates by fermentation. *Carbohydr. Res.* **2003**, *338*, 1127–1132. [CrossRef]
56. Gámbaro, A.; Giménez, A.; Ares, G.; Gilardi, V. Influence of Enzymes on the Texture of Brown Pan Bread. *J. Texture Stud.* **2006**, *37*, 300–314. [CrossRef]
57. Giménez, A.; Ares, G.; Gámbaro, A. Survival Analysis to Estimate Sensory Shelf Life Using Acceptability Scores. *J. Sens. Stud.* **2008**, *23*, 571–582. [CrossRef]

58. Giménez, A.; Varela, P.; Salvador, A.; Ares, G.; Fiszman, S.; Garitta, L. Shelf life estimation of brown pan bread: A consumer approach. *Food Qual. Prefer.* **2007**, *18*, 196–204. [CrossRef]
59. TISI. *Thai Community Product Standard, Bread*; Thai Industrial Standards Institute: Bangkok, Thailand, 2012.
60. Behera, S.S.; Ray, R.C.; Zdolec, N. *Lactobacillus plantarum* with Functional Properties: An Approach to Increase Safety and Shelf-Life of Fermented Foods. *BioMed Res. Int.* **2018**, *2018*, 9361614. [CrossRef] [PubMed]
61. Nikoo, M.; Regenstein, J.M.; Ahmadi Gavlighi, H. Antioxidant and Antimicrobial Activities of (−)-Epigallocatechin-3-gallate (EGCG) and its Potential to Preserve the Quality and Safety of Foods. *Compr. Rev. Food Sci. Food Saf.* **2018**, *17*, 732–753. [CrossRef] [PubMed]
62. Sae-Lee, N.; Kerdchoechuen, O.; Laohakunjit, N. Composition of epigallocatechin gallate (EGCG) and catechins derivatives in Chinese and assam tea. *J. Agric. Sci.* **2009**, *40*, 9–12.
63. Axel, C.; Brosnan, B.; Zannini, E.; Peyer, L.C.; Furey, A.; Coffey, A.; Arendt, E.K. Antifungal activities of three different *Lactobacillus* species and their production of antifungal carboxylic acids in wheat sourdough. *Appl. Microbiol. Biotechnol.* **2016**, *100*, 1701–1711. [CrossRef] [PubMed]
64. Ananingsih, V.K.; Sharma, A.; Zhou, W. Green tea catechins during food processing and storage: A review on stability and detection. *Food Res. Int.* **2013**, *50*, 469–479. [CrossRef]

foods

MDPI

Article

Metagenetic Analysis for Microbial Characterization of *Focaccia* Doughs Obtained by Using Two Different Starters: Traditional Baker's Yeast and a Selected *Leuconostoc citreum* Strain

Massimo Ferrara, Angelo Sisto, Giuseppina Mulè, Paola Lavermicocca and Palmira De Bellis *

Institute of Sciences of Food Production (ISPA), National Research Council (CNR), Via G. Amendola 122/O, 70126 Bari, Italy; massimo.ferrara@ispa.cnr.it (M.F.); angelo.sisto@ispa.cnr.it (A.S.); giuseppina.mule@ispa.cnr.it (G.M.); paola.lavermicocca@ispa.cnr.it (P.L.)
* Correspondence: mirella.debellis@ispa.cnr.it; Tel.: +39-080-5929459

Abstract: Lactic acid bacteria (LAB) decisively influence the technological, nutritional, organoleptic and preservation properties of bakery products. Therefore, their use has long been considered an excellent strategy to improve the characteristics of those goods. The aim of this study was the evaluation of microbial diversity in different doughs used for the production of a typical Apulian flatbread, named *focaccia*. Leavening of the analyzed doughs was obtained with baker's yeast or by applying an innovative "yeast-free" protocol based on a liquid sourdough obtained by using *Leuconostoc citreum* strain C2.27 as a starter. The microbial populations of the doughs were studied by both a culture-dependent approach and metagenetic analyses. The flours used for dough preparation were also subjected to the same analyses. The metagenetic analyses were performed by sequencing the V5–V6 hypervariable regions of the 16S rRNA gene and the V9 hypervariable region of the 18S rRNA gene. The results indicate that these hypervariable regions were suitable for studying the microbiota of doughs, highlighting a significant difference between the microbial community of *focaccia* dough with baker's yeast and that of the dough inoculated with the bacterial starter. In particular, the dough made with baker's yeast contained a microbiota with a high abundance of *Proteobacteria* (82% of the bacterial population), known to be negatively correlated with the biochemical properties of the doughs, while the *Proteobacteria* in dough produced with the *L. citreum* starter were about 43.5% lower than those in flour and dough prepared using baker's yeast. Moreover, the results show that the *L. citreum* C2.27 starter was able to dominate the microbial environment and also reveal the absence of the genus *Saccharomyces* in the dough used for the production of the "yeast-free" *focaccia*. This result is particularly important because it highlights the suitability of the starter strain for obtaining an innovative "yeast-free" product.

Keywords: yeast-free dough; baker's yeast; liquid sourdough; metagenetic analysis; lactic acid bacteria; microbiota

Citation: Ferrara, M.; Sisto, A.; Mulè, G.; Lavermicocca, P.; De Bellis, P. Metagenetic Analysis for Microbial Characterization of *Focaccia* Doughs Obtained by Using Two Different Starters: Traditional Baker's Yeast and a Selected *Leuconostoc citreum* Strain. *Foods* **2021**, *10*, 1189. https://doi.org/10.3390/foods10061189

Academic Editors: Barbara Speranza and Arun K. Bhunia

Received: 14 April 2021
Accepted: 21 May 2021
Published: 25 May 2021

1. Introduction

Traditionally, dough leavening of bakery products has been obtained using sourdough, but with the development of industrial baking at the beginning of the 20th century, baker's yeast (*Saccharomyces cerevisiae*) almost completely replaced sourdough as a more rapid leavening agent, which also occurred in several traditional productions [1]. In fact, baker's yeast is a ready-to-use starter able to produce high amounts of CO_2 and flavor compounds [2,3]. The typical Apulian *focaccia* flatbread has recently been included in the Italian list of unprotected typical/local products (PAT, Prodotti Agroalimentari Tradizionali) [4]. Its production is characterized by a fermentation process using baker's yeast on a blend of soft and durum wheat flours and water, extra-virgin olive oil and salt [5]. The nutritional values of a commercial Apulian *focaccia* (100 g) are approximately the following:

305 kcal, carbohydrates 44.28 g of which sugars 2.39 g, fats 11 g of which saturates 1.36 g, proteins 6.9 g, salt 0,74 g [6].

Although the use of baker's yeast as a starter shortens the leavening time, it does not provide products with high-quality standards [7,8]. Therefore, the use of sourdough is once again increasing, especially in retail or artisanal bakeries, due to its characteristics [9,10]. In fact, sourdough fermentation may improve the functional/nutritional features of leavened baked goods by lowering the glycemic index, increasing mineral bioavailability, decreasing the gluten content, masking the decreased salt content and/or enriching bakery products with functional antihypertensive compounds [11,12], in addition to determining a better flavor, an extended shelf life and a reduction in additives [13]. The abundant studies in the literature show the features of sourdough fermentation and the need to develop new products based on the use of selected or novel sourdoughs and on targeted processes in relation to the desired characteristics of food products [12,14,15]. In fact, sourdough is a mixture of flour and water, fermented with lactic acid bacteria (LAB) and yeast, and the traditional sourdough fermentation is generally a long process involving interactions among many factors such as microbiota and related metabolic activities, fermentation parameters and raw materials [7,16]. Sourdough standardization is difficult due to its complexity, and this constitutes a problem for the bakery industry, which requires product stability and reproducibility, and short leavening times [15]. To overcome these limitations, liquid sourdough (type II) fermentation with single or mixed starter cultures has recently been introduced in bakeries [13,15,17,18]. Liquid sourdough offers several advantages compared to traditional sourdough, such as great flexibility of use, easier fermentation control, easier management and reproducibility [7]. Moreover, tailored biotechnological protocols can be used to modify the nutritional and/or functional features of the products. In fact, the growing consumer interest in foods with improved nutritional and functional quality has prompted food producers to adapt protocols for upgrading product attributes, including those of traditional products [19]. Product labels often include nutritional claims, such as "low fat", "high fiber", "light", "sugar-free", "source of protein" and "low salt". Similarly, there is an increasing interest of the market in "yeast-free" leavened foods, due to the recent concern about the role of *S. cerevisiae* (particularly its anti-antibodies) in several pathologies, including Crohn's disease, gastro-esophageal reflux, acne inversa and gut fermentation syndrome [20–23]. In this context, processes to develop "yeast-free" bakery products have been formulated and applied at the industry level [24–26]. In particular, we developed a protocol based on the use of a liquid sourdough started with a single bacterial strain for "yeast-free" bread and *focaccia* production, obtaining products with improved functional/nutritional characteristics compared to those made with baker's yeast [24,27]. The strain *L. citreum* C2.27 was selected for its technological characteristics, particularly its leavening capacity, thus allowing the production of doughs without using baker's yeast.

The aim of this study was to characterize the bacterial and fungal microbiota of doughs used for *focaccia* production made with liquid sourdough inoculated with *L. citreum* C2.27 or with baker's yeast. The microbial community was investigated by using culture-dependent and culture-independent approaches. Several studies have investigated the bacterial microbiota associated with traditional sourdough [28,29], but only a few metagenetic studies have been concerned with the characterization of the fungal microbiota of sourdoughs [30–32]. The absence of metagenetic studies on fungal communities is probably due to non-standardized metagenetic analysis of yeast, including selection of the optimal region to amplify [33,34], or to a lower general interest in the yeast composition [35]. To our knowledge, this is the first study that explores, also using an in-depth metagenetic approach, the bacterial and fungal microbiota of doughs leavened with baker's yeast or with a type II sourdough (started with a selected bacterial strain) used to obtain an innovative "yeast-free" product.

2. Materials and Methods

2.1. Bacterial Starter Strain and Culture Conditions

The strain *L. citreum* C2.27 (ITEM 17404) belongs to the ITEM Culture Collection of the Institute of Sciences of Food Production of the National Research Council (ISPA-CNR) and was isolated from Italian durum wheat semolina [36]. The cell culture was routinely propagated (2% *v*/*v*) in de Man Rogosa Sharpe (MRS) broth (Oxoid Ltd., Basingstoke, UK) and incubated at 30 °C for 24 h.

2.2. Liquid Sourdough and Dough Fermentation

Two different doughs were used for the production of *focaccia*: (a) dough using the liquid sourdough started with lactic acid bacteria; (b) dough using baker's yeast as described below. *L. citreum* C2.27, which belongs to a species included in the list of biological agents with Qualified Presumption of Safety (QPS) [37], was used as starter to produce the liquid sourdough as previously described [24].

Liquid sourdough (S) (500 g) was prepared by mixing wheat flour (17% *w*/*w*), sterile tap water (58% *v*/*w*) and bacterial suspension (25% *v*/*w*). The cell density in the sourdough was ca. 8 log cfu/g and dough yield (DY) was 600. The mixture was incubated at 30 °C for 16 h (Figure 1).

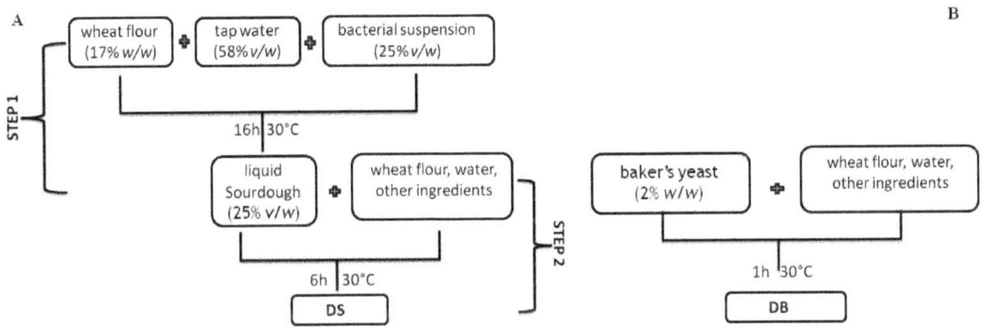

Figure 1. Schematic representation of the production of two different doughs: (**A**) dough with the liquid sourdough started with *L. citreum* C2.27 (DS), and (**B**) dough with baker's yeast (DB).

Focaccia dough (DY ca. 172) with liquid sourdough (DS) was prepared by mixing durum wheat semolina (27% *w*/*w*, Divella, Rutigliano, ITA), soft wheat flour type "00" (27% *w*/*w*; Casillo, Corato, ITA), tap water (18.36% *v*/*w*), extra-virgin olive oil (2.1% *v*/*w*; Agridè, Bitonto, ITA), malt barley flour (0.54% *w*/*w*; Antico Molino Rosso, Buttapietra, ITA) and S (25% *v*/*w*). DS was compared with *focaccia* dough (DB) prepared using baker's yeast (2% *w*/*w*, corresponding to a final yeast density of ca. 8 log cfu/g) and without liquid sourdough. Three independent tests were carried out. Dough portions of 180 g were placed in round non-stick pans and fermented at 30 °C for 6 h for DS or 1 h for DB. After fermentation, the microbial communities were analyzed, together with the pH drop (ΔpH, pH units) and total titratable acidity (TTA, ml di NaOH 0.1 N/10 g) for S, DS and DB [30].

2.3. Microbiological Analyses of Doughs

After fermentation, liquid sourdough (S) was immediately subjected to decimal dilutions and plating, while 20 g aliquots of *focaccia* doughs (DS and DB) or of a mixture (1:1 ratio) of durum wheat semolina and soft wheat flour type "00" (F) were each homogenized with 180 mL of sterile NaCl solution (0.85%, *w*/*v*) in a Stomacher (Seward, London, UK) for 2 min. After serial dilutions, the microbial suspensions were plated on an mMRS agar (Oxoid, UK) [38] supplemented with 100 mg/l of cycloheximide (EMD Millipore Corp., Billerica, MA, USA) for the determination of lactic acid bacteria (LAB), and on a Sabouraud Dextrose Agar (SDA, Oxoid, UK) supplemented with 200 mg/l chloramphenicol (Sigma, Milan, Italy) for the enumeration of yeasts and molds. Moreover, an aliquot of each microbial suspension was heat treated for 20 min at 90 °C, plated on a Plate Count Agar (PCA, Difco, Franklin Lakes, NJ, USA) and incubated for 24 h at 30 °C for spore-forming bacteria counts. A total of 20% of the colonies from the countable mMRS agar and SDA plates (incubated at 30 °C for 48 h and at 25 °C for 72 h, respectively) were randomly taken, purified and stored at −80 °C.

2.4. Characterization and Identification of LAB and Yeasts

Bacterial DNA was extracted from overnight cultures grown in mMRS broth (Oxoid, UK) at 30 °C, using a Clonsaver Card Kit (Whatman, Maidstone, UK) and analyzed by Repetitive Extragenic Palindromic-PCR (REP-PCR) [39]. The identification of *L. citreum* C2.27 was based on the comparison of its strain-specific REP-PCR profile with that of each LAB isolate from liquid sourdough and doughs. Bacterial isolates showing an REP-PCR profile different from that of the starter strain were identified by the sequencing of the almost complete 16S rRNA gene as previously described [24,39], using an ABI Prism 3730 × 1 DNA Analyzer (Thermo Fisher Scientific, Waltham, MA, USA). The species *Lactiplantibacillus paraplantarum* (basonym, *Lactobacillus paraplantarum*) [40] and *Lacticaseibacillus paracasei* (basonym *Lactobacillus paracasei*) were also identified by multiplex-PCR methods as described by Torriani et al. [41] and Ventura et al. [42], respectively.

The DNA of yeast isolates was extracted from 1.5 mL cultures grown in YEPG (Yeast Extract 1% *w*/*v*, Peptone 1% *w*/*v* and Dextrose 2% *w*/*v*) at 25 °C for 24 h, using the Wizard Genomic DNA Purification kit (Promega Corporation, Madison, WI, USA), and amplified by the oligonucleotide $(GTG)_5$ [43]. The isolates were identified by amplification and sequencing of the D1/D2 domain of the 26S rDNA using the primers NL1 and NL4 [44]. Bacterial and yeast strains were assigned to the species on the basis of the highest scores of alignment and percentage of identity (>99%) between their 16S rRNA/26S rRNA gene sequences and those of type strains in the NCBI Nucleotide database [24].

2.5. Culture-Independent Community Identifications

Ninety milliliters of saline solution was added to 10 g of flour mixture (F) or doughs DS and DB and homogenized for 3 min. Homogenates were centrifuged (1000× *g* for 5 min at 4 °C) and the supernatants were recovered and centrifuged (5000× *g* for 15 min at 4 °C). Each pellet was suspended in 0.5 mL of saline solution and the suspension was subjected to DNA extraction. Total genomic DNA was extracted using the FastDNA Pro Soil-Direct kit (MPBiomedicals, Santa Ana, CA, USA) coupled to the Retsch MM301 instrument (Retsch Gmbh, Germany), according to the manufacturer's instructions. Quality and quantity of DNA extracts were estimated using NanoDrop ND1000 (NanoDrop Technologies, Inc., Wilmington, NC, USA) and by 1% (*w*/*v*) agarose gel electrophoresis in TAE buffer. DNA extraction was carried out in triplicate on each sample.

2.6. Library Preparation and Sequencing

The total DNA extracted from the flour mixture and dough samples was used as template for 16S and 18S metagenetic analyses. All DNA samples were equalized at a final concentration of 10 ng/µL for NGS library preparation. Library preparation was performed using a bidirectional fusion primer set to specifically amplify the target regions. This was accomplished by combining primers targeting the regions of interest to the Ion Torrent sequencing adapters Ion A and truncated P1 (trP1). Adapter A included unique Ion Xpress Barcode sequences for sample multiplexing. Bidirectional sequencing was achieved by swapping adapter sequences A and trP1. Therefore, the V5–V6 hypervariable region of the16S rRNA gene was amplified with primers 785F (GGATTAGATACCCTGGTA) and 1100R (GGGTTGCGCTCGTTG). For 18S libraries, the V9 hypervariable region was amplified with primers 1380F (CCCTGCCHTTTGTACACAC) and 1510R (CCTTCYGCAGGTTCACCTAC). The PCR reactions were carried out in triplicates using Platinum SuperFi DNA Polymerase (Invitrogen, USA) with the following conditions:

98 °C for 30 min, 25 cycles of 98 °C for 10 s, 58 °C for 20 s and 72 °C for 30 s and a final extension of 72 °C for 5 min. Amplified libraries were verified on 2% agarose gel, and PCR products were purified using Agencourt AMPure XP magnetic beads (Agencourt Bioscience, Beverly, MA, USA). Purified libraries were quantified with Qubit dsDNA HS Assay Kit (Invitrogen, Carlsbad, CA, USA) and pooled at a final concentration of 100 pM. Libraries were sequenced on an Ion S5 Sequencing System (Thermofisher, Waltham, MA, USA). Following the sequencing run, a FASTQ file was generated for each sample and demultiplexed by using the built-in software of Torrent Suite.

2.7. Bioinformatics

Quality-filtered reads were aligned using MALT (v0.4.1) [45], in BlastN mode, against the SILVA ribosomal RNA sequence database (SSURef_NR99, release 128), with default settings and alignment type set to "SemiGlobal". For taxonomic binning, MEGAN software (MEGAN 6 v6.18.1) [46] was used. For 16S libraries, the LCA algorithm was set with the following settings: minScore = 500, maxExpected = 1.0, minIdentity= 0.01, topPercent = 10 and minSupport = 10. For 18S libraries, the LCA algorithm was set with the following settings: minScore = 250, maxExpected = 1.0, minIdentity= 0.01, topPercent = 10 and minSupport = 10. Sequences were assigned at the taxonomic level with at least 97% identity for genus and 99% for species in the reference database.

2.8. Statistics

Analysis of variance (ANOVA) combined with the Tukey–Kramer method as a post hoc test was applied for microbiological analysis. Significant differences ($p < 0.05$) among *focaccia* dough samples are marked with different letters. Rank abundance and alpha and beta diversity indexes were estimated using Microbiome Analyst [47]. Moreover, principal coordinate analysis (PCoA) was calculated using the Bray–Curtis index. Hypothesis testing was conducted by the analysis of molecular variance (AMOVA) test ($p < 0.05$) [48]. Statistical comparison among taxonomic categories was performed using STAMP software [49]. Differences between groups were analyzed using Welch's t-test ($p < 0.05$) [50], and the Benjamini–Hochberg false discovery rate (FDR) method (q-value < 0.05) was used [51].

2.9. Data Availability

The sequence data are available at NCBI SRA under BioProject ID: PRJNA668040.

3. Results

3.1. Culture-Dependent Microbiological Analyses

Durum wheat semolina and soft wheat flour type "00" in a 1:1 ratio (F) and, at the end of fermentation, S, DS and DB doughs were subjected to microbiological analyses to assess the cell load of LAB, yeasts, molds and spore-forming bacteria (Table 1). The use of sourdough influenced the pH values of DS, which, after fermentation, showed pH 4.2 and ΔpH 1.3, while DB presented a pH value (5.38) that was significantly higher ($p < 0.05$). Moreover, S showed pH and TTA values lower than DS. This apparently contrasting result is due to the higher value of DY in S than in DS. In fact, it is known that a low DY value, as in DS, increases the buffering capacity of the flour, lowering the rate of acidification, even if organic acids are present at higher levels [52].

S and DS showed cell densities of LAB (9 log cfu/g) that were significantly higher than DB (4 log cfu/g). LAB isolates from liquid sourdough and doughs (20% of the colonies from countable plates) were characterized by REP-PCR. The analysis of their electrophoretic profiles revealed the presence of only the starter strain in S and DS, while four LAB strains were detected in DB, belonging to *Leuconostoc mesenteroides*, *Furfurilactobacillus rossiae* (basonym *Lactobacillus rossiae*), *Lp. paraplantarum* and *Lc. paracasei*. These species are typically contained in sourdough [17] and are among the species found in doughs made with different flours from the same geographical area [24]. In addition to the absence of LAB strains isolated from DB, the only strain isolated directly from the flours, belonging to

Loigolactobacillus coryniformis (basonym *Lactobacillus coryniformis*), was also absent in DS. It is probable that this strain was no longer found in the doughs because it was dominated during fermentation by the starter strain or by the other strains, the latter evidently present in the flours in quantities lower than the detection limit.

The presence of spore-forming bacteria in the doughs was also evaluated. Bacterial counts were very low (<1 log cfu/g) and therefore below the quantities which could cause product alteration (ropy bread) (≥2 log cfu/g) or constitute a risk for consumer health (>5 log cfu/g) [53].

Yeasts were absent in S and DS, while molds were present in very low quantities. On the contrary, DB contained a high density of yeasts (8.25 log cfu/g), while molds were not detected. The REP-PCR analysis of the yeasts showed the presence of only one strain belonging to *S. cerevisiae*. Finally, only one yeast strain belonging to *Cryptococcus victoriae* was isolated from the flours but was not subsequently found in the doughs, probably due to the characteristics of the ecosystem and the presence of the starters. Overall, the results show the capacity of the *L. citreum* C2.27 strain to dominate the microbiota in DS, and, at the same time, the ability of *S. cerevisiae* to dominate the mycobiota in DB.

3.2. Bacterial and Fungal Microbiota in Flour and Doughs

After quality filtering, a total of 3,201,028 and 7,194,310 reads were generated and 3,055,906 (95.5%) and 6,648,063 (92.4%) were assigned to a taxon for 16S and 18S, respectively.

The abundance-based coverage estimators (ACE) and the alpha diversity index Chao1 indicated that the richness of the samples was DB > F > DS (Figure 2A). The inoculum of *L. citreum* reduced the abundance of OTUs, if compared with DB and F samples. The loss of OTUs also reduced both richness (Chao1 and ACE) and diversity (Shannon and Simpson) indexes in DS samples, indicating that the use of *L. citreum* as a bacterial starter had a strong effect on microbial community diversity. Considering the same population indexes, even the mycobiota diversity was strongly influenced by the presence of *S. cerevisiae* in the DB sample (Figure 2B).

Table 1. Acidification (ΔpH, pH units), total titratable acidity (TTA, ml di NaOH 0.1 N/10 g) and microbiological characteristics of durum wheat semolina and soft wheat flour type "00" in a 1:1 ratio (F), liquid sourdough (S), doughs started with the liquid sourdough (DS) and dough made with baker's yeast (DB).

Sample	pH	ΔpH	TTA	LAB (log cfu/g)	Other LAB Species Isolates	Spore-Forming Bacteria (log cfu/g)	Yeasts (log cfu/g)	Molds (log cfu/g)	Yeast Species
F	-	-	-	0.46 ± 0.28	*Ll. coryniformis*	0.8 ± 0.72	0.7 ± 1.15	2.59 ± 0.12	*Cr. victoriae*
S	3.53 ± 0.07	2.42 ± 0.14	4.15 ± 0.21	9.47 ± 0.13	nd	nd	nd	1.57 ± 0.04	nd
DS	4.23 ± 0.04 ^{b}	1.35 ± 0.09 ^{a}	8.17 ± 0.55 ^{a}	9.09 ± 0.01 ^{a}	nd	0.54 ± 0.08	nd	2.70 ± 0.36	nd
DB	5.38 ± 0.02 ^{a}	0.25 ± 0.01 ^{b}	3.55 ± 0.21 ^{b}	4.48 ± 0.08 ^{b}	*Fl. rossiae* *L. mesenteroides* *Lp. paraplantarum* *Lc. paracasei*	nd	8.25 ± 0.01	nd	*S. cerevisiae*

Data represent means of three independent experiments ± standard error. $^{a–b}$ Values referring to *focaccia* dough in the same column with different letters differ significantly ($p < 0.05$). *Ll., Loigolactobacillus; Fl., Furfurilactobacillus; L., Leuconostoc; Lp., Lactiplantibacillus; Lc., Lacticaseibacillus; Cr., Cryptococcus; S., Saccharomyces*. nd: not detected (<LOD, limit of detection).

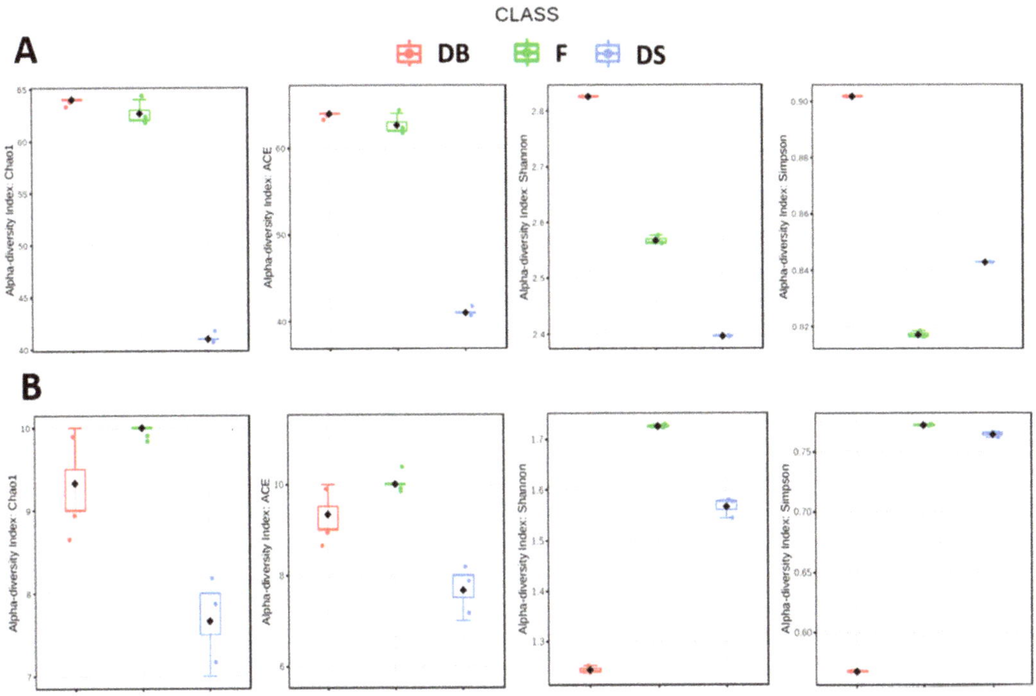

Figure 2. Alpha diversity indexes (Chao1, ACE, Shannon, Simpson) of bacterial (**A**) and fungal (**B**) microbiota in dough with baker's yeast (DB), dough with sourdough (DS) and flour (F) samples.

The PCoA analysis showed that the microbial communities in DB, DS and F samples were distinct from each other. The sample triplicates resemble each other very closely (Figure 3). The results were also validated with analysis of molecular variance (AMOVA). This confirmed that the inoculum of *L. citreum* or *S. cerevisiae* significantly affected microbial consortia in DS and DB samples, respectively.

Considering the absolute number of assigned reads, the phyla of *Proteobacteria* (1,508,548 assigned reads) and *Firmicutes* (1,322,874 assigned reads) were the most representative in all the samples. The DB microbial consortium was composed mainly of ten different genera belonging to the *Proteobacteria* phylum. *Stenotrophomonas* (relative abundance of 30.0%) was predominant, followed by *Pseudomonas* (16.6%), *Acinetobacter* (15.5%), *Sphingomonas* (5.0%), *Agrobacterium* (2.5%), *Siccibacter* (2.3%), *Cronobacter* (1.4%), *Enterobacter* (1.3%), *Gluconobacter* (1.2%) and *Brevundimonas* (1.0%). Within the *Firmicutes* phylum, the relative abundance was mainly represented by *Lactococcus* (6.8%), *Lactobacillus* (2.9%), *Paenibacillus* (2.7%), *Pediococcus* (2.0%) and *Leuconostoc* (1.9%). From the DS sample, the genera belonging to the *Proteobacteria* phylum were *Stenotrophomonas* (11.5%), *Acinetobacter* (10.7%), *Pseudomonas* (10.3%), *Sphingomonas* (3.2%), *Siccibacter* (2.0%), *Cronobacter* (1.7%), *Gluconobacter* (1.2%), *Enterobacter* (1.1%) and *Xanthomonas* (1.1%). Within *Firmicutes*, the genus *Leuconostoc* (41.4%) was dominant, followed by *Lactococcus* (7.1%), *Pediococcus* (3.7%), *Paenibacillus* (1.0%) and *Lactobacillus* (<1%). From the F sample, the genera belonging to the *Proteobacteria* phylum were *Pseudomonas* (49.1%), *Rosenbergiella* (7.9%), *Xanthomonas* (6.4%), *Sphingomonas* (5.4%), *Stenotrophomonas* (2.5%), *Enterobacter* (2.3%), *Novosphingobium* (2.0%) and *Siccibacter* (1.3%). Within *Firmicutes*, *Paenibacillus* (8.0%), *Staphylococcus* (4.4%), *Lactobacillus* (2.1%) and *Leuconostoc* (1.1%) were detected (Figure 4).

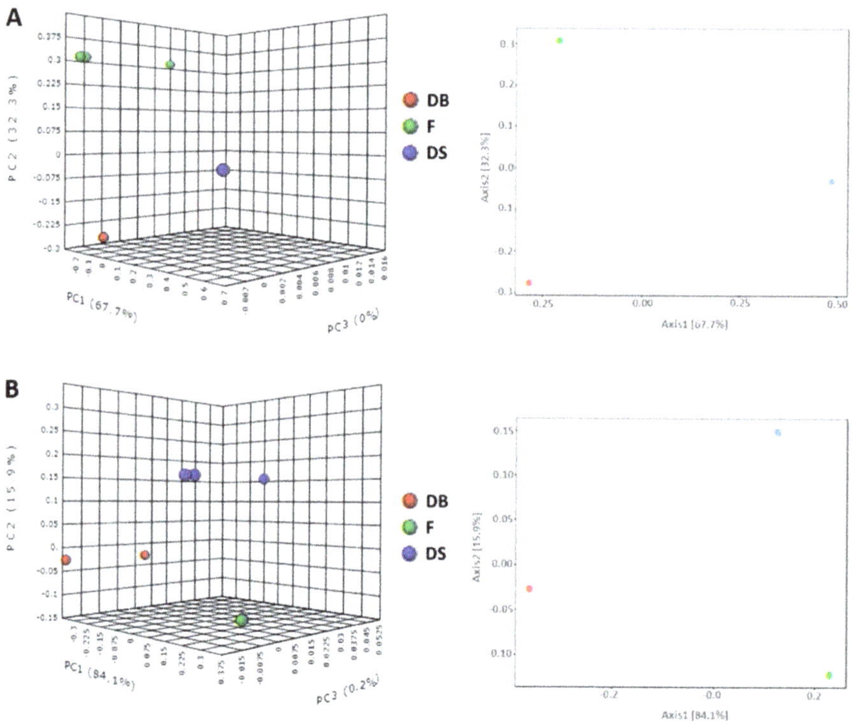

Figure 3. Principal coordinate analysis (PCoA) as 2D (right) and 3D (left) plots of bacterial (**A**) and fungal (**B**) communities of dough with baker's yeast (DB), dough with sourdough (DS) and flour (F) samples. PCoA was calculated using the Bray–Curtis index to compute dissimilarities among different samples. Hypothesis testing was conducted by analysis of molecular variance (AMOVA) test ($p < 0.05$).

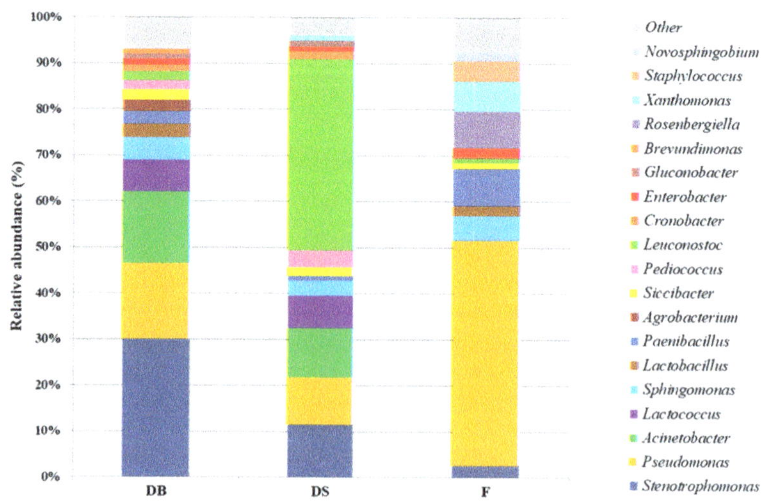

Figure 4. Relative abundance and taxonomic assignments of microbial flora at genus level of dough with baker's yeast (DB), dough with sourdough (DS) and flour (F) samples. Minor genera with <0.1% relative abundance are represented as "Other".

The relative abundances of identified genera were further compared between sample groups, as shown in Figures 5 and 6.

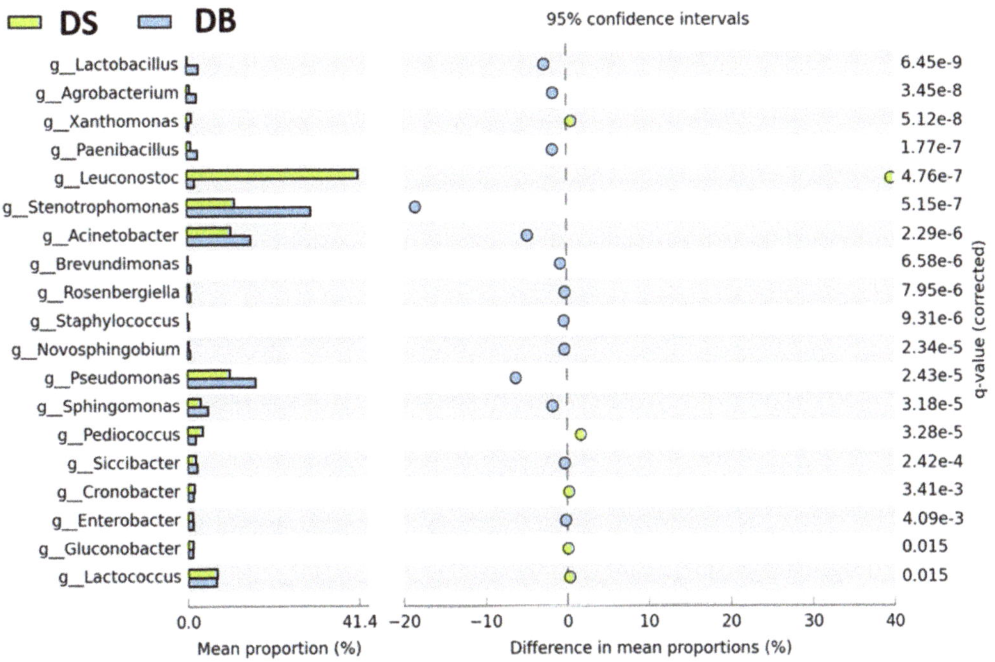

Figure 5. Genera with a significantly different abundance between dough with sourdough (DS) and dough with baker's yeast (DB) samples. The statistical cutoffs of $p < 0.05$ (Welch test) and *q-value* < 0.05 (FDR) were set as the significance threshold.

In addition to the dominance of the genus *Leuconostoc* (relative abundance 39.6% higher than in DB), DS showed a reduction in abundance of 18.5% for *Stenotrophomonas*, 6.3% for *Pseudomonas*, 4.8% for *Acinetobacter* and 1.7% for *Paenibacillus*, compared to DB. Furthermore, within LAB, *Lactobacillus* was significantly lower and *Pediococcus* was significantly higher in DS than in DB (Figure 5). In terms of relative abundance, flour (F) had significantly higher percentages of genera *Pseudomonas* (up to 39%), *Paenibacillus*, *Enterobacter* and *Xanthomonas* than both doughs (Figure 6), while *Sphingomonas* and *Lactobacillus* were higher in F than in DS. *Novosphingobium*, *Rosenbergiella* and *Staphylococcus* genera were present only in F.

The metagenetic analysis of mycobiota revealed that the *Saccharomyces* genus dominated the mycobiota in DB and was not detected in other samples (Figures 7 and 8). Considering the absolute number of reads assigned to a taxon, in DB, the overall number of reads assigned to the *Ascomycota* phylum was 6805, while it was 933 in DS and 992 in F. In detail, in the DB samples, 91.5% of reads were assigned to the *Saccharomyces* genus, followed by *Saturnispora* (3.7%), *Geotrichum* (2.0%), *Wickerhamomyces* (2.0%) and *Penicillium* (0.7%). In DS, the reads were predominately assigned to *Saturnispora* (69.3%), followed by *Wickerhamomyces* (27.6%) and *Geotrichum* (3.1%), while in F, the reads were assigned to genera belonging to *Aspergillus* (45.4%), *Hyphopichia* (30.2%), *Cryptococcus* (10.2%), *Penicillium* (9.3%) and *Cladosporium* (4.9%).

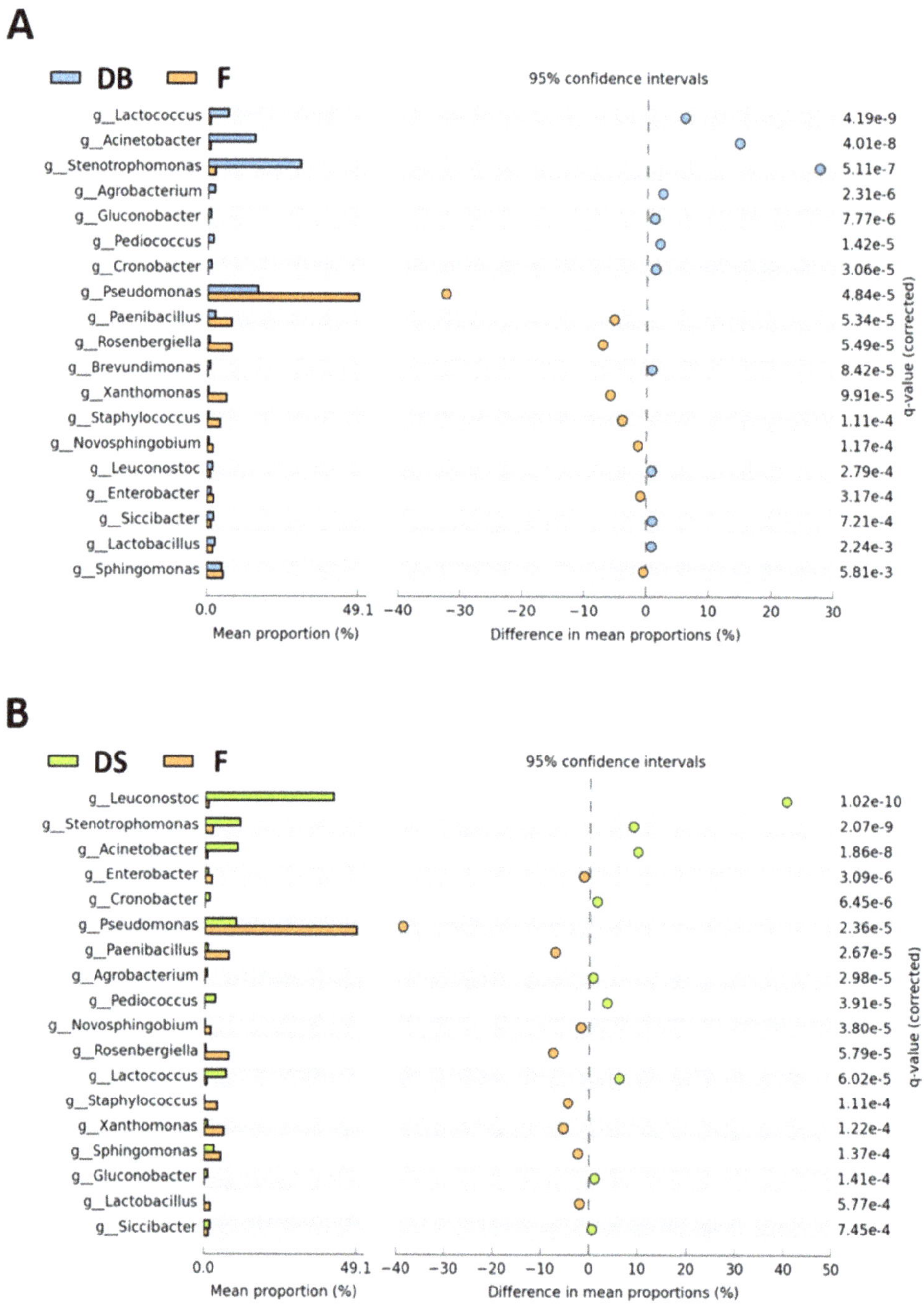

Figure 6. Genera with a significantly different abundance between dough with baker's yeast (DB) and flour (F) samples (**A**) and dough with sourdough (DS) and flour (F) samples (**B**). The statistical cutoffs of $p < 0.05$ (Welch test) and *q-value* < 0.05 (FDR) were set as the significance threshold.

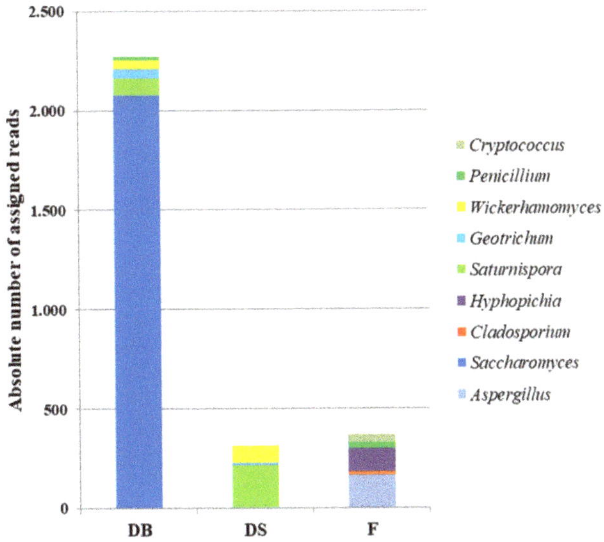

Figure 7. Absolute number of assigned reads to genera belonging to Ascomycota and Basidiomycota phyla for dough with baker's yeast (DB), dough with sourdough (DS) and flour (F) samples.

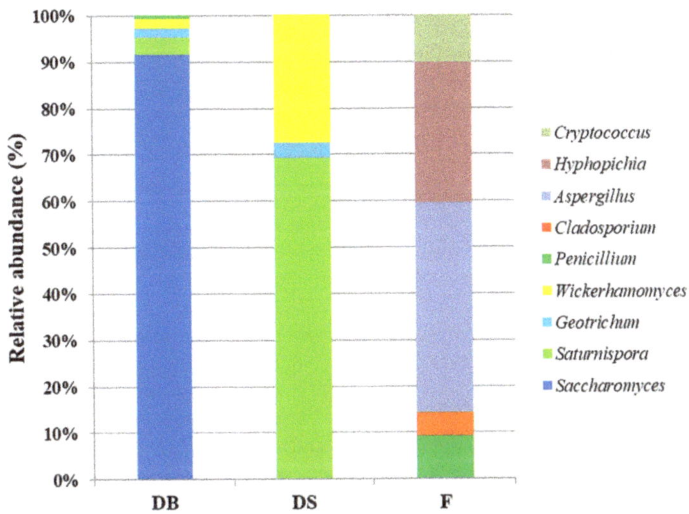

Figure 8. Relative abundance of genera belonging to *Ascomycota* and *Basidiomycota* phyla for dough with baker's yeast (DB), dough with sourdough (DS) and flour (F) samples. Data are represented as a normalized percentage of the assigned genera.

4. Discussion

The use of selected starter cultures for dough fermentation is useful to obtain products with determined characteristics and to standardize the production processes [12,13,54]. Starters are known to have a strong impact on the microbial populations of doughs, consequently affecting product features. Therefore, the present work used the metagenetic method and the culture-dependent approach to analyze the biodiversity of doughs started with sourdough (type II) or baker's yeast. It is well known that during traditional sourdough (type I) fermentation, the bacterial evolution is characterized by a decrease in

Gram-negative and an increase in Gram-positive biota [55]. Ercolini et al. [29] observed flours with great diversity, mainly contaminated by genera (*Acinetobacter*, *Pantoea*, *Pseudomonas*, *Comamonas*, *Enterobacter*, *Erwinia* and *Sphingomonas*) belonging to the phylum *Proteobacteria*, although this population was almost completely inhibited after 1 day of sourdough propagation. On the contrary, although the phylum *Firmicutes* (mainly represented by LAB) was present at low relative abundances in the flours, it already became dominant after the first fermentation [29,55]. These microbial dynamics from flour to mature sourdough are well known and are in accordance with the microbial ecology of other fermented foods [16]. Flours analyzed in the present study were also found to be strongly contaminated by *Proteobacteria* (ca. 82%), and *Pseudomonas* dominated (49%) this microbiota. Contaminating bacteria originate from grain milling or are epiphytic and endophytic populations of wheat. In this study, sourdough type II (25%) or baker's yeast (2%) was used as a single starter for dough leavening. The metagenetic analysis highlighted that, after fermentation, the *Proteobacteria* in DS were about 43.5% lower than in F and DB. At the same time, the flour population was replaced mainly by the starter strain. In fact, the metagenetic analysis showed that the *Leuconostoc* genus constituted 78.6% of LAB in DS. Moreover, as supported by the identification of cultivable LAB, *L. citreum* C2.27 was the only LAB strain isolated from DS.

On the contrary, *Proteobacteria* was the dominant phylum in F and DB samples. This phylum represents metabolically active populations, and the abundance of species belonging to the phylum *Proteobacteria* was mostly negatively correlated with dough and sourdough qualities [29,56]. It can be hypothesized that in dough with baker's yeast, the fermentation conditions and the autochthonous microbiota of wheat flour, characterized, in particular, by the low LAB density, did not cause the reduction in *Proteobacteria*, which constituted, as in F, 82% of the bacterial population. In fact, the activity of the acid-producing bacterial starter during fermentation caused a rapid increase in the acidity of DS. Such an increase, together with other ecological parameters [16], may influence the microbial succession favoring LAB. On the other hand, LAB showed an increase within *Firmicutes* also in DB; in particular, *Lactococcus*, *Lactobacillus*, *Pediococcus* and *Leuconostoc* increased compared with F, but this was not sufficient to decrease the pH.

As described in other studies, *Staphylococcus*, *Streptococcus* and *Paenibacillus* had low ability to survive and colonize the doughs, especially DS. In fact, these genera are not generally found in doughs [17,31]. Regarding the presence of the genus *Leuconostoc* in F and DB detected by the metagenetic approach, it should be noted that this is probably due to the species *L. mesenteroides* isolated from DB together with *Fb. rossiae*, *Lp. paraplantarum* and *Lc. paracasei.* Metagenetic analysis also showed the presence of *Lactococcus* in the doughs, although it was not isolated from DS, probably due to the high starter load, or from DB, due to competition with other genera. In this regard, it should be considered that differences can be noted between metagenetic analysis and plating, especially for those LAB species that are generally difficult to cultivate [35].

Therefore, the flour plays a key role in establishing microbial consortia, but only species and/or strains adapted to the sourdough environment in relation to the nutrient availability and physic-chemical parameters of the process will grow and dominate in this ecosystem [17]. In type II sourdough, fermentation occurs after the inoculation of a starter culture. This starter culture can dominate and inhibit the growth of autochthonous dough microbiota because it is added in a high density [13]. However, the starter strain should be well adapted to the cereal environment in order to compete with the endogenous microbiota and be suitable for the process [57]. *L. citreum* C2.27, which was selected for its good leavening capacity, was isolated from durum wheat semolina and was suitable for the production process of yeast-free bread in the bakery [24]. In fact, it was the only strain isolated from liquid sourdough and DS, and the present study also confirmed its dominance by culture-independent analysis. This result is fundamental to guarantee the technological characteristics of the product made without using baker's yeast, resulting, at the same time, in positive chemical-physical and nutritional characteristics of the dough and final product, as determined in a previous work [27].

The results show that yeasts were present in very low quantities in DS and F. They are normally associated with flours up to 3.3 log cfu/g [57], and the genera were those typically found in flours and doughs [17,24,58]. *Ascomycota* was the only phylum present in doughs, but not in the flours evaluated. The mycobiota changed from flour to doughs; in fact, the genera *Hyphopichia* and *Cryptococcus* were found only in F, and the latter was also the only genus isolated directly from F. Considering the mycobiota of flours, molds were dominant in F, but, overall, in very low quantities. Therefore, the metagenetic approach made it possible to obtain more information on the mycobiota composition even if present in small numbers.

The absence of *Saccharomyces* in F and DS was interesting. *S. cerevisiae* is the species of yeast most frequently isolated in sourdoughs from central and southern Italy [52]. However, several studies have shown different compositions of yeast species between artisanal bakery and spontaneous laboratory sourdoughs and, in particular, have hypothesized the presence of *S. cerevisiae* in bakery sourdoughs due to contamination of the bakery environment with commercial baker's yeast [58,59]. Conversely, as observed in the present study, yeasts detected in sourdough and doughs made in the laboratory could only come from flours [24,60].

Finally, the results for DS, which was dominated by *L. citreum* C2.27, also show a reduction in the fungal diversity, probably related to the antifungal properties of the starter strain [36], while DB, with a *S. cerevisiae*-dominated mycobiota, presented a greater bacterial diversity.

5. Conclusions

This study showed how the use of a microbial starter deeply affects the composition of the dough microbiota, which is directly responsible for the quality of the product. To our knowledge, this work is the first that analyzes, also using the metagenetic method, the bacterial and fungal microbiota of doughs leavened with baker's yeast or type II sourdough (started with a selected bacterial strain), also making it possible to verify the absence of *S. cerevisiae* in the latter used to obtain an innovative "yeast-free" product.

Metagenetic analyses indicated that the V5–V6 hypervariable regions of the 16S rRNA gene and the V9 hypervariable region of the 18S rRNA gene were suitable for studying the microbiota of doughs, providing a comprehensive overview of the microbial community. The culture-independent approach allowed gaining deeper and wider knowledge of the starter impact on the microbial populations of the doughs, even if the association with the culture-dependent analysis is still very useful to obtain more precise information at the species level on certain microbial categories. This study highlighted that, differently from doughs produced with the *L. citreum* starter, the dough made with baker's yeast contained a microbiota with a high abundance of *Proteobacteria* (82% of the bacterial population), which were negatively correlated with the biochemical properties of the doughs [27,56]. Furthermore, the analyses showed the ability of the *L. citreum* C2.27 starter to dominate the microbiota, also inhibiting the growth of *S. cerevisiae*. This result is particularly important because *L. citreum* C2.27 has been adopted for its leavening abilities in a biotechnological protocol for the production of "yeast-free" bakery products.

Author Contributions: Conceptualization, P.D.B.; formal analysis, M.F.; investigation, M.F. and P.D.B.; supervision, P.D.B.; visualization, M.F.; writing—original draft, M.F. and P.D.B.; writing—review and editing, A.S., G.M., P.L. and P.D.B. All authors have read and agreed to the published version of the manuscript.

Funding: This research was funded by CNR project NUTR-AGE (FOE-2019, DSB.AD004.271).

Data Availability Statement: The sequence data are available at NCBI SRA under BioProject ID: PRJNA668040.

Acknowledgments: The authors thank Lisa Lonigro for her technical assistance in REP-PCR analysis, Domenico Genchi for his IT support and Sarah Jane Christopher for the English editing of the manuscript.

Conflicts of Interest: The authors declare no conflict of interest.

References

1. Catzeddu, P. Sourdough Breads. In *Flour and Breads and Their Fortification in Health and Disease Prevention*; Preedy, V.R., Watson, R.R., Patel, V.B., Eds.; Academic Press: Cambridge, MA, USA; Elsevier: London, UK; Burlington, VT, USA; San Diego, CA, USA 2011; pp. 37–46. [CrossRef]
2. Liu, T.; Li, Y.; Sadiq, F.A.; Yang, H.; Gu, J.; Yuan, L.; Lee, Y.K.; He, G. Predominant yeasts in Chinese traditional sourdough and their influence on aroma formation in Chinese steamed bread. *Food Chem.* **2018**, *242*, 404–411. [CrossRef] [PubMed]
3. Suo, B.; Nie, W.; Wang, Y.; Ma, J.; Xing, X.; Huang, Z.; Xu, C.; Li, Z.; Ai, Z. Microbial diversity of fermented dough and volatile compounds in steamed bread prepared with traditional Chinese starters. *LWT Food Sci. Technol.* **2020**, *126*, 109350. [CrossRef]
4. Decreto Ministeriale 7 Febbraio 2019. Aggiornamento Dell'elenco Nazionale Dei Prodotti Agroalimentari Tradizionali Ai Sensi Dell'articolo 12, Comma 1, Della Legge 12 Dicembre 2016, N. 238. Gazzetta Ufficiale N. 60 Del 12 Marzo 2019–Supplemento Ordinario N. 9. Available online: http://www.gazzettaufficiale.it/eli/gu/2019/03/12/60/so/9/sg/pdf (accessed on 3 May 2021).
5. Pasqualone, A.; Delcuratolo, D.; Gomes, T. Focaccia Italian flat fatty bread. In *Flour and Breads and Their Fortification in Health and Disease Prevention*; Preedy, V.R., Watson, R.R., Patel, V.B., Eds.; Academic Press: Cambridge, MA, USA; Elsevier: London, UK, Burlington, VT, USA; San Diego, CA, USA, 2011; pp. 47–58. [CrossRef]
6. Calorie in Focaccia Pugliese e Valori Nutrizionali. Available online: https://www.fatsecret.it/calorie-nutrizione/generico/focaccia-pugliese (accessed on 3 May 2021).
7. Carnevali, P.; Ciati, R.; Leporati, A.; Paese, M. Liquid sourdough fermentation: Industrial application perspectives. *Food Microbiol.* **2007**, *24*, 150–154. [CrossRef] [PubMed]
8. Li, Z.; Song, K.; Li, H.; Ma, R.; Cui, M. Effect of mixed *Saccharomyces cerevisiae* Y10 and *Torulaspora delbrueckii* Y22 on dough fermentation for steamed bread making. *Int. J. Food Microbiol.* **2019**, *303*, 58–64. [CrossRef]
9. Lattanzi, A.; Minervini, F.; Di Cagno, R.; Diviccaro, A.; Antonielli, L.; Cardinali, G.; Cappelle, S.; De Angelis, M.; Gobbetti, M. The lactic acid bacteria and yeast microbiota of eighteen sourdoughs used for the manufacture of traditional Italian sweet leavened baked goods. *Int. J. Food Microbiol.* **2013**, *163*, 71–79. [CrossRef] [PubMed]
10. Reale, A.; Di Renzo, T.; Boscaino, F.; Nazzaro, F.; Fratianni, F.; Aponte, M. Lactic Acid Bacteria Biota and Aroma Profile of Italian Traditional Sourdoughs from the Irpinian Area in Italy. *Front. Microbiol.* **2019**, *10*, 1621. [CrossRef] [PubMed]
11. Gobbetti, M.; Rizzello, C.G.; Di Cagno, R.; De Angelis, M. How the sourdough may affect the functional features of leavened baked goods. *Food Microbiol.* **2014**, *37*, 30–40. [CrossRef]
12. Gobbetti, M.; De Angelis, M.; Di Cagno, R.; Calasso, M.; Archetti, G.; Rizzello, C.G. Novel insights on the functional/nutritional features of the sourdough fermentation. *Int. J. Food Microbiol.* **2019**, *302*, 103–113. [CrossRef]
13. Siepmann, F.B.; Ripari, V.; Waszczynskyj, N.; Spier, M.R. Overview of sourdough technology: From production to marketing. *Food Bioprocess Technol.* **2018**, *11*, 242–270. [CrossRef]
14. Belz, M.C.E.; Axel, C.; Arendt, E.K.; Lynch, K.M.; Brosnan, B.; Sheehan, E.M.; Coffey, A.; Zannini, E. Improvement of taste and shelf life of yeasted low-salt bread containing functional sourdoughs using *Lactobacillus amylovorus* DSM 19280 and *Weissella cibaria* MG1. *Int. J. Food Microbiol.* **2019**, *302*, 69–79. [CrossRef]
15. Mantzourani, I.; Plessas, S.; Odatzidou, M.; Alexopoulos, A.; Galanis, A.; Bezirtzoglou, E.; Bekatorou, A. Effect of a novel Lactobacillus paracasei starter on sourdough bread quality. *Food Chem.* **2019**, *271*, 259–265. [CrossRef]
16. Minervini, F.; De Angelis, M.; Di Cagno, R.; Gobbetti, M. Ecological parameters influencing microbial diversity and stability of traditional sourdough. *Int. J. Food Microbiol.* **2014**, *171*, 136–146. [CrossRef] [PubMed]
17. De Vuyst, L.; Van Kerrebroeck, S.; Harth, H.; Huys, G.; Daniel, H.M.; Weckx, S. Microbial ecology of sourdough fermentations: Diverse or uniform? *Food Microbiol.* **2014**, *37*, 11–29. [CrossRef] [PubMed]
18. Valerio, F.; Bavaro, A.R.; Di Biase, M.; Lonigro, S.L.; Logrieco, A.F.; Lavermicocca, P. Effect of Amaranth and Quinoa Flours on Exopolysaccharide Production and Protein Profile of Liquid Sourdough Fermented by *Weissella cibaria* and *Lactobacillus plantarum*. *Front. Microbiol.* **2020**, *11*, 967. [CrossRef]
19. Papadimitriou, K.; Zoumpopoulou, G.; Georgalaki, M.; Alexandraki, V.; Kazou, M.; Anastasiou, R.; Tsakalidou, E. Sourdough Bread. In *Innovations in Traditional Foods*; Galanakis, C.M., Ed.; Woodhead Publishing: Cambridge, UK; Elsevier: Amsterdam, The Netherlands, 2019; pp. 127–158. [CrossRef]
20. Zhang, Z.; Li, C.; Zhao, X.; Lv, C.; He, Q.; Lei, S.; Guo, Y.; Zhi, F. Anti-Saccharomyces cerevisiae Antibodies Associate with Phenotypes and Higher Risk for Surgery in Crohn's Disease: A Meta-Analysis. *Dig. Dis. Sci.* **2012**, *57*, 2944–2954. [CrossRef]
21. Caselli, M.; Lo Cascio, N.; Rabitti, S.; Eusebi, L.H.; Zeni, E.; Soavi, C.; Cassol, F.; Zuliani, G.; Zagari, R.M. Pattern of food intolerance in patients with gastro-esophageal reflux symptoms. *Minerva Med.* **2017**, *108*, 496–501. [CrossRef] [PubMed]
22. Cannistrà, C.; Finocchi, V.; Trivisonno, A.; Tambasco, D. New perspectives in the treatment of hidradenitis suppurativa: Surgery and brewer's yeast–exclusion diet. *Surgery* **2013**, *154*, 1126–1130. [CrossRef]
23. Cordell, B.; Kanodia, A. Auto-Brewery as an Emerging Syndrome: Three Representative Case Studies. *J. Clin. Med. Case Rep.* **2015**, *2*, 5.

24. De Bellis, P.; Rizzello, C.G.; Sisto, A.; Valerio, F.; Lonigro, S.L.; Conte, A.; Lorusso, V.; Lavermicocca, P. Use of a selected *Leuconostoc citreum* strain as a starter for making a "yeast-free" bread. *Foods* **2019**, *8*, 70. [CrossRef]
25. Musatti, A.; Cappa, C.; Mapelli, C.; Alamprese, C.; Rollini, M. *Zymomonas mobilis* in Bread Dough: Characterization of Dough Leavening Performance in Presence of Sucrose. *Foods* **2020**, *9*, 89. [CrossRef]
26. Bottani, M.; Brasca, M.; Ferraretto, A.; Cardone, G.; Casiraghi, M.C.; Lombardi, G.; De Noni, I.; Cattaneo, S.; Silvetti, T. Chemical and nutritional properties of white bread leavened by lactic acid bacteria. *J. Funct. Foods* **2018**, *45*, 330–338. [CrossRef]
27. De Bellis, P.; Montemurro, M.; D'Imperio, M.; Rizzello, C.G.; Sisto, A.; Lavermicocca, P. Production of a yeast-free focaccia with reduced salt content using a selected *Leuconostoc citreum* strain and seawater. *LWT Food Sci. Technol.* **2020**, *134*, 109918. [CrossRef]
28. De Angelis, M.; Minervini, F.; Siragusa, S.; Rizzello, C.G.; Gobbetti, M. Wholemeal wheat flours drive the microbiome and functional features of wheat sourdoughs. *Int. J. Food Microbiol.* **2019**, *302*, 35–46. [CrossRef] [PubMed]
29. Ercolini, D.; Pontonio, E.; De Filippis, F.; Minervini, F.; La Storia, A.; Gobbetti, M.; Di Cagno, R. Microbial ecology dynamics during rye and wheat sourdough preparation. *Appl. Environ. Microbiol.* **2013**, *79*, 7827–7836. [CrossRef]
30. Coda, R.; Kianjam, M.; Pontonio, E.; Verni, M.; Di Cagno, R.; Katina, K.; Rizzello, C.G.; Gobbetti, M. Sourdough-type propagation of faba bean flour: Dynamics of microbial consortia and biochemical implications. *Int. J. Food Microbiol.* **2017**, *248*, 10–21. [CrossRef]
31. Minervini, F.; Lattanzi, A.; De Angelis, M.; Celano, G.; Gobbetti, M. House microbiotas as sources of lactic acid bacteria and yeasts in traditional Italian sourdoughs. *Food Microbiol.* **2015**, *52*, 66–76. [CrossRef] [PubMed]
32. Raimondi, S.; Amaretti, A.; Rossi, M.; Fall, P.A.; Tabanelli, G.; Gardini, F.; Montanari, C. Evolution of microbial community and chemical properties of a sourdough during the production of Colomba, an Italian sweet leavened baked product. *LWT Food Sci. Technol.* **2017**, *86*, 31–39. [CrossRef]
33. Bokulich, N.A.; Mills, D.A. Improved selection of internal transcribed spacer-specific primers enables quantitative, ultra-high-throughput profiling of fungal communities. *Appl. Environ. Microbiol.* **2013**, *79*, 2519–2526. [CrossRef]
34. De Filippis, F.; Laiola, M.; Blaiotta, G.; Ercolini, D. Different amplicon targets for sequencing-based studies of fungal diversity. *Appl. Environ. Microbiol.* **2017**, *83*, e00905-17. [CrossRef]
35. Weckx, S.; Van Kerrebroeck, S.; De Vuyst, L. Omics approaches to understand sourdough fermentation processes. *Int. J. Food Microbiol.* **2019**, *302*, 90–102. [CrossRef]
36. Valerio, F.; Favilla, M.; De Bellis, P.; Sisto, A.; De Candia, S.; Lavermicocca, P. Antifungal activity of strains of lactic acid bacteria isolated from a semolina ecosystem against *Penicillium roqueforti, Aspergillus niger* and *Endomyces fibuliger* contaminating bakery products. *Syst. Appl. Microbiol.* **2009**, *32*, 438–448. [CrossRef]
37. EFSA BIOHAZ Panel (EFSA Panel on Biological Hazards); Koutsoumanis, K.; Allende, A.; Alvarez-Ordóñez, A.; Bolton, D.; Bover-Cid, S.; Chemaly, M.; Davies, R.; De Cesare, A.; Hilbert, F.; et al. Scientific Opinion on the update of the list of QPS-recommended biological agents intentionally added to food or feed as notified to EFSA (2017–2019). *EFSA J.* **2020**, *18*, e05966. [CrossRef] [PubMed]
38. Corsetti, A.; Lavermicocca, P.; Morea, M.; Baruzzi, F.; Tosti, N.; Gobbetti, M. Phenotypic and molecular identification and clustering of lactic acid bacteria and yeasts from wheat (species *Triticum durum* and *Triticum aestivum*) sourdoughs of Southern Italy. *Int. J. Food Microbiol.* **2001**, *64*, 95–104. [CrossRef]
39. De Bellis, P.; Valerio, F.; Sisto, A.; Lonigro, S.L.; Lavermicocca, P. Probiotic table olives: Microbial populations adhering on olive surface in fermentation sets inoculated with the probiotic strain *Lactobacillus paracasei* IMPC2.1 in an industrial plant. *Int. J. Food Microbiol.* **2010**, *140*, 6–13. [CrossRef] [PubMed]
40. Zheng, J.; Wittouck, S.; Salvetti, E.; Franz, C.M.A.P.; Harris, H.M.B.; Mattarelli, P.; O'Toole, P.W.; Pot, B.; Vandamme, P.; Walter, J.; et al. A taxonomic note on the genus *Lactobacillus*: Description of 23 novel genera, emended description of the genus *Lactobacillus* Beijerinck 1901, and union of *Lactobacillaceae* and *Leuconostocaceae*. *Int. J. Syst. Evol. Microbiol.* **2020**, *70*, 2782–2858. [CrossRef]
41. Torriani, S.; Felis, G.E.; Dellaglio, F. Differentiation of *Lactobacillus plantarum*, L. pentosus, and L. paraplantarum by recA gene sequence analysis and multiplex PCR assay with recA gene-derived primers. *Appl. Environ. Microbiol.* **2001**, *67*, 3450–3454. [CrossRef]
42. Ventura, M.; Canchaya, C.; Meylan, V.; Klaenhammer, T.R.; Zink, R. Analysis, characterization, and *loci* of the *tuf* genes in *Lactobacillus* and *Bifidobacterium* species and their direct application for species identification. *Appl. Environ. Microbiol.* **2003**, *69*, 6908–6922. [CrossRef]
43. Gadanho, M.; Almeida, J.M.G.C.F.; Sampaio, J.P. Assessment of yeast diversity in a marine environment in the south of Portugal by microsatellite-primed PCR. *Ant. Van Leeuwenhoek* **2003**, *84*, 217–227. [CrossRef] [PubMed]
44. Kurtzman, C.P.; Robnett, C.J. Identification and phylogeny of ascomycetous yeasts from analysis of nuclear large subunit (26S) ribosomal DNA partial sequences. *Ant. Van Leeuwenhoek* **1998**, *73*, 331–371. [CrossRef]
45. Herbig, A.; Maixner, F.; Bos, K.I.; Zink, A.; Krause, J.; Huson, D.H. MALT: Fast alignment and analysis of metagenomic DNA sequence data applied to the Tyrolean Iceman. *BioRxiv* **2016**. [CrossRef]
46. Huson, D.H.; Beier, S.; Flade, I.; Górska, A.; El-Hadidi, M.; Mitra, S.; Ruscheweyh, H.J.; Tappu, R. MEGAN Community Edition-Interactive Exploration and Analysis of Large-Scale Microbiome Sequencing Data. *PLoS Comput. Biol.* **2016**, *12*, e1004957. [CrossRef] [PubMed]
47. Dhariwal, A.; Chong, J.; Habib, S.; King, I.L.; Agellon, L.B.; Xia, J. MicrobiomeAnalyst: A web-based tool for comprehensive statistical, visual and meta-analysis of microbiome data. *Nucl. Acids Res.* **2017**, *45*, W180–W188. [CrossRef]

48. Mengoni, A.; Bazzicalupo, M. The statistical treatment of data and the analysis of MOlecular VAriance (AMOVA) in molecular microbial ecology. *Ann. Microbiol.* **2002**, *52*, 95–101.
49. Parks, D.H.; Beiko, R.G. Identifying biologically relevant differences between metagenomic communities. *Bioinformatics* **2010**, *26*, 715–721. [CrossRef]
50. White, J.R.; Nagarajan, N.; Pop, M. Statistical methods for detecting differentially abundant features in clinical metagenomic samples. *PLoS Comput. Biol.* **2009**, *5*, e1000352. [CrossRef]
51. Benjamini, Y.; Hochberg, Y. Controlling the false discovery rate: A practical and powerful approach to multiple testing. *J. R. Stat. Soc. Ser. B Methodol.* **1995**, *57*, 289–300. [CrossRef]
52. Di Cagno, R.; Pontonio, E.; Buchin, S.; De Angelis, M.; Lattanzi, A.; Valerio, F.; Gobbetti, M.; Calasso, M. Diversity of the lactic acid bacterium and yeast microbiota in the switch from firm- to liquid-sourdough fermentation. *Appl. Environ. Microbiol.* **2014**, *10*, 3161–3172. [CrossRef]
53. Valerio, F.; De Bellis, P.; Di Biase, M.; Lonigro, S.L.; Giussani, B.; Visconti, A.; Lavermicocca, P.; Sisto, A. Diversity of spore-forming bacteria and identification of *Bacillus amyloliquefaciens* as a species frequently associated with the ropy spoilage of bread. *Int. J. Food Microbiol.* **2012**, *156*, 278–285. [CrossRef]
54. Montemurro, M.; Pontonio, E.; Gobbetti, M.; Rizzello, C.G. Investigation of the nutritional, functional and technological effects of the sourdough fermentation of sprouted flours. *Int. J. Food Microbiol.* **2019**, *302*, 47–58. [CrossRef] [PubMed]
55. Gobbetti, M.; Minervini, F.; Pontonio, E.; Di Cagno, R.; De Angelis, M. Drivers for the establishment and composition of the sourdough lactic acid bacteria biota. *Int. J. Food Microbiol.* **2016**, *239*, 3–18. [CrossRef] [PubMed]
56. Rizzello, C.G.; Cavoski, I.; Turk, J.; Ercolini, D.; Nionelli, L.; Pontonio, E.; De Angelis, M.; De Filippis, F.; Gobbetti, M.; Di Cagno, R. Organic cultivation of Triticum turgidum subsp. durum is reflected in the flour-sourdough fermentation-bread axis. *Appl. Environ. Microbiol.* **2015**, *81*, 3192–3204. [CrossRef] [PubMed]
57. De Vuyst, L.; Neysens, P. The sourdough microflora: Biodiversity and metabolic interactions. *Trends Food Sci. Technol.* **2005**, *16*, 43–56. [CrossRef]
58. De Vuyst, L.; Harth, H.; Kerrebroeck, S.V.; Leroy, F. Yeast diversity of sourdoughs and associated metabolic properties and functionalities. *Int. J. Food Microbiol.* **2016**, *239*, 26–34. [CrossRef]
59. Vrancken, G.; De Vuyst, L.; Van der Meulen, R.; Huys, G.; Vandamme, P.; Daniel, H.M. Yeast species composition differs between artisan bakery and spontaneous laboratory sourdoughs. *FEMS Yeast Res.* **2010**, *10*, 471–481. [CrossRef]
60. Minervini, F.; Lattanzi, A.; De Angelis, M.; Di Cagno, R.; Gobbetti, M. Influence of artisan bakery- or laboratory-propagated sourdoughs on the diversity of lactic acid bacterium and yeast microbiotas. *Appl. Environ. Microbiol.* **2012**, *78*, 5328–5340. [CrossRef] [PubMed]

foods

MDPI

Article

Effects of Five Different Lactic Acid Bacteria on Bioactive Components and Volatile Compounds of Oat

Zhishu He [1], Hao Zhang [1], Tao Wang [1], Ren Wang [1,*] and Xiaohu Luo [2,*]

[1] National Engineering Research Center of Cereal Fermentation and Food Biomanufacturing, Jiangnan University, Wuxi 214122, China
[2] College of Food and Pharmaceutical Sciences, Ningbo University, Ningbo 315211, China
* Correspondence: nedved_wr@jiangnan.edu.cn (R.W.); xh06326@gmail.com (X.L.)

Abstract: In this research, oats were fermented with *Lactobacillus plantarum*, *Lactobacillus acidophilus*, *Lactobacillus casei*, *Lactobacillus bulgaricus* and *Streptococcus thermophilus* for 48 h at 37 °C. The purpose of this work was to compare the growth capacities of the five lactic acid bacteria (LAB) in the oat matrix and the effects of fermentation on the contents of the bioactive components of oat, such as β-glucan, polyphenols, flavonoids and volatile compounds at different time (0, 4, 8, 12, 24, 36 and 48 h). After 48 h of fermentation, the number of living *L. acidophilus* in oat reached 7.05×10^9 cfu/mL, much higher than that of other strains. *S. thermophilus* retained the greatest β-glucan content, and *L. casei* had increased total polyphenol and total flavonoid contents. The proportion of free and bound polyphenols and flavonoids in all samples was changed by microbial action, indicating that forms of polyphenols and flavonoids can be transformed during the fermentation process, and the changes varied with different strains. The samples with *L. plantarum*, *L. acidophilus*, and *L. casei* fermentation contained more alcohols, whereas those with *S. thermophilus* and *L. bulgaricus* fermentation had more aldehydes, which revealed that the composition of volatile components was related to strains. The results indicate that oat substrate is a good medium for LAB growth. This study provides a reference for the use of different strains to achieve different fermentation purposes and a theoretical basis for the further processing of oat and fermented oat beverages.

Keywords: fermentation; lactic acid bacteria; oats; polyphenols; flavonoids

Citation: He, Z.; Zhang, H.; Wang, T.; Wang, R.; Luo, X. Effects of Five Different Lactic Acid Bacteria on Bioactive Components and Volatile Compounds of Oat. *Foods* **2022**, *11*, 3230. https://doi.org/10.3390/foods11203230

Academic Editors: Carlo Giuseppe Rizzello and Palmira De Bellis

Received: 16 September 2022
Accepted: 8 October 2022
Published: 16 October 2022

1. Introduction

Oats are grown all over the world, rank seventh in global production behind corn, wheat, rice, barley, sorghum, and millet [1], and are rich in carbohydrates, balanced protein, essential fatty acids, vitamins, and other nutrients [2]. In addition, oats contain many bioactive substances, such as β-glucan, polyphenols, and flavonoids. Among them, β-glucan is a dietary fiber with hypoglycemic and lipid-lowering effects, which can reduce the risk of obesity, diabetes, and cardiovascular diseases and is of great benefit to human health [3,4]. New research shows that polyphenols can improve intestinal health and plasma inflammation and participate in cell signal transduction pathways owing to their anti-inflammatory, antithrombotic, and antioxidant activities [5]. The oat-based food industry has been promoted by the U.S. Food and Drug Administration's recommendation to eat more oats because of their high nutritional benefits and the growing emphasis on healthy food [6]. Thus, exploring more oat processing technologies, enriching oat products, and improving economic benefits are the directions of future development.

Ways to improve the nutritional value of grains include cooking, grinding, and fermentation [7]. Bioactive substances in the grain bran tend to bind to the complex structure of the cell wall, which resists traditional crushing processes; however, bioprocessing techniques can effectively solve this problem [8]. Fermentation may be most economical and the simplest way to increase the nutritional value and functional quality of oats [9]. Lactic acid bacteria (LAB) are probiotics commonly used in food production, especially in dairy fermentation. However, exposure to lactose intolerance, milk allergy, and cholesterol content, and the increase in vegetarianism, have led people to pay more attention to the development of fermented products from plant sources [4,10]. In fact, the enzymes produced by LAB, as well as their metabolic capacity, are vital for the synthesis of some beneficial substances and could facilitate their application in the oat beverage market [11]. In the process of fermentation, the glycosidic bonds of some substrates such as polyphenols and flavonoids are hydrolyzed by the β-glucosidase of LAB, which may release and increase their concentration, thus improving the potential value of oat the fermented beverage. At the same time, the activation of the peptidase system increases the digestibility of proteins and the level of limiting amino acids. In addition, phytase activity has been increased in low pH environment for better hydrolysis of phytates and enhanced mineral bioaccessibility [12]. There have been some reports of oat-based fermented products, such as a beverage containing 25% of oats fermented with *Lactobacillus plantarum* LP09, which improved polyphenol utilization, antioxidant activity and flavor [13]. Bocchi et al. [14] found that co-fermentation of oat milk by *Lactobacillus* and *Bifidobacterium* decreased phytic acid content and increased the bioavailability of amino acids, polyphenols and vitamins Martensson et al. [15] combined the prebiotic β-glucan of oats with the potential benefits of probiotics to produce a functional product that could lower cholesterol levels.

There is much literature on the development of beverages produced from grains, but studies on the selection of appropriate fermentation conditions and strains are few [13]. In this paper, oats were fermented by different LAB strains, and the growth ability of strains and changes in oat bioactive substances and volatile compounds were monitored to provide guidance for the development of fermented oat beverages.

2. Materials and Methods

2.1. Materials

Whole oats were bought from a local supermarket (Wuxi, China). *Lactobacillus plantarum* 22158, *Lactobacillus acidophilus* 6089, and *Lactobacillus casei* 6117 were obtained from the China Center of Industrial Culture Collection (Beijing, China). *Lactobacillus delbrueckii* subsp. *bulgaricus* 57004 and *Streptococcus thermophilus* 58013 were obtained from Hubei Provincial Center for the Preservation and Research of Industrial Microbial Strains (Wuhan, China). α-Amylase (3000 U/mL) and glucoamylase (260,000 U/mL) were provided by Jiangsu Boli Biological Products Co., Ltd. (Taizhou, China). de Man–Rogosa–Sharpe (MRS) medium was obtained from Guangdong Huankai Microbial Sci. & Tech. Co., Ltd. (Guangzhou, China). Other analytical chemicals were obtained from China National Medicines Co., Ltd. (Beijing, China).

2.2. Preparation of Oat Substrate

Whole oats were cleaned, soaked (1 h), dried, roasted at 170 °C for 15 min, ground by ball mill (SJM-5L, Mitr Instrument Equipment Co., Ltd., Changsha, China) at 4000 rpm for 20 min, added with deionized water according to the solid–liquid ratio of 1:4 (m/v), liquefied for 30 min (20 U/g α-amylase, 60 °C, pH 6.0), saccharified for 30 min (130 U/g glucoamylase, 60 °C, pH 4.2), homogenized, adjusted to pH 7.0, and pasteurized for reserve use.

2.3. Fermentation of Oat Substrates with LAB Strains

L. plantarum, *L. acidophilus*, *L. casei*, *L. bulgaricus*, and *S. thermophilus* were activated on the optimal medium (MRS) for two generations. Afterward, 1% LAB (v/v) was inoculated in 20 mL of oat substrate, then mixed and incubated at 37 °C for 48 h. All samples were shaken gently in the bottle before being tested to achieve a homogeneous sample. The samples fermented with *L. plantarum*, *L. acidophilus*, *L. casei*, *L. bulgaricus*, and *S. thermophilus* were named O-Lp, O-La, O-Lc, O-Lb, and O-St, respectively.

2.4. Determination of Viable Counts and pH Value

The bacterial counts of different strains fermented in oat substrate for 0, 4, 8, 12, 24, 36, and 48 h were determined according to the standard plate count method. Sterile saline was used for gradient dilution, and the bacterial solution was inoculated separately onto MRS agar for 48 h and then counted [16].

Additionally, the pH of the samples was measured with a pH meter (PB-10, Sartorius AG, Goettingen, Germany).

2.5. Determination of β-Glucan Content

β-Glucan content was evaluated by a β-glucan assay kit (mixed linkage) (Megazyme, Wicklow, Ireland). A volume of 3 mL of sample was added to 9 mL of 95% ethanol and stood for 5 min. After the supernatant was discarded (1000× *g*, 10 min; Sigma, Lower Saxony, Germany), 10 mL of 50% ethanol was added, dispersed again and the precipitate reserved (1000× *g*, 10 min). Volumes of 0.2 mL of 50% ethanol and 4 mL of sodium phosphate (20 mM, pH 6.5) were added to the precipitate followed by heating in a boiling water bath for 3 min. The tube at was incubated at 50 °C for 5 min and 0.2 mL of lichenase solution (50 U/mL) was added. The reaction was carried out for 1 h with vigorous stirring (3–4 times). After that, 5 mL sodium acetate buffer (200 mM, pH 4.0) was added and the tube allowed to equilibrate to room temperature. The supernatant after centrifugation (0.1 mL; 1000× *g*, 10 min) was hydrolyzed with 0.1 mL β-glucosidase solution, and the obtained solution was used for color reaction.

2.6. Extraction of Phenolic Compounds

The extraction of phenolic compounds was performed according to the method described by Zhang et al. [17] with some modifications. O-Lp, O-La, O-Lc, O-Lb, and O-St samples (4 mL) with different fermentation time were collected, mixed with 20 mL of 80% ethanol, and placed in an ultrasonic cleaner (KQ-50E, Kun Shan Ultrasonic Instruments Co., Ltd., Kun Shan, China) at 25 °C for 20 min. The obtained liquid was centrifuged at 4000 rpm for 10 min at 4 °C and the above steps repeated three times. The supernatant was collected and concentrated with a vacuum rotary evaporator (RV 10 digital V, IKA, Baden-Wuerttemberg, Germany) at 40 °C. The liquid, containing free phenolics, was kept at a constant volume of 10 mL with methanol and stored under dark condition. The remaining precipitate was added to hexane to remove lipids. Then, the precipitate was hydrolyzed by addition of 20 mL of 4 M NaOH, shaken for 1 h, and adjusted to pH 2.0–3.0 with 6 M HCl. The mixture was extracted with 20 mL of ethyl acetate, followed by ultrasonication for 20 min and centrifugation at 4000 rpm for 10 min. The operation was repeated three times to collect the supernatant, which was vacuum-evaporated at 40 °C. The liquid, containing the bound phenolics, was kept at a constant volume of 10 mL with methanol and stored under dark conditions.

2.7. Determination of Phenolic Content

Phenolic content was determined using an adapted and validated method [18] with slight modifications. Briefly, 0.25 mL of extract was mixed with 1 mL of distilled water and 0.25 mL of Folin–Ciocalteu's phenol reagent was added for reaction for 6 min. Then, 2.5 mL of 7% Na_2CO_3 and methanol were added to a total volume of 10 mL. The obtained liquid was incubated for 90 min in darkness at room temperature. Subsequently, absorbance was measured on a spectrophotometer (UV-2100, UNICO, Shanghai, China) at 760 nm. The phenolic compounds were quantified using a gallic acid standard calibration curve. The results were expressed in milligram of gallic acid equivalent (GAE) per 1 L of sample.

2.8. Determination of Flavonoid Content

Flavonoid content was measured with reference to the method proposed by Kim et al. [19]. A 1 mL of sample was diluted to 5 mL with 70% ethanol and mixed with 0.3 mL of 5% $NaNO_2$ for 5 min. Afterward, 0.3 mL of 10% $AlCl_3 \cdot 6H_2O$ was added to react for 6 min.

Finally, 2 mL of 1 M NaOH and 2.4 mL of 70% ethanol were added, and the obtained liquid was incubated for 15 min. The absorbance was read at 420 nm. A rutin standard calibration curve was used for the quantification of flavonoids, and the results were expressed as milligram of rutin equivalent (RE) per 1 L of sample.

2.9. Determination of Volatile Components

Samples were incubated at 50 °C for 10 min in a headspace flask (20 mL), extracted continuously with fiber (DVB/CAR/PDMS, 50/30 μm) at 50 °C for 30 min, and desorbed at 250 °C for 5 min.

Two-dimensional gas chromatography–time-of-flight mass spectrometry analysis was conducted using a Pegasus GC-HRT+ 4D high-performance mass spectrometer (LECO Corp., San Jose, CA, USA). Separation was carried out on a MAT-WAX column (30 m × 0.25 mm × 0.25 μm film thickness; Restek, Bellefonte, PA, USA) with a helium (purity > 99.999%) carrier gas, maintaining a constant flow rate of 1 mL/min. The splitless mode was operated. The inlet temperature was set at 250 °C. The oven temperature was initially set at 40 °C for 3 min, then programmed to increase to 230 °C at a rate of 10 °C/min, and held at 230 °C for 6 min. The transmission line temperature and ion source temperature were set to 250 °C. Mass spectra were measured over a range of 33–400 m/z utilizing an electron energy of 70 eV.

2.10. Statistical Analysis

All experiments were conducted in triplicate unless specified. Diagrams were drawn using Origin 2022b (Origin Statistical Software, Northampton, MA, USA). ANOVA with Duncan's test ($p < 0.05$) was used to analyse the differences between samples in SPSS 24.0 statistics software (SPSS Inc., Chicago, IL, USA).

3. Results and Discussion

3.1. LAB Growth Curve

The purpose of this work was to monitor and compare the cell viabilities of *L. plantarum*, *L. acidophilus*, *L. casei*, *S. thermophilus*, and *L. bulgaricus* inoculated in oats. As shown in Figure 1, the five strains showed remarkable differences in growth activity during the first 12 h of the fermentation. The differences were caused by the strains' utilization of carbon and nitrogen sources and adaptability to substrates [20]. *L. acidophilus* and *L. plantarum* had the fastest growth rates. The maximum concentrations of all strains except *L. bulgaricus* were determined after inoculation for 24 h. At this time, *L. acidophilus* entered the stable stage, and the number of viable bacteria reached 1.36×10^{10} colony forming units (cfu)/mL. The results showed that *L. acidophilus* and *L. plantarum* had stronger growth abilities in oat substrate than the other strains. The viable counts of the five strains in the lag phase were all higher than the recommended minimum of 10^6 cfu/mL for probiotic products [21], indicating that all strains could use the nutrients in oats, and the oat substrate was a suitable fermentation substrate. Similar results were shown by Rathore et al. [22]. Who found that *L. plantarum* and *L. acidophilus* grew well in barley and malt substrates. In fact, *L. plantarum*, *L. acidophilus*, and *L. casei* are considered common strains for cereal fermentation [12]. LAB are heterotrophic organisms that lack some biosynthetic processes and have complex nutritional needs [23]. Different LAB can metabolize different carbon sources, and each microorganism shows a specific preference for one or more sugars [24]. This is also why *L. bulgaricus* and *S. thermophilus* have poorer growth abilities in oat compared with other strains, and prefer to use lactose.

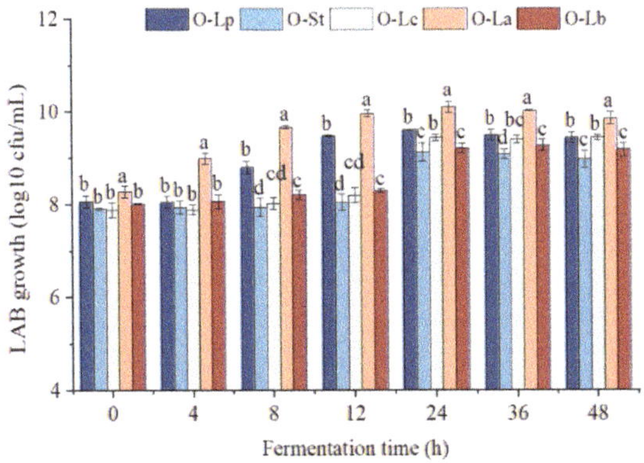

Figure 1. Growth curves of five strains at different incubation time in oat substrate. Different letters indicate significant differences among different strains a given fermentation time ($p < 0.05$). O-Lp, sample fermented with *L. plantarum*; O-La, sample fermented with *L. acidophilus*; O-Lc, sample fermented with *L. casei*; O-Lb, sample fermented with *L. bulgaricus*; O-St, sample fermented with *S. thermophilus*. h, hours; cfu, colony forming units; LAB, Lactic acid bacteria.

3.2. Effect of Fermentation on pH

The ability of LAB to produce acid during fermentation is related to whether the growth of miscellaneous bacteria can be inhibited during product storage, and also affects the sensory characteristics of the final product. The results revealed that the five strains could reduce the pH of the substrate to varying degrees during fermentation (Table 1) and finally reach a pH of 3.15–4.25. After 8 h of fermentation, the pH values of *L. plantarum* and *L. acidophilus* decreased rapidly by 27.39% and 27.67%, respectively, which were distinctly different from those of other strains ($p < 0.05$). The maximum decrease in the pH value of the two strains occurred at the same time with the index growth period of the strain. By contrast, *S. thermophilus* has the weakest acid-producing capacity, reaching a pH of 4.25 after 48 h of fermentation, which may be attributed to *L. acidophilus* preferring lactose to glucose as its main energy source [25]. LAB can produce lactic acid through the carbohydrate metabolism pathway. The production of various organic acids may be the main reason for the decrease in pH [26]. The above results are supported by the study of Mirmohammadi et al. [27]. They concluded that the pH values of different substrates decrease rapidly within 12 h after fermentation by LAB. The strains used in the production of fermented cereal beverages and fermentation time affect the pH of the product.

Table 1. Changes of pH of oat substrate during 48 h fermentation by different lactic acid bacteria.

Time (h)	O-Lp	O-La	O-St	O-Lb	O-Lc
0	7.01 ± 0.01 a	7.01 ± 0.03 a	7.01 ± 0.03 a	7.01 ± 0.03 a	7.01 ± 0.03 a
4	6.23 ± 0.02 d	6.24 ± 0.01 d	6.76 ± 0.01 a	6.54 ± 0.01 b	6.35 ± 0.04 c
8	5.09 ± 0.08 d	5.07 ± 0.03 d	6.27 ± 0.02 a	5.95 ± 0.04 b	5.85 ± 0.02 c
12	4.27 ± 0.05 d	4.12 ± 0.03 e	5.85 ± 0.06 a	5.46 ± 0.07 b	5.36 ± 0.05 c
24	3.54 ± 0.05 c	3.5 ± 0.06 c	5.17 ± 0.06 a	4.48 ± 0.01 b	4.56 ± 0 b
36	3.3 ± 0.06 d	3.23 ± 0.01 e	4.48 ± 0.02 a	3.79 ± 0.01 b	3.55 ± 0.02 c
48	3.15 ± 0.01 d	3.14 ± 0.03 d	4.25 ± 0.01 a	3.76 ± 0.04 b	3.22 ± 0.03 c

Different lowercase letters on the same line indicate significant differences among samples ($p < 0.05$). O-Lp, sample fermented with *L. plantarum*; O-La, sample fermented with *L. acidophilus*; O-Lc, sample fermented with *L. casei*; O-Lb, sample fermented with *L. bulgaricus*; O-St, sample fermented with *S. thermophilus*. h, hours.

3.3. Effect of Fermentation on β-Glucan Content

Our work aimed to study the effect of LAB on β-glucan content during oat fermentation. The β-glucan contents of the five strains shown in Figure 2 increased at the initial fermentation stage, which may be related to the fact that insoluble β-glucan was degraded to soluble β-glucanase by LAB [28], and then LAB consumed a large amount of carbohydrates for proliferation. β-Glucan can also provide growth substrates (prebiotics) for some LAB. Therefore, the contents of all samples were significantly decreased at 8–12 h of fermentation ($p < 0.05$), but the decreases were different. The sample fermented with *S. thermophilus* for 24 h decreased the β-glucan content by 5.09% compared with the unfermented sample; thus, O-St had the most β-glucan content. Evidence shows that *S. thermophilus* TKM3 KKP2030p does not grow well in oat–banana matrix and does not utilize β-glucan [29]. This finding was supported by the results of the viable count experiment and β-glucan content in the present study. Studies showed that the change in β-glucan content during fermentation is related to the strain type. Interestingly, Sims et al. [30] concluded that β-glucan oligosaccharide supported *L. rhamnosus* growth, but *B. lactis* and *L. acidophilus* did not grow on this substrate. Therefore, the different utilization of β-glucan by different strains in the oat matrix may be the reason for the inconsistent decrease in β-glucan content. These data provide guidance for the development of fermented oat beverages. Strains that do not ferment β-glucan can be selected to maximize the potential of probiotics.

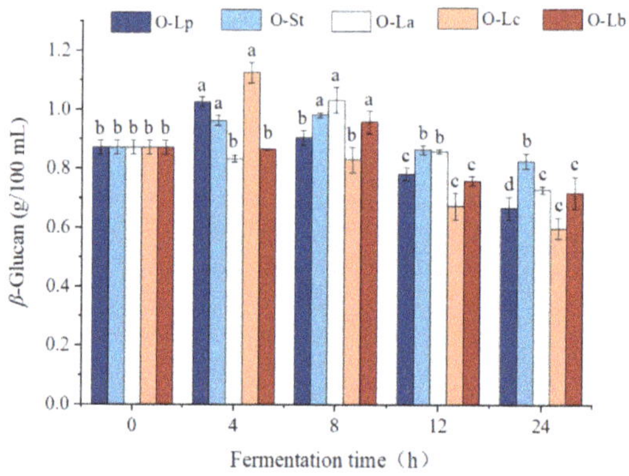

Figure 2. Changes of β-glucan content in oats fermented by different strains at different times. Different letters indicate significant differences among different fermentation time a given fermentation strain ($p < 0.05$). O-Lp, sample fermented with *L. plantarum*; O-La, sample fermented with *L. acidophilus*; O-Lc, sample fermented with *L. casei*; O-Lb, sample fermented with *L. bulgaricus*; O-St, sample fermented with *S. thermophilus*. h, hours.

3.4. *Effect of Fermentation on Phenolic Content*

In fermented products, the glycoside form of polyphenols can be transformed into the aglycone form by microorganisms to improve their bioavailability in the intestine and perform their beneficial functions better. Changes in total phenolic content (TPC), bound phenolic content (BPC), and free phenolic content (FPC) during oat fermentation by LAB are depicted in Figure 3a–c, respectively. The results indicate that a strain has strict specificity in phenolic acid metabolism/degradation/hydrolysis as reported in the literature [31]. For example, as shown in the figure, the TPCs of some strains were distinctly different at the given fermentation time ($p < 0.05$). Compared with the unfermented samples, the TPCs of all fermented samples showed a decreasing trend at the later stage of fermentation, particularly after 12 h, except for O-Lc. However, after 48 h of fermentation, only the TPC of O-Lc increased by 3.14%. Similar results were found by Li et al. [32], who fermented jujube juice with *L. plantarum* and *L. casei* to increase TPC. Other results showed that fermentation could also reduce phenolic content. Moreover, extractability was reduced by the self-polymerization of phenolic compounds and/or the interaction with other macromolecules (such as amino acids and starch) [33]. In addition, compounds are also be transformed and degraded into other healthy monomers [8].

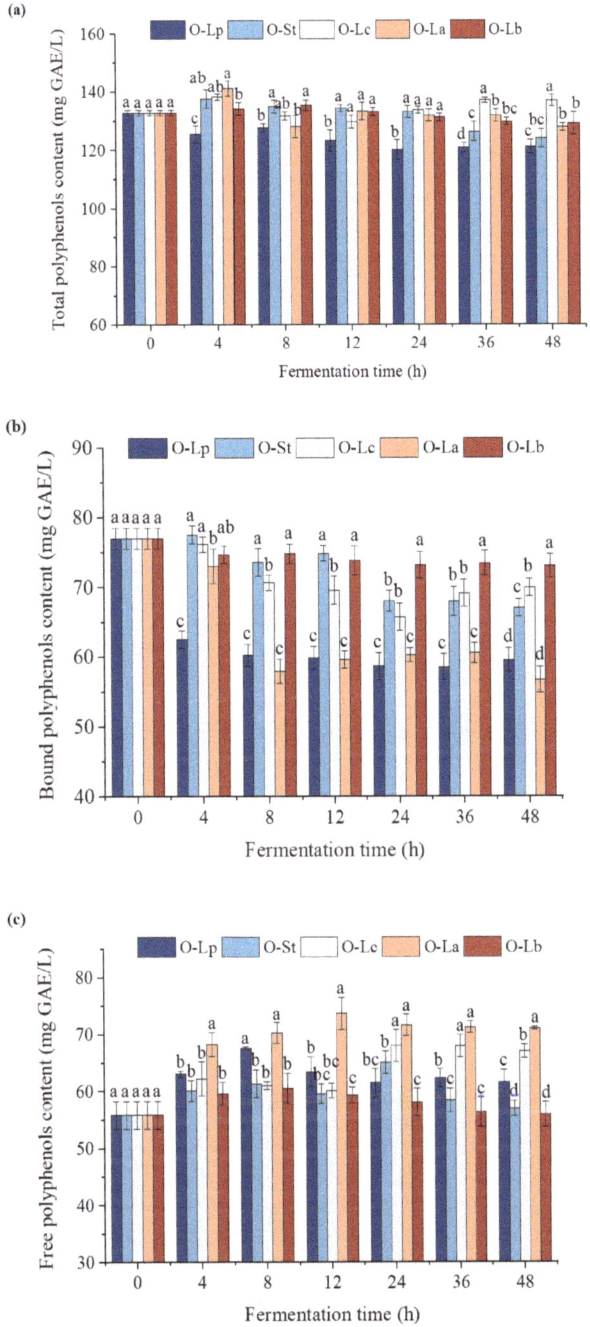

Figure 3. Contents of total polyphenols (**a**), bound polyphenols (**b**) and free polyphenols (**c**) in samples of different strains at different fermentation times. Different letters indicate significant differences among different strains a given fermentation time ($p < 0.05$). O-Lp, sample fermented with *L. plantarum*; O-La, sample fermented with *L. acidophilus*; O-Lc, sample fermented with *L. casei*; O-Lb, sample fermented with *L. bulgaricus*; O-St, sample fermented with *S. thermophilus*. h, hours.

Figure 3b,c show that the FPCs of the five strains were higher than that of the control sample within 24 h of fermentation, whereas the BPCs had the opposite trend. Additionally, an increase in the FPC of each sample was often accompanied by a decrease in BPC. This phenomenon can be explained by the transformation of BPC into FPC due to the metabolic activities of microorganisms. LAB grow rapidly in the early stage of fermentation and can use sugar and protein to partially release BPC. In addition, esterase, decarboxylase, and β-glucanase are produced in the proliferation process. Among them, β-glucanase hydrolyzes the β-glycosidic bond of conjugated phenolic compounds [34], resulting in the release of conjugated phenolic compounds and an increase in FPC [35]. Ferulic acid esterase also releases BPC from the grain cellulose matrix into a free form. At different fermentation times, the FPC of O-Lc was always higher than those of other samples ($p < 0.05$), especially at 12 h, when it could reach 73.70 mg GAE/L, which was 1.32 times higher than that of the control samples. The BPC in O-Lc was remarkably lower than those of O-St, O-Lc, and O-Lb. In addition to glucanase activity, the ability of bacterial strains to degrade phenol esters or tannins, different induced phenol decarboxylase activities, and different substrate acidity may also contribute to this result [36]. Overall, fermentation can change the phenolic contents in the samples. However, whether the compositions of free and bound monomeric phenolic compounds change needs further verification.

3.5. *Effect of Fermentation on Flavonoid Content*

Flavonoids are important natural organic compounds that exist widely in nature. Most flavonoids have strong biological activities, which has aroused research interest. Flavonoids occupy an increasingly important position in daily diet and disease treatment owing to their extensive pharmacological effects and low toxicity. The content change of flavonoids during oat fermentation is shown in Figure 4. The content of total flavonoids in oats fermented by different strains showed a downward trend within 48 h as shown in Figure 4a. This result may be related to the large reduction of conjugated flavonoids. According to previous reports, flavonoids can be transformed into free forms in the fermentation process of soybean and tea, but a decrease in the total amount may not indicate a decrease in biological activity. Bound and free flavonoids decreased during the fermentation of corn and other grains [37]. Researchers attributed this to the metabolism of these compounds during microbial fermentation, such as the degradation/polymerization of flavonoids into dihydroxy flavonoids analogues and the activity changes of enzymes, including glycosidase, glycosyltransferase, tannase, esterase, and hydrolase [34]. In addition, some reactions may induce flavonoids to transform into other metabolites through flavonoid methylation, glycosylation, flavonoid deglycosylation, and flavonoid–sulfuric acid conjugation.

Figure 4b,c shows that after 4 h of fermentation, the free flavonoids in the samples of the other four bacteria except O-Lp were increased, and the bound flavonoids were decreased. O-Lp had the opposite trends initially but gradually had the same trends in the later fermentation time. This outcome could be related to the depolymerization of bound flavonoids and the formation of soluble free flavonoids. Interestingly, we found that at each fermentation time point, the free flavonoid content of O-Lc was distinctly higher than in those of other strains ($p < 0.05$), reaching 202.60 mg RE/L at 48 h, an increase of 20.47% compared with the unfermented sample. Flavonoid glycosides may be consumed during fermentation and released in the form of aglycones. LAB can convert flavonoid glycosides in Cudrania tricuspidate leaves into flavonols, quercetin, and kaempferol [38].

Current studies on the biotransformation of flavonoids in food fermentation processes are few. Xu et al. [39] fermented milk containing *Scutellaria baicalensis* with *Lactobacillus brevis* and found that baicalin and wogonoside were converted into their aglycone forms, baicalein and wogonin, respectively, which have higher biological activities. The biotransformation mechanism of flavonoids in LAB-fermented food remains to be further studied.

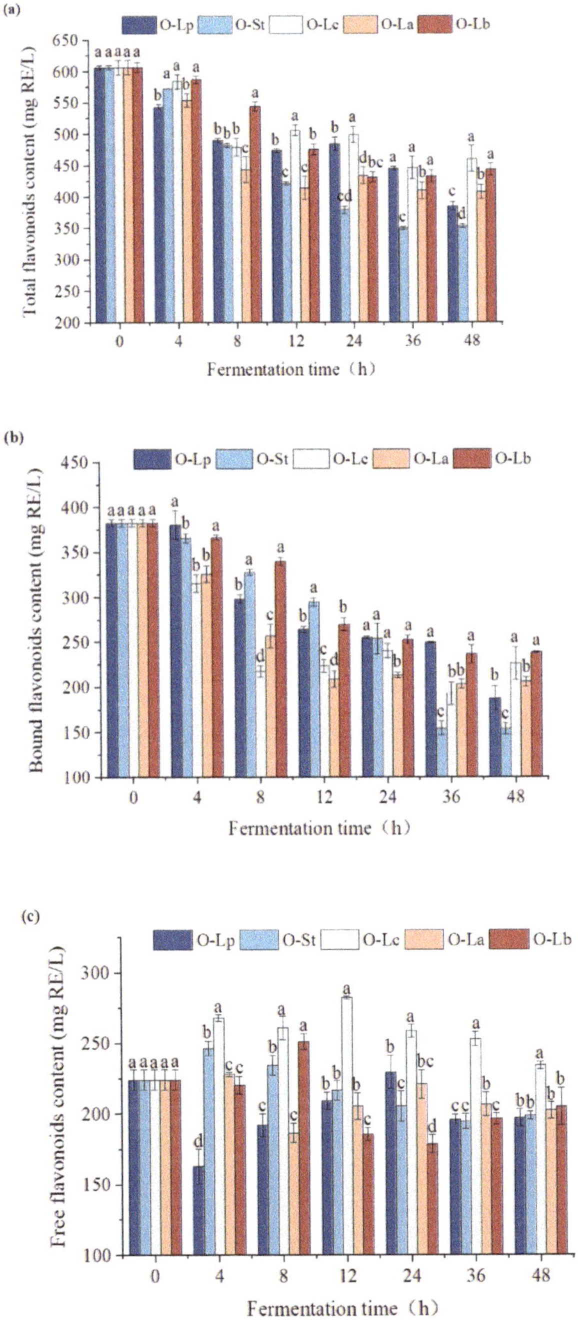

Figure 4. Contents of total flavonoids (**a**), bound flavonoids (**b**) and free flavonoids (**c**) in samples of different strains at different fermentation time. Different letters indicate significant differences among different strains a given fermentation time ($p < 0.05$). O-Lp, sample fermented with *L. plantarum*; O-La, sample fermented with *L. acidophilus*; O-Lc, sample fermented with *L. casei*; O-Lb, sample fermented with *L. bulgaricus*; O-St, sample fermented with *S. thermophilus*. H, hours.

3.6. Effect of Fermentation on Volatile Components

The volatile components of oat fermented by different LAB strains for 48 h are shown in Table 2. The volatile components included 10 alcohols, 10 aldehydes, 7 acids, 15 ketones, 7 esters, 11 furan derivatives, 8 hydrocarbons, and 1 terpene. Oat is prone to oxidative rancidity and deterioration during processing, storage, and circulation, which is related to the high fat content of oat, especially the high percentage of unsaturated fatty acids and the large amount of lipase with high activity in the endosperm. The oxidative cleavage of oleic or linoleic acid during the contact between unsaturated fatty acids and lipase in oat crushing or milling produces hexanal and nonanal. Nonanal production may be caused by the loss of a hydrogen from the 10th carbon of the oleic acid chain, followed by the absorption of a hydrogen peroxyfree group (OOH) and subsequent fracture. The cleavage of the 13-hydroperoxide after the oxygenation of linoleic acid chain may lead to the formation of hexaldehyde. The fatty oxygenase-specific oxidation of 9-hydroperoxides can form 2-pentylfurans, which have greeny, beany, and buttery aromas, whereas 1-octen-3-ol may arise from the 10-hydroperoxide of linoleic acid. This finding was confirmed by the fact that 1-octene-3-ol, hexanal, and nonanal had high concentrations in the unfermented samples as presented in Table 2, revealing that partial oxidation occurred before fermentation. At the same time, we observed that 2-pentylfuran, as one of the important volatile compounds, was detected in each fermented sample, and was significantly higher than that in the fermented sample ($p < 0.05$). This indicates that oxidation also occurred during fermentation.

No pentanal, hexanal and heptanal were detected in O-Lc and O-La. This result was very similar to the conclusion of Lee et al. [40], who fermented oats with *L. paracasei* and found that hexanal content decreased considerably within the first 2 h and was completely undetectable at 24 h. This outcome was due to the fact that microbial action can convert aldehydes into alcohols and acids. In our experiment we observed the fact that 1-pentanol, 1-hexanol, 1-heptanol, acetic acid and hexanoic acid increased correspondingly. Overall, after LAB fermentation, the contents of aldehydes in the sample decreased, and the contents of alcohols, acids, and ketones increased. However, there were distinct differences in flavor components among fermented samples with different strains. As can be seen from Table 2, the contents of alcohols and acids in O-Lp, O-Lc and O-La were higher than those in O-St and O-Lb by up to 5.86 and 2.55 times (O-La compared to O-Lb). In contrast, the contents of aldehydes and esters in O-St and O-Lb were higher than those of O-Lp, O-Lc and O-La, up to 2.85 and 9.61 times (O- St compared to O-La).This difference may be closely related to carbohydrate and amino acid metabolism. The taste characteristics of amino acids and their catabolic by-products during LAB fermentation make an important contribution to the flavor of cereal-based foods [41]. O-St and O-Lb contained lesser 1-hexanol, benzyl alcohol (fruity and slightly floral), phenylethanol (floral and slightly rose), acetic acid, hexanoic acid (sweaty and cheesy), valeric acid, and nonanoic acid and more abundant hexanal (grassy, tallow, fatty), heptanal octanal (fatty, greeny), nonanal (fatty, citrus, greeny), and esters (floral and fruit flavor). Alcohols were predominant in O-Lp, O-Lc, and O-La. 2,3-Butanedione and 3-hydroxy-2-butanone were only present in O-Lc and O-La, which showed a buttery odor. Limonene, which has a citrus-like aroma, was detected in all samples but at low concentrations.

Fermented oats contain volatile components, such as 1-hexanol, hexanal, nonanal, acetic acid and 2-pentylfuran. These components are also considered the key flavor compounds in Lactobacillus-fermented foods and influence the organoleptic properties of fermented products.

Table 2. Changes of volatile compounds in oat samples fermented by different lactic acid bacteria for 48 h.

Volatile Compounds	RT (s)	Control (%)	O-Lp (%)	O-Lc (%)	O-La (%)	O-St (%)	O-Lb (%)
Alcohols							
2-Butanol	411.88	ND	ND	ND	ND	0.33 ± 0.02 a	ND
1-Pentanol	603.74	ND	0.49 ± 0.03 b	0.46 ± 0.04 b	0.59 ± 0.02 a	0.29 ± 0.02 d	0.34 ± 0.02 c
1-Hexanol	696.64	0.09 ± 0.01 d	4.81 ± 0.32 a	1.96 ± 0.23 b	5.22 ± 0.41 a	0.63 ± 0.06 c	0.10 ± 0.03 d
1-Octen-3-ol	778.63	0.46 ± 0.03 b	0.39 ± 0.02 c	0.44 ± 0.03 bc	0.75 ± 0.03 a	0.38 ± 0.05 c	0.43 ± 0.03 bc
1-Heptanol	782.84	0.25 ± 0.02 c	0.69 ± 0.03 a	0.35 ± 0.05 b	0.76 ± 0.08 a	0.18 ± 0.03 c	0.19 ± 0.04 c
1-Octanol	863.64	0.46 ± 0.03 d	0.95 ± 0.04 b	0.58 ± 0.02 c	1.08 ± 0.02 a	0.40 ± 0.04 e	0.47 ± 0.03 d
2-Octen-1-ol, (Z)-	908.51	0.05 ± 0.01 b	0.04 ± 0.00 b	0.03 ± 0.01 b	0.11 ± 0.01 a	0.03 ± 0.00 b	0.03 ± 0.02 b
1-Nonanol	939.49	ND	0.99 ± 0.12 a	0.32 ± 0.02 b	0.96 ± 0.17 a	ND	0.04 ± 0.02 c
Benzenemethanol	1098.00	0.02 ± 0.00 b	0.12 ± 0.05 a	0.09 ± 0.03 a	0.12 ± 0.02 a	0.02 ± 0.01 b	0.03 ± 0.01 b
Benzeneethanol	1121.35	0.02 ± 0.00 b	0.09 ± 0.01 a	0.08 ± 0.01 a	0.09 ± 0.02 a	0.02 ± 0.00 b	0.02 ± 0.00 b
Aldehydes							
Pentanal	321.59	0.35 ± 0.09 a	ND	ND	ND	0.25 ± 0.05 a	0.28 ± 0.02 a
Hexanal	431.62	4.63 ± 0.34 a	0.06 ± 0.01 c	ND	ND	3.34 ± 0.37 b	3.84 ± 0.30 b
Heptanal	542.25	0.60 ± 0.04 b	ND	ND	ND	0.42 ± 0.02 c	0.74 ± 0.02 a
2-Hexenal, (E)-	578.54	0.10 ± 0.02 a	0.04 ± 0.00 b	ND	0.04 ± 0.01 b	0.09 ± 0.02 a	0.11 ± 0.01 a
Octanal	644.40	1.81 ± 0.25 a	ND	ND	0.36 ± 0.04 c	0.95 ± 0.03 b	0.88 ± 0.05 b
2-Heptenal, (E)-	679.51	0.41 ± 0.02 a	0.27 ± 0.04 c	0.14 ± 0.01 d	0.35 ± 0.02 b	0.26 ± 0.05 c	0.30 ± 0.04 bc
Nonanal	738.364	2.32 ± 0.32 a	0.95 ± 0.04 c	0.95 ± 0.03 c	1.10 ± 0.05 c	1.96 ± 0.12 b	2.48 ± 0.31 a
2 Octenal	771.40	0.19 ± 0.02 a	0.14 ± 0.01 c	0.06 ± 0.02 d	0.16 ± 0.02 bc	0.02 ± 0.00 e	0.17 ± 0.01 ab
Benzaldehyde	852.46	ND	ND	ND	0.51 ± 0.03 b	0.56 ± 0.04 a	ND
2-Nonenal, (E)-	856.75	0.55 ± 0.03 a	0.44 ± 0.13 b	0.05 ± 0.04 c	0.40 ± 0.01 b	0.47 ± 0.03 ab	0.56 ± 0.01 a
Acids							
Acetic acid	788.20	ND	2.31 ± 0.34 a	1.62 ± 0.12 b	2.54 ± 0.24 a	0.39 ± 0.09 c	0.56 ± 0.11 c
Hexanoic acid	1073.15	0.41 ± 0.02 d	0.73 ± 0.05 c	1.37 ± 0.12 a	0.88 ± 0.03 b	0.50 ± 0.01 d	0.70 ± 0.02 c
Pentanoic acid	1073.79	ND	0.25 ± 0.01 b	0.08 ± 0.01 c	0.88 ± 0.07 a	ND	0.05 ± 0.02 c
Heptanoic acid	1141.51	0.31 ± 0.02 a	0.28 ± 0.03 ab	0.24 ± 0.01 bc	0.23 ± 0.03 c	0.21 ± 0.02 c	0.23 ± 0.02 c
Octanoic acid	1206.61	0.13 ± 0.02 e	0.65 ± 0.03 a	0.44 ± 0.03 c	0.53 ± 0.02 b	0.28 ± 0.02 d	0.41 ± 0.01 c
Nonanoic acid	1268.57	ND	0.12 ± 0.01 a	0.11 ± 0.01 a	0.11 ± 0.00 a	0.06 ± 0.00 b	ND
Decanoic acid	1327.65	ND	0.01 ± 0.00 b	ND	ND	ND	0.07 ± 0.01 a
Ketones							
2,4-Pentanedione	148.80	ND	ND	ND	ND	0.24 ± 0.04 a	ND
2-Propanone	188.36	ND	0.07 ± 0.00 d	0.09 ± 0.01 c	0.09 ± 0.01 c	0.17 ± 0.00 b	0.19 ± 0.01 a
2- butanone	245.99	ND	ND	0.03 ± 0.00 a	ND	ND	ND
2-Pentadecanone	260.50	ND	ND	0.05 ± 0.00 a	ND	ND	ND
2,3-Butanedione	322.60	ND	ND	0.39 ± 0.02 a	0.11 ± 0.01 b	ND	ND
2,3-Pentanedione	409.87	ND	ND	ND	ND	0.10 ± 0.01 a	ND
2-Heptanone	539.31	0.58 ± 0.04 b	0.54 ± 0.07 b	0.61 ± 0.03 ab	0.69 ± 0.07 a	0.54 ± 0.01 b	0.59 ± 0.03 b
3-Octanone	610.63	ND	0.04 ± 0.01 ab	0.03 ± 0.00 b	0.05 ± 0.01 a	0.03 ± 0.00 b	ND
2-Octanone	640.36	ND	ND	0.15 ± 0.01 a	0.16 ± 0.02 a	ND	0.14 ± 0.01 a
Methoxy-1-phenyl-2-propanone	641.12	ND	ND	0.71 ± 0.02 a	0.16 ± 0.03 b	ND	ND
2-Butanone, 3-hydroxy-	644.82	ND	ND	0.56 ± 0.03 a	0.36 ± 0.02 b	ND	ND
1-Octen-3-one	656.24	0.09 ± 0.00 c	0.13 ± 0.01 b	0.07 ± 0.00 d	0.19 ± 0.01 a	0.06 ± 0.00 d	ND
2-Propanone, 1-hydroxy-	659.94	0.01 ± 0.00 b	0.02 ± 0.01 ab	0.02 ± 0.00 ab	0.03 ± 0.01 a	ND	0.01 ± 0.00 b
3,5-octadiene-2-one	843.73	0.09 ± 0.02 a	0.10 ± 0.03 a	0.09 ± 0.02 a	0.10 ± 0.01 a	0.01 ± 0.01 b	ND
6-Undecanone	845.96	0.02 ± 0.01 a	ND	0.02 ± 0.00 a	ND	ND	ND
Esters							
Formic acid, pentyl ester	473.96	0.07 ± 0.02 a	ND	0.05 ± 0.01 a	ND	ND	0.06 ± 0.01 a
Heptanoic acid, methyl ester	639.94	1.08 ± 0.04 a	0.32 ± 0.09 b	ND	ND	0.32 ± 0.05 b	ND
1-Cyclopropylpentyl acetate	688.58	ND	0.23 ± 0.03 a	0.19 ± 0.03 a	0.19 ± 0.02 a	0.19 ± 0.01 a	0.22 ± 0.05 a
2-etynyl-3-ethyl-2-buten-4-olide	738.56	ND	ND	ND	ND	1.97 ± 0.32 b	2.48 ± 0.51 a
Hexanoic acid, pentyl ester	831.13	ND	0.02 ± 0.00 a	0.01 ± 0.00 a	0.02 ± 0.01 a	0.01 ± 0.00 a	0.02 ± 0.01 a
2-methylpropyl heptanoate	845.64	ND	0.03 ± 0.01 a	ND	0.03 ± 0.00 a	0.02 ± 0.00 a	0.02 ± 0.01 a
Octadecanoic acid, methyl ester	1296.47	0.03 ± 0.01 ab	0.05 ± 0.02 a	0.04 ± 0.00 ab	0.03 ± 0.00 ab	0.02 ± 0.00 b	0.04 ± 0.01 ab
Furan derivatives							
Furan, 2,3-dihydro-	124.44	ND	0.88 ± 0.06 a	0.81 ± 0.09 a	ND	ND	ND
Furan, 2-methyl-	222.88	ND	ND	ND	0.08 ± 0.02 a	ND	ND
Furan, 2-ethyl-	295.38	ND	0.08 ± 0.01 a	0.02 ± 0.00 b	0.09 ± 0.01 a	0.07 ± 0.00 a	0.08 ± 0.02 a

Table 2. *Cont.*

Volatile Compounds	RT (s)	Control (%)	O-Lp (%)	O-Lc (%)	O-La (%)	O-St (%)	O-Lb (%)
2-Propylfuran	377.78	ND	0.07 ± 0.02 a	ND	0.08 ± 0.01 a	0.06 ± 0.01 a	0.07 ± 0.00 a
2-Butylfuran	483.20	0.10 ± 0.00 a	ND	0.07 ± 0.01 b	0.11 ± 0.01 a	0.07 ± 0.00 b	0.11 ± 0.02 a
2-tert-Butoxytetra-hydrofuran	515.08	0.11 ± 0.02 b	ND	0.19 ± 0.05 a	ND	ND	ND
Furan, 2-pentyl-	585.34	ND	2.20 ± 0.09 b	2.74 ± 0.13 a	2.27 ± 0.31 b	2.08 ± 0.08 b	2.66 ± 0.22 a
Furan, 2-(1-pentenyl)-, (E)-	739.45	ND	ND	0.04 ± 0.01 b	0.08 ± 0.02 b	ND	2.48 ± 0.27 a
2-Heptyl furan	769.09	0.01 ± 0.01 a	0.01 ± 0.00 a	0.000 ± 0.00 a	0.01 ± 0.01 a	0.01 ± 0.00 a	0.01 ± 0.00 a
Furan, 2-(methoxymethyl)	981.74	0.14 ± 0.04 a	0.11 ± 0.03 ab	0.05 ± 0.01 c	0.08 ± 0.01 bc	0.13 ± 0.02 ab	0.14 ± 0.03 a
2-Phenylfuran	1087.83	0.06 ± 0.02 a	0.07 ± 0.01 a	0.05 ± 0.02 a	0.06 ± 0.01 a	0.06 ± 0.01 a	0.06 ± 0.02 a
Hydrocarbons							
Pentane	127.57	ND	0.48 ± 0.03 a	0.45 ± 0.03 a	0.43 ± 0.02 a	0.49 ± 0.05 a	0.31 ± 0.04 b
Heptane	176.43	0.07 ± 0.02 b	0.41 ± 0.12 a	ND	ND	0.06 ± 0.02 b	0.09 ± 0.01 b
Undecane	338.38	0.01 ± 0.00 b	0.01 ± 0.00 b	0.01 ± 0.00 b	0.09 ± 0.02 a	0.01 ± 0.00 b	0.01 ± 0.00 b
Dodecane	543.61	ND	0.25 ± 0.04 a	ND	0.16 ± 0.03 b	ND	ND
Pentadecane	546.45	0.17 ± 0.02 ab	ND	0.14 ± 0.01 b	ND	0.19 ± 0.03 a	0.16 ± 0.02 ab
Tetradecane	643.72	ND	0.36 ± 0.03 a	0.01 ± 0.01 b	0.36 ± 0.02 a	0.04 ± 0.00 b	ND
Nonadecane	893.62	0.05 ± 0.01 a	0.02 ± 0.01 c	0.05 ± 0.00 a	0.05 ± 0.01 a	0.03 ± 0.00 bc	0.04 ± 0.01 ab
Terpenes							
Limonene	548.57	0.01 ± 0.00 c	0.03 ± 0.01 b	0.05 ± 0.02 ab	0.03 ± 0.01 b	0.06 ± 0.01 a	0.05 ± 0.00 ab

Different lowercase letters indicate significant differences between different lactic acid bacteria fermented samples ($p < 0.05$). O-Lp, sample fermented with *L. plantarum*; O-La, sample fermented with *L. acidophilus*; O-Lc, sample fermented with *L. casei*; O-Lb, sample fermented with *L. bulgaricus*; O-St, sample fermented with *S. thermophilus*. ND, not detected; RT: retention time; h, hours.

4. Conclusions

In summary, oat substrate is a suitable fermentation substrate. *L. plantarum*, *L. casei*, *L. acidophilus*, *L. bulgaricus*, and *S. thermophilus* could grow well in oat substrate, and the number of viable bacteria could reach 10^6 cfu/mL even at the late fermentation stage. *L. acidophilus* showed the strongest growth ability in oat substrate, and the number of live bacteria was the highest. Soaking oats for 1 h prior to fermentation made them more conducive to subsequent cleaning and absorption of water without significantly affecting the bioactive components. *L. plantarum* and *L. acidophilus* had the strongest acid-producing capacity during the fermentation process, and *S. thermophilus* retained the most β-glucan content after 48 h of fermentation. Moreover, the contents of total polyphenols and total flavonoids in oats varied with different strains, among which O-Lc was the highest. Aldehydes were predominant in O-Lb and O-St, but alcohols were predominant in O-Lp, O-Lc, and O-La. In conclusion, different strains used in oat fermentation have different effects. Therefore, suitable LAB can be selected in future research to improve beneficial ingredients through microbial-mediated biotransformation, providing guidance for the research and development of cereal fermentation products and other options for vegetarians and lactose-intolerant people.

Author Contributions: Conceptualization, Z.H.; methodology, Z.H.; software, Z.H., H.Z. and T.W.; validation, H.Z. and T.W.; formal analysis, Z.H.; investigation, Z.H.; resources, H.Z., R.W. and X.L.; data curation, R.W. and T.W.; writing—original draft preparation, Z.H.; writing—review and editing, Z.H. and H.Z.; visualization, R.W.; supervision, X.L.; project administration, R.W. and X.L.; funding acquisition, R.W. and X.L. All authors have read and agreed to the published version of the manuscript.

Funding: This research was funded by the Science and Technology Key Program of Inner Mongolia, grant number 2021SZD0017.

Data Availability Statement: The data presented in this study are available on request from the corresponding author.

Conflicts of Interest: The author declared no potential conflict of interest of this article.

References

1. Bai, X.; Zhang, M.L.; Zhang, Y.Y.; Zhang, J.; Wang, C.; Zhang, Y.K. Effect of steam, microwave, and hot-air drying on antioxidant capacity and in vitro digestion properties of polyphenols in oat bran. *J. Food Process. Preserv.* **2021**, *45*, 12. [CrossRef]
2. Paudel, D.; Dhungana, B.; Caffe, M.; Krishnan, P. A Review of Health-Beneficial Properties of Oats. *Foods* **2021**, *10*, 2591. [CrossRef] [PubMed]
3. Nguyen, T.T.L.; Gilbert, R.G.; Gidley, M.J.; Fox, G.P. The contribution of beta-glucan and starch fine structure to texture of oat-fortified wheat noodles. *Food Chem.* **2020**, *324*, 8. [CrossRef] [PubMed]
4. Salmeron, I. Fermented cereal beverages: From probiotic, prebiotic and synbiotic towards Nanoscience designed healthy drinks. *Lett. Appl. Microbiol.* **2017**, *65*, 114–124. [CrossRef]
5. Nignpense, B.E.; Francis, N.; Blanchard, C.; Santhakumar, A.B. Bioaccessibility and Bioactivity of Cereal Polyphenols: A Review. *Foods* **2021**, *10*, 1595. [CrossRef]
6. Hozlar, P.; Gregusova, V.; Nemecek, P.; Slikova, S.; Havrlentova, M. Study of Dynamic Accumulation in beta-D-Glucan in Oat (*Avena sativa* L.) during Plant Development. *Polymers* **2022**, *14*, 2668. [CrossRef] [PubMed]
7. Karademir, E.; Yalcin, S.K.; Yalcin, E. Effect of fermentation process on nutritional properties and bioactive compounds of cereal and legume based foods. *Gida* **2018**, *43*, 163–173. [CrossRef]
8. Reque, P.M.; Pinilla, C.M.B.; Tinello, F.; Corich, V.; Lante, A.; Giacomini, A.; Brandelli, A. Biochemical and functional properties of wheat middlings bioprocessed by lactic acid bacteria. *J. Food Biochem.* **2020**, *44*, 11. [CrossRef]
9. Blandino, A.; Al-Aseeri, M.E.; Pandiella, S.S.; Cantero, D.; Webb, C. Cereal-based fermented foods and beverages. *Food Res. Int.* **2003**, *36*, 527–543. [CrossRef]
10. Nematollahi, A.; Sohrabvandi, S.; Mortazavian, A.M.; Jazaeri, S. Viability of probiotic bacteria and some chemical and sensory characteristics in cornelian cherry juice during cold storage. *Electron. J. Biotechnol.* **2016**, *21*, 49–53. [CrossRef]
11. Yepez, A.; Russo, P.; Spano, G.; Khomenko, I.; Biasioli, F.; Capozzi, V.; Aznar, R. In situ riboflavin fortification of different kefir-like cereal-based beverages using selected Andean LAB strains. *Food Microbiol.* **2019**, *77*, 61–68. [CrossRef]
12. Ziarno, M.; Cichonska, P. Lactic Acid Bacteria-Fermentable Cereal- and Pseudocereal-Based Beverages. *Microorganisms* **2021**, *9*, 2532. [CrossRef] [PubMed]
13. Nionelli, L.; Coda, R.; Curiel, J.A.; Poutanen, K.; Marco, G.; Rizzello, C.G. Manufacture and characterization of a yogurt-like beverage made with oat flakes fermented by selected lactic acid bacteria. *Int. J. Food Microbiol.* **2014**, *185*, 17–26. [CrossRef]
14. Bocchi, S.; Rocchetti, G.; Elli, M.; Lucini, L.; Lim, C.Y.; Morelli, L. The combined effect of fermentation of lactic acid bacteria and in vitro digestion on metabolomic and oligosaccharide profile of oat beverage. *Food Res. Int.* **2021**, *142*, 12. [CrossRef] [PubMed]
15. Martensson, O.; Biorklund, M.; Lambo, A.M.; Duenas-Chasco, M.; Irastorza, A.; Holst, O.; Norin, E.; Welling, G.; Oste, R.; Onning, G. Fermented, ropy, oat-based products reduce cholesterol levels and stimulate the bifidobacteria flora in humans. *Nutr. Res.* **2005**, *25*, 429–442. [CrossRef]
16. Zhai, Y.H.; Pan, L.H.; Luo, X.H.; Zhang, Y.W.; Wang, R.; Chen, Z.X. Effect of electron beam irradiation on storage, moisture and eating properties of high-moisture rice during storage. *J. Cereal Sci.* **2022**, *103*, 6. [CrossRef]
17. Zhang, J.G.; Li, M.Y.; Cheng, J.H.; Zhang, X.H.; Li, K.X.; Li, B.; Wang, C.Y.; Liu, X.M. Viscozyme L hydrolysis and *Lactobacillus* fermentation increase the phenolic compound content and antioxidant properties of aqueous solutions of quinoa pretreated by steaming with alpha-amylase. *J. Food Sci.* **2021**, *86*, 1726–1736. [CrossRef]
18. Rekha, C.R.; Vijayalakshmi, G. Biomolecules and Nutritional Quality of Soymilk Fermented with Probiotic Yeast and Bacteria. *Appl. Biochem. Biotechnol.* **2008**, *151*, 452–463. [CrossRef]
19. Kim, M.Y.; Jang, G.Y.; Lee, Y.; Li, M.; Ji, Y.M.; Yoon, N.; Lee, S.H.; Kim, K.M.; Lee, J.; Jeong, H.S. Free and bound form bioactive compound profiles in germinated black soybean (*Glycine max* L.). *Food Sci. Biotechnol.* **2016**, *25*, 1551–1559. [CrossRef]
20. Peyer, L.C.; Zannini, E.; Jacob, F.; Arendt, E.K. Growth Study, Metabolite Development, and Organoleptic Profile of a Malt-Based Substrate Fermented by Lactic Acid Bacteria. *J. Am. Soc. Brew. Chem.* **2015**, *73*, 303–313. [CrossRef]
21. Angelov, A.; Gotcheva, V.; Hristozova, T.; Gargova, S. Application of pure and mixed probiotic lactic acid bacteria and yeast cultures for oat fermentation. *J. Sci. Food Agric.* **2005**, *85*, 2134–2141. [CrossRef]
22. Rathore, S.; Salmeron, I.; Pandiella, S.S. Production of potentially probiotic beverages using single and mixed cereal substrates fermented with lactic acid bacteria cultures. *Food Microbiol.* **2012**, *30*, 239–244. [CrossRef] [PubMed]
23. Reddy, G.; Altaf, M.; Naveena, B.J.; Venkateshwar, M.; Kumar, E.V. Amylolytic bacterial lactic acid fermentation—A review. *Biotechnol. Adv.* **2008**, *26*, 22–34. [CrossRef] [PubMed]
24. Charalampopoulos, D.; Pandiella, S.S.; Webb, C. Growth studies of potentially probiotic lactic acid bacteria in cereal-based substrates. *J. Appl. Microbiol.* **2002**, *92*, 851–859. [CrossRef]
25. Cui, Y.H.; Xu, T.T.; Qu, X.J.; Hu, T.; Jiang, X.; Zhao, C.Y. New Insights into Various Production Characteristics of *Streptococcus thermophilus* Strains. *Int. J. Mol. Sci.* **2016**, *17*, 1701. [CrossRef]
26. Degrain, A.; Manhivi, V.; Remize, F.; Garcia, C.; Sivakumar, D. Effect of Lactic Acid Fermentation on Color, Phenolic Compounds and Antioxidant Activity in African Nightshade. *Microorganisms* **2020**, *8*, 1324. [CrossRef]
27. Mirmohammadi, R.; Zamindar, N.; Razavi, S.H.; Mirmohammadi, M.; Paidari, S. Investigation of the possibility of fermentation of red grape juice and rice flour by *Lactobacillus plantarum* and *Lactobacillus casei*. *Food Sci. Nutr.* **2021**, *9*, 5370–5378. [CrossRef]

28. Lu, J.; Shan, L.K.; Xie, Y.T.; Min, F.F.; Gao, J.Y.; Guo, L.C.; Ren, C.Z.; Yuan, J.L.; Gilissen, L.; Chen, H.B. Effect of fermentation on content, molecule weight distribution and viscosity of beta-glucans in oat sourdough. *Int. J. Food Sci. Technol.* **2019**, *54*, 62–67 [CrossRef]
29. Goncerzewicz, A.; Misiewicz, A.; Owczarek, L.; Jasinska, U.; Skapska, S. The Effect of a Newly Developed Oat-Banana Fermented Beverage with a Beta-glucan Additive on ldhL Gene Expression in *Streptococcus thermophilus* TKM3 KKP 2030p. *Curr. Microbiol.* **2016**, *73*, 773–780. [CrossRef]
30. Sims, I.M.; Ryan, J.L.J.; Kim, S.H. In vitro fermentation of prebiotic oligosaccharides by *Bifidobacterium lactis* HN019 and *Lactobacillus* spp. *Anaerobe* **2014**, *25*, 11–17. [CrossRef]
31. Ferri, M.; Serrazanetti, D.I.; Tassoni, A.; Baldissarri, M.; Gianotti, A. Improving the functional and sensorial profile of cereal-based fermented foods by selecting *Lactobacillus plantarum* strains via a metabolomics approach. *Food Res. Int.* **2016**, *89*, 1095–1105. [CrossRef]
32. Li, T.; Jiang, T.; Liu, N.; Wu, C.; Xu, H.; Lei, H. Biotransformation of phenolic profiles and improvement of antioxidant capacities in jujube juice by select lactic acid bacteria. *Food Chem.* **2021**, *339*, 10. [CrossRef]
33. Adebo, O.A.; Njobeh, P.B.; Adebiyi, J.A.; Kayitesi, E. Co-influence of fermentation time and temperature on physicochemical properties, bioactive components and microstructure of ting (a Southern African food) from whole grain sorghum. *Food Biosci.* **2018**, *25*, 118–127. [CrossRef]
34. Huynh, N.T.; Van Camp, J.; Smagghe, G.; Raes, K. Improved Release and Metabolism of Flavonoids by Steered Fermentation Processes: A Review. *Int. J. Mol. Sci.* **2014**, *15*, 19369–19388. [CrossRef] [PubMed]
35. Georgetti, S.R.; Vicentini, F.; Yokoyama, C.Y.; Borin, M.F.; Spadaro, A.C.C.; Fonseca, M.J.V. Enhanced in vitro and in vivo antioxidant activity and mobilization of free phenolic compounds of soybean flour fermented with different beta-glucosidase-producing fungi. *J. Appl. Microbiol.* **2009**, *106*, 459–466. [CrossRef]
36. Matejcekova, Z.; Mikulajova, A.; Vlkova, E.; Liptakova, D.; Mosovska, S.; Hybenova, E.; Valik, L. Design of Bacterial Cultures in Fermented Functional Maize Product Formulation. *Pol. J. Food Nutr. Sci.* **2019**, *69*, 417–426. [CrossRef]
37. Adebo, O.A.; Oyedeji, A.B.; Adebiyi, J.A.; Chinma, C.E.; Oyeyinka, S.A.; Olatunde, O.O.; Green, E.; Njobeh, P.B.; Kondiah, K. Kinetics of Phenolic Compounds Modification during Maize Flour Fermentation. *Molecules* **2021**, *26*, 6702. [CrossRef]
38. Lee, Y.; Oh, J.; Jeong, Y.S. *Lactobacillus plantarum*-mediated conversion of flavonoid glycosides into flavonols, quercetin, and kaempferol in *Cudrania tricuspidata* leaves. *Food Sci. Biotechnol.* **2015**, *24*, 1817–1821. [CrossRef]
39. Xu, C.; Ji, G.E. Bioconversion of Flavones during Fermentation in Milk Containing Scutellaria baicalensis Extract by *Lactobacillus brevis*. *J. Microbiol. Biotechnol.* **2013**, *23*, 1422–1427. [CrossRef]
40. Lee, S.M.; Oh, J.; Hurh, B.S.; Jeong, G.H.; Shin, Y.K.; Kim, Y.S. Volatile Compounds Produced by *Lactobacillus paracasei* during Oat Fermentation. *J. Food Sci.* **2016**, *81*, C2915–C2922. [CrossRef]
41. Peyer, L.C.; Zannini, E.; Arendt, E.K. Lactic acid bacteria as sensory biomodulators for fermented cereal-based beverages. *Trends Food Sci. Tech.* **2016**, *54*, 17–25. [CrossRef]

foods MDPI

Article

Use of Selected Lactic Acid Bacteria for the Fermentation of Legume-Based Water Extracts

Chiara Demarinis [1], Michela Verni [1], Loris Pinto [2], Carlo Giuseppe Rizzello [3] and Federico Baruzzi [2,*]

1 Department of Soil, Plant and Food Sciences, University of Bari Aldo Moro, 70126 Bari, Italy
2 Institute of Sciences of Food Production, National Research Council of Italy (CNR-ISPA), 70126 Bari, Italy
3 Department of Environmental Biology, Sapienza University of Rome, Piazzale Aldo Moro 5, 00185 Rome, Italy
* Correspondence: federico.baruzzi@ispa.cnr.it; Tel.: +39-080-592-9319

Abstract: In this study, the effect of selected *Lactobacillus acidophilus* ATCC 4356, *Limosilactobacillus fermentum* DSM 20052, and *Lacticaseibacillus paracasei* subsp. *paracasei* DSM 20312 strains on the sensory characteristics, and protein and amino acid content of fermented water extracts derived from lupin, pea, and bean grains is reported. Even though all strains were able to grow over 7 log cfu mL^{-1} and to decrease pH in the range of −0.52 to −1.25 within 24 h, the release of an unpleasant ferric-sulfurous off-odor from the fermented bean water extract prohibited further characterization. Lupin and pea grain-based beverages underwent an in-depth sensory evaluation using a simplified check-all-that-apply (CATA) method, finding new and appreciable sensory notes such as cooked ham, almonds, and sandalwood. Fermented lupin water extract showed higher total protein content (on average, 0.93 mg mL^{-1}) in comparison to that of pea grains (on average, 0.08 mg mL^{-1}), and a free amino acid content (on average, 3.9 mg mL^{-1}) close to that of cow milk. The concentrations of these nutrients decreased during refrigerated storage, when the lactic acid bacteria load was always higher than 7 log cfu mL^{-1}. The results of this study indicated that lactic fermentation improves the sensory characteristics of these innovative legume-based beverages, which sustained high loads of viable lactobacilli up to the end of cold storage.

Keywords: lactic acid bacteria; legumes; bioprocessing; non-dairy alternatives; sensory properties; free amino acids and peptides

Citation: Demarinis, C.; Verni, M.; Pinto, L.; Rizzello, C.G.; Baruzzi, F. Use of Selected Lactic Acid Bacteria for the Fermentation of Legume-Based Water Extracts. *Foods* **2022**, *11*, 3346. https://doi.org/10.3390/foods11213346

Academic Editors: Albert Bordons and Dae-Kyung Kang

Received: 14 September 2022
Accepted: 20 October 2022
Published: 25 October 2022

1. Introduction

Although milk is essential for the body development of mammals, thanks to its content of high-biological-value nutrients, its consumption is decreasing due to lactose intolerance, cow's milk protein allergy, and to the increase in people following a vegan or vegetarian diet [1]. In addition, part of the reduction in milk and dairy product consumption arises from a new consumer awareness towards animal welfare [2] and the reduction of the carbon footprint of animal farms [3]. Consequently, the world demand for non-dairy alternatives is estimated to grow by around 250% in the period 2020–2028 [4].

Plant-based products can be a good alternative. In agreement with Chaturvedi and Chakraborty [5], plant-based milk substitutes can be described as water-soluble extracts of legumes, oilseeds, cereals, or pseudo-cereals that resemble bovine milk in appearance and can be produced by different methods. Lactic acid bacteria (LAB) fermentation of legume grains, flours, and extracts is a sustainable approach to increase their (i) mineral bioavailability thanks to microbial and/or endogenous phytases, able to hydrolyze phytic salt into free inositol, metal ions (mainly calcium, iron, magnesium, and zinc) and phosphates, and (ii) protein and amino acid availability resulting from microbial proteolytic activity against structural seed proteins and specific enzymes responsible for the inhibition of trypsin and chymotrypsin under gastric environmental conditions [6,7]. In addition, LAB decrease the concentration of fermentable oligo-, di-, monosaccharides, and polyols in fermented legumes, with particular concern regarding the raffinose family of oligosaccharides, considered anti-nutritional factors responsible for flatulence [6,7].

Soy (*Glycine max*, L.) grains are the most studied and employed legumes for the production of fermented plant-based beverages, and the effect of lactic acid fermentation on the quality of soy milk was reported more than 40 years ago [8]. However, other legume grains, such as lupins [9], beans [10–12], chickpeas [13–15], or faba beans [16], have been also fermented by LAB. The characterization of a novel lentil-based beverage fermented with different LAB strains showed a strong reduction in both phytic acid and raffinose oligosaccharide, well known as anti-nutritional factors, and a high concentration of soluble and highly digestible proteins [17]. In addition, a yogurt-like beverage including chickpea and lentil flours has been fermented with *Lactobacillus* strains, increasing the antioxidant activity and sustaining the survival of a commercial probiotic strain [18].

Recently, the application of different technologies allowed the realization of new beverages from pea, chickpea, and lupin, incorporating a large amount of seed components with low release of by-products [19].

It is well known that the sensory profiles of legume-based beverages are the main key restraints that prevent their larger diffusion among consumers. In particular, volatile compounds such as n-hexanal and n-hexanol, which originate from lipid oxidation, are mainly responsible for the beany off-flavor [20], whereas tannins and saponins, terpenes, glucosinolates, and flavonoids impart bitter or astringent tastes [21]. In addition, greenish, greyish, or brownish colors, and a chalky or sandy texture, due to the presence of insoluble particles, are the main sensory features negatively affecting the consumer acceptability of legume-based beverages [20].

Some of these drawbacks can be reduced by LAB fermentation thanks to the production of several metabolites, such as lactic and acetic acid or acetoin and acetaldehyde, able to reduce the beany flavor, to enhance the fruity flavor, or to mask the "green-note" off-flavor [22–24]. However, their excess concentration, as demonstrated for acetic acid in beer, could also be detrimental to the organoleptic acceptance of the beverage [25].

Thus, the aim of this work was to ferment different legume grain watery extracts to obtain a legume-based beverage with acceptable sensory traits, rich in soluble peptides and amino acids, that could be a source of probiotic lactobacilli at the end of the cold storage period.

2. Materials and Methods

2.1. Legume-Based Water Extract Preparation and Enumeration of Their Autochthonous Microbial Populations

Pea (*Pisum sativum* Asch. et Gr.), bean (*Phaseolus vulgaris* L.), and lupin (*Lupinus albus* L.) were kindly provided by Terre di Altamura S.r.l. (Altamura, Bari, Italy) and used to prepare legume-based fermented beverages.

Grains were dipped in tap water for around 16 h using the ratio of dry grain/water of 1:10 (*w*/*w*). Since grains absorbed different amounts of water, the ratio of dry grain/water was maintained at 1:10 by adding fresh water to soaked grains. Grain suspensions were homogenized by a hand blender and filtered using cotton gauze [17].

Legume- based water extracts were characterized for their main microbial populations before and after incubation at 24 h at 37 °C. The extracts were serially diluted in sterile 0.1% *w*/*v* buffered peptone water. Then, dilutions were plated in triplicate on (i) Plate Count Agar (PCA) for counting of total aerobic bacteria (24 h at 30 °C, as required by the ISO standard n° 4833 [26]), (ii) acidified (pH 5.4) de Man, Rogosa, and Sharpe (MRS Agar ISO formulation) for mesophilic lactic acid bacteria (anaerobic incubation for 48 h at 30 °C, as already reported [27]); (iii) Potato Dextrose Agar (PDA) supplemented with chloramphenicol (0.1 g/L) for yeasts and molds (incubation for 3 days at 25 °C), and (iv) Reinforced Clostridial Medium (RCM) immediately after thermal treatment (80 °C for 10 min) for aerobic spore-forming bacteria (incubation at 30 °C for 3 days [28]).

All media were purchased from Biolife Italiana S.r.l., Milan, Italy. Microbial loads were expressed as log colony forming unit (cfu) mL^{-1}.

2.2. Lactic Acid Fermentation of Legume-Based Water Extracts

Strains of the former *Lactobacillus* genus used in this work were included in the Agro-Food Microbial Culture Collection (ITEM) at the Institute of Sciences of Food Production of Bari, Italy (http://server.ispa.cnr.it/ITEM/Collection/ (accessed on 13 September 2022) (Table 1). Fresh microbial cultures of *Lactobacillus* spp. strains from frozen cultures (−80 °C) were routinely grown in MRS (MRS Broth ISO Formulation, Biolife Italiana S.r.l., Milan, Italy) for 48 h at 37 °C under anaerobic conditions (Anaerogen, AN0025, Oxoid S.p.A., Milan, Italy).

Table 1. *Lactobacillus* spp. strains used for the fermentation of legume-based water extracts [17].

Specie	Strain
Lactobacillus acidophilus	ATCC 4356
Limosilactobacillus fermentum	DSM 20052
Lactobacillus gasseri	ITEM 13541
Lactobacillus helveticus	ATCC 15009
Lactobacillus johnsonii	NCC533
Lacticaseibacillus paracaseis subsp. *paracasei*	DSM 20312
Lacticaseibacillus rhamnosus	ATCC 53103

ATCC: American Type Culture Collection, Manassas, Virginia, USA. DSM: Deutsche Sammlung von Mikrorganismen und Zellkulturen GmBH, Braunschweig, Germany. ITEM: Agro-Food Microbial Culture Collection of the Institute of Sciences of Food Production, Bari, Italy. NCC: Nestlé Culture Collection, Lausanne, Switzerland.

When used for fermentation, fresh lactobacilli cultures were centrifuged (10,000 rpm for 3 min, centrifuge model Sigma 3–30 KS, Sigma Laborzentrifugen GmbH, Germany), removing the supernatant and re-suspending the cell pellet in saline solution until an absorbance reading at 600 nm of 0.3 ± 0.05 (ca. 8 log cfu mL^{-1}) was obtained. Before lactic acid fermentation, legume-based water extracts were sterilized in an autoclave at 110 °C for 10 min, as previously reported [17]. The three sterilized legume-based water extracts were singly inoculated with 1% of cell suspension (final cell density of around 5–6 log cfu mL^{-1}) and incubated for up to 48 h at 37 °C under anaerobiosis [17].

Before and after fermentation, viable cell counting of lactic acid bacteria was carried out as reported in Section 2.1, incubating plates at 37 °C under anaerobiosis, and the pH was measured (Model pH50 Lab pH Meter XS-Instrument, Concordia, Italy). In order to understand the ability of different strains to grow in legume-based water extracts, as well as their ability to acidify them, the values of Δlog cfu mL^{-1} and ΔpH were calculated. These differences were achieved by subtracting the values obtained at 24 h or 48 h from those of the previous sampling time, e.g., the beginning of fermentation or 24 h.

2.3. Microbial, Biochemical, and Nutritional Characterization of the Legume-Based Beverages

2.3.1. Shelf-Life Evaluation

Legume-based beverages, fermented with the three best-fermenting *Lactobacillus* strains selected during fermentation assays, were stored at 4 °C for 28 days. Changes in microbial populations and acidification rate capabilities were evaluated during cold storage (4 °C) every 7 days.

2.3.2. Total Proteins, Peptides, and Free Amino Acids (FAA)

Total protein content and concentrations of peptides and free amino acids were analyzed at 7-day intervals during cold storage. The protein concentration was evaluated using the Bradford method [29]. Standard solutions of bovine serum albumin (BSA, Sigma No. A-7030) in the range of 0.05–1 mg mL^{-1} were used to build the calibration curve. Subsequently, the absorbance of unknown samples was measured at 595 nm by an automatic spectrofluorometer (Varioskan Flash, ThermoFischer Scientific, Milan, Italy) in multiwell plates. For each well, 5 readings were taken after incubation for 5 min at 25 °C and brief stirring (5 min at 120 rpm).

The concentration of free peptides and free amino acids was performed through the o-phtaldialdehyde method (OPA) [30]. A standard curve of casamino acid mixture (BD BactoCasamino Acids, BD Biosciences, San Jose, CA, USA) was used as a reference in the range of 0.1–2.0 mg mL^{-1}.

For the measurement of free amino acid concentrations, a Biochrom 30 series amino acid analyzer (Biochrom Ltd., Cambridge, UK) equipped with a Li cation exchange column (20 cm × 0.46 cm of internal diameter) was used [31].

2.3.3. Sensory Analysis

The sensory analyses of the fermented legume-based beverages were carried out during 28 days of cold storage at regular 7-day intervals. The check-all-that-apply (CATA) method [32] was simplified to describe the visual, olfactory, and taste characteristics of the beverages. In particular, the group of panelists was composed of 10 untrained judges in the recognition of different sensory properties, as differently reported [33,34]. They evaluated the different fermented beverages, describing their main perceptions simply as they were perceived.

Since these beverages were produced at lab scale for the first time, at the first instance, it was necessary to collect preliminary information about their acceptance. Thus, main traits were scored for the macro-descriptors "Appearance", "Odor", and "Taste" as 0—unexpected and/or unwanted feature, or 1—expected and/or welcome feature. The scores were then normalized for the number of panelists with the result of expressing the three macro-descriptors in a 0–1 interval. Even though this approach was less informative in comparison to other descriptive tests based on a hedonic scale for each descriptor, it was considered more useful to distinguish among samples by considering only the main, "good" or "bad", trait perceived.

Participants were recruited among people answering a short survey sent by email to students, researchers, and professors, as well as members of their families, already in contact with the authors and their colleagues. Then, panelists were selected among those habitually consuming fermented beverages (different from wine and beer) and legumes. During the first session of sensory analysis, a moderator explained the aim of the study, the characteristics of the products (liquid and ready-to-drink), and instructions on how to complete the questionnaire, and also answered technical questions and queries.

2.4. *Statistical Analysis*

The average lactic acid strain population enumerated during the cold refrigerated period for each legume-based beverage was analyzed by two-way analysis of variance (ANOVA); the significance of differences ($p < 0.05$) between mean values was evaluated by Fisher's least significant difference test. Statistical differences in protein concentration and free amino acid and peptide concentration were evaluated by the T test. Statistical analyses were performed with Microsoft Excel software, implemented with the statistical analysis tool add-in (Microsoft Corporation, Redmond, WA, USA).

3. Results and Discussion

3.1. *Autochthonous Microbial Populations of Legume-Based Water Extracts*

Viable cell counts of autochthonous microbial populations of legume-based water extracts, before and after 24 h of fermentation at 37 °C, are shown in Figure 1. The legume-based water extracts showed the presence of different types of autochthonous microbial populations, mainly belonging to presumptive lactic acid rods and cocci. In particular, LAB counts, presumptively represented by lactic acid rods, ranged from 2.39 ± 0.36 log cfu mL^{-1} to 4.62 ± 0.22 log cfu mL^{-1}, with lupin and peas having the lowest and highest density, respectively (Figure 1). Aerobic endospore-forming bacteria were found before incubation only in peas and beans. However, after the incubation, they were detected in all samples, reaching 5.85 ± 0.01 log cfu mL^{-1} in beans, 6.10 ± 0.09 log cfu mL^{-1} in lupin, and approximately 1.88 ± 0.09 cfu mL^{-1} in pea water extracts (Figure 1). Yeast and molds were found only in the bean extracts, but, after incubation, they were not further detected (data not shown).

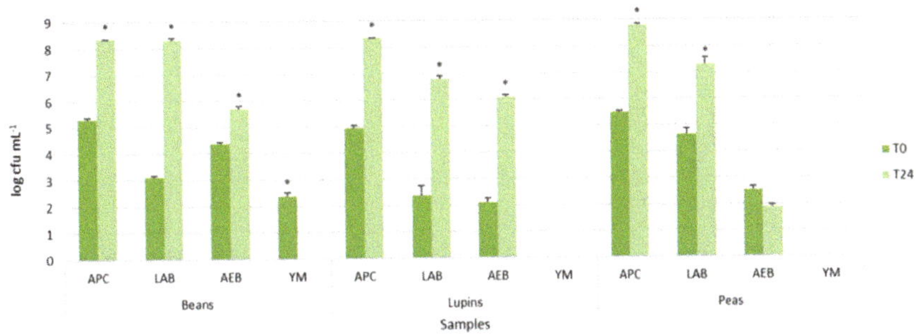

Figure 1. Viable cell counts of main microbial populations of legume grains in water before and after 24 h of incubation at 37 °C. Histograms represent microbial average population values ± standard deviation (error bars). The asterisks indicate statistical differences ($p < 0.05$) after T test comparison. Abbreviations: Aerobic Plate Count (APC), Lactic Acid Bacteria (LAB), Aerobic Endospore-Forming Bacteria (AEB), Yeasts and Molds (YM).

The comparison of the means of each population before and after incubation within the same extract resulted in statistically significant values for all microbial populations, as shown by the asterisks in Figure 1, excluding aerobic endospore-forming bacteria of peas.

These results confirmed that the legume grains were characterized by autochthonous and potentially useful lactic acid bacteria, able to grow during incubation, and spoilage bacteria, such as the endospore-forming ones.

As far as the lactic acid microbial population is concerned, the viable loads here reported are in agreement with those found in several fruit and vegetables [35–38], belonging to species such as *Lactiplantibacillus plantarum*, *Lactiplantibacillus pentosus*, *Lm. fermentum*, *Latilactobacillus curvatus*, *Levilactobacillus brevis*, *Leuconostoc mesenteroides*, *Weissella* spp., and *Enterococcus* spp.

As found for lactic acid bacteria, endospore-forming bacterial loads also increased during incubation. This result, in agreement with the occurrence of *Bacillus cereus*, *B. nitratireducens*, B. pumilus, B. safensis, and *B. australimaris* in different types of fermented food [39,40], forced us to heat-treat extracts before fermentation.

3.2. Fermentation Assays of Legume-Based Water Extracts

In order to select the LAB strains for the production of fermented legume-based beverages, legume-based water extracts, after sterilization (110 °C for 10 min), were first inoculated with each strain at 5–6 log cfu mL^{-1} and incubated for 48 h at 37 °C. As shown in Tables 2 and 3, *L. acidophilus* ATCC 4356, *Lm. fermentum* DSM 20052, and *Lc. paracasei* DSM 20312 showed the highest increase in average viable cell counts and a marked decrease in pH values within 48 h of fermentation in comparison to the beginning of incubation.

All LAB strains grew in legume extracts, but only *L. acidophilus* ATCC 4356, *Lm. fermentum* DSM 20052, and *Lc. paracasei* DSM 20312 showed Δlog cfu mL^{-1} higher than one in their viable loads in the three legume extracts after 48 h at 37 °C (Table 2).

The fermentation assays employing these strains were repeated, lowering the initial inoculum level to 4–5 log cfu mL^{-1} and evaluating pH changes and cell viability at 24 h and 48 h (Tables 4 and 5).

Table 2. Differences in cell densities, reported as Δlog cfu mL^{-1}, after 48 h of incubation at 37 °C, for legume grain water extracts inoculated with selected LAB strains.

	Beans	Lupins	Peas	Strain Average Values
L. acidophilus ATCC 4356	1.77 ± 0.25^{Aa}	1.21 ± 0.13^{Bc}	0.19 ± 0.03^{Cd}	1.06 ± 0.13
Lm. fermentum DSM 20052	1.10 ± 0.10^{Ab}	3.04 ± 0.10^{Ba}	1.15 ± 0.07^{Ac}	1.77 ± 0.09
L. gasseri ITEM 13541	0.76 ± 0.10^{Bc}	0.16 ± 0.08^{Ce}	0.96 ± 0.06^{Ac}	0.63 ± 0.08
L. helveticus ATCC 15009	0.31 ± 0.03^{Bd}	0.04 ± 0.02^{Ce}	2.52 ± 0.10^{Aa}	0.96 ± 0.05
L. johnsonii NCC533	1.06 ± 0.06^{Ab}	0.74 ± 0.11^{Bd}	-0.40 ± 0.13^{Ce}	0.47 ± 0.10
Lc. paracasei DSM 20312	1.64 ± 0.26^{Aa}	1.54 ± 0.22^{Ab}	1.66 ± 0.12^{Ab}	1.61 ± 0.2
Lc. rhamnosus ATCC 53103	0.32 ± 0.19^{Bd}	0.88 ± 0.15^{Ad}	-0.25 ± 0.08^{Ce}	0.32 ± 0.14

Two-way ANOVA analysis was applied to estimate the effect of legume and starter strain on cell density. The least significant difference comparison values (LSD, 95% confidence interval) were calculated. Cell densities: legume, 0.16 Δlog cfu g^{-1}; strain, 0.25 Δlog cfu g^{-1}. Superscript letters indicate significant differences within rows, while lowercase letters indicate significant differences within columns.

Table 3. Differences in pH values, reported as ΔpH values, after 48 h of incubation at 37 °C, for legume grain water extracts inoculated with selected LAB strains.

	Beans	Lupins	Peas	Strain Average Values
L. acidophilus ATCC 4356	-2.58 ± 0.20^{Bd}	-0.58 ± 0.16^{Aa}	-3.57 ± 0.08^{Cc}	-2.24 ± 0.14
Lm. fermentum DSM 20052	-1.51 ± 0.06^{Bb}	-1.12 ± 0.14^{Ab}	-3.51 ± 0.21^{Cc}	-2.05 ± 0.14
L. gasseri ITEM 13541	-1.48 ± 0.19^{Ab}	-1.72 ± 0.17^{Bc}	-3.24 ± 0.12^{Cb}	-2.15 ± 0.16
L. helveticus ATCC 15009	-1.81 ± 0.24^{Bc}	-1.17 ± 0.07^{Ab}	-3.44 ± 0.11^{Cc}	-2.14 ± 0.14
L. johnsonii NCC533	-1.56 ± 0.22^{Ab}	-2.06 ± 0.15^{Bd}	-3.02 ± 0.11^{Cb}	-2.21 ± 0.16
Lc. paracasei DSM 20312	-1.56 ± 0.22^{Bb}	-1.24 ± 0.14^{Ab}	-3.67 ± 0.09^{Cc}	-2.15 ± 0.15
Lc. rhamnosus ATCC 53103	-0.59 ± 0.14^{Aa}	-0.98 ± 0.05^{Bb}	-2.12 ± 0.18^{Ca}	-1.23 ± 0.12

Two-way ANOVA analysis was applied to estimate the effect of legume and starter strain on pH values. The least significant difference comparison values (LSD, 95% confidence interval) were calculated. pH values: legume 0.19 ΔpH; strain, 0.29 ΔpH. Superscript letters indicate significant differences within rows, while lowercase letters indicate significant differences within columns. Initial pH values: beans 6.62, lupins 5.78, peas 7.02. Negative values represent the unit of pH reduction of fresh extracts that occurred during fermentation.

Table 4. Differences in cell densities, reported as Δlog cfu mL^{-1} after 24 and 48 h of incubation at 37 °C, of legume grain water extracts inoculated with selected LAB strains.

	Beans		Lupins		Peas	
	t24	t48	t24	t48	t24	t48
L. acidophilus ATCC 4356	2.81 ± 0.21^{Aa}	3.24 ± 0.07^{Aa}	2.76 ± 0.09^{Aa}	3.15 ± 0.06^{Aa}	1.87 ± 0.02^{Ba}	2.42 ± 0.07^{Ba}
Lm. fermentum DSM 20052	2.97 ± 0.04^{Aa}	3.43 ± 0.06^{Aa}	2.23 ± 0.10^{Bb}	2.64 ± 0.21^{Bb}	1.85 ± 0.07^{Ca}	2.63 ± 0.11^{Ba}
Lc. paracasei DSM 20312	2.17 ± 0.07^{Bb}	3.19 ± 0.11^{Ab}	2.89 ± 0.10^{Aa}	3.08 ± 0.07^{Aa}	1.68 ± 0.05^{Ca}	2.18 ± 0.11^{Bb}

Two-way ANOVA analysis was applied to estimate the effect of legume and starter strain on cell density. The least significant difference comparison values (LSD, 95% confidence interval) were calculated. Cell densities: legume, 0.19 Δlog cfu g^{-1}; strain, 0.21 Δlog cfu g^{-1}. At each sampling time, superscript letters indicate significant differences within rows, while lowercase letters indicate significant differences within columns.

Two-way ANOVA analysis showed that legume-based beverage, starter strain, and their interaction significantly ($p \leq 0.05$) affected cell density and pH values at each sampling time. The effect of the strain on cell density values was not significant at 48 h. All strains confirmed their ability to grow in all legume-based water extracts already after 24 h, without excessive acidification, except in the case of *L. acidophilus* ATCC 4356 in the pea extract. After 48 h of incubation, the pea extract showed the lowest viable cell increase. The same legume-based beverage showed the highest reduction in pH values for all three inoculated strains after 48 h of incubation. Generally, the pH decreased differently during fermentation, depending on the legume extract and the strain used.

Table 5. Differences in pH values, reported as ΔpH values after 24 and 48 h of incubation at 37 °C, of legume grain water extracts inoculated with selected LAB strains.

	Beans		Lupins		Peas	
	t24	t48	t24	t48	t24	t48
L. acidophilus ATCC 4356	-1.01 ± 0.12 ^{Bc}	-2.19 ± 0.10 ^{Bc}	-0.28 ± 0.06 ^{Aa}	-1.19 ± 0.04 ^{Aa}	-2.45 ± 0.05 ^{Cc}	-3.42 ± 0.17 ^{Cc}
Lm. fermentum DSM 20052	-0.76 ± 0.11 ^{Bb}	-0.56 ± 1.79 ^{Aa}	-0.95 ± 0.07 ^{Cb}	-1.15 ± 0.06 ^{Ba}	-0.46 ± 0.04 ^{Aa}	-2.86 ± 0.06 ^{Ca}
Lc. paracasei DSM 20312	-0.16 ± 0.05 ^{Aa}	-1.87 ± 0.02 ^{Bb}	-0.38 ± 0.03 ^{Ba}	-1.14 ± 0.06 ^{Aa}	-1.01 ± 0.12 ^{Cb}	-3.05 ± 0.07 ^{Cb}

Two-way ANOVA analysis was applied to estimate the effect of legume and starter strain on pH values. The least significant difference comparison values (LSD, 95% confidence interval) were calculated. pH values: legume 0.15 ΔpH; strain, 0.16 ΔpH. At each sampling time, superscript letters indicate significant differences within rows, while lowercase letters indicate significant differences within columns. Initial pH values: beans 6.59, lupins 5.81, peas 6.98. Negative values represent the unit of pH reduction of fresh extracts after fermentation.

The extension in incubation time from 24 to 48 h increased the average viable cell count only by 0.53 Δlog cfu mL^{-1} (as calculated from values reported in Table 4), with a drop in the pH values from 5.57 ± 0.49 to 4.76 ± 0.19 (Table 5). Results confirmed the ability of these strains to grow and reduce the pH in fermented legumes, as reported in soy for *L. acidophilus* ATCC 4356 [41], and for these three strains in lentil grains [17].

Thus, the fermentation step of the legume-based beverage for these strains was set up following incubation for 24 h at 37 °C, avoiding the increase in the concentration of acids from sugar fermentation, which could be unpleasant for the consumer.

As far as the bean water extract is concerned, the release of off-odors at the end of fermentation, probably due to volatile sulfur compounds, prevented its use for the production of a bean-based fermented beverage. As a consequence, bean water extracts were not further characterized.

3.3. *Evaluation of the Shelf-Life during Cold Storage*

Based on previous results, the fermented legume-based beverages were prepared from lupin and pea water extracts singly fermented for 24 h at 37 °C with *L. acidophilus* ATCC 4356, *Lm. fermentum* DSM 20052, and *Lc. paracasei* DSM 20312.

The microbial viability during 28 days of cold storage of these strains is reported in Table 6. During 28 days of cold storage, the cell density of all strains reached values close to 8 log cfu mL^{-1}, remaining stable or, in some cases, increasing by one order of magnitude in comparison to the beginning of incubation.

Table 6. Cell density (log cfu mL^{-1}) of selected lactobacilli in legume-based beverages during 28 days of storage at 4 °C.

Beverage	Days of Storage	*L. acidophilus* ATCC 4356	*Lm. fermentum* DSM 20052	*Lc. paracasei* DSM 20312
Lupins	0	7.60 ± 0.06 ^{Bd}	7.57 ± 0.07 ^{Be}	7.72 ± 0.05 ^{Ac}
	7	8.40 ± 0.05 ^{Bb}	7.71 ± 0.07 ^{Cd}	8.63 ± 0.06 ^{Ab}
	14	8.11 ± 0.04 ^{Cc}	8.92 ± 0.06 ^{Aa}	8.69 ± 0.04 ^{Bb}
	21	8.83 ± 0.06 ^{Aa}	7.89 ± 0.09 ^{Bc}	8.80 ± 0.10 ^{Aa}
	28	8.77 ± 0.07 ^{Aa}	8.30 ± 0.06 ^{Bb}	8.86 ± 0.07 ^{Aa}
	Average values	8.67 ± 0.08	8.36 ± 0.05	8.65 ± 0.06
Peas	0	7.54 ± 0.08 ^{Bd}	7.59 ± 0.06 ^{Ac}	7.67 ± 0.08 ^{Ac}
	7	7.87 ± 0.07 ^{Ac}	7.76 ± 0.09 ^{Ab}	7.66 ± 0.07 ^{Bc}
	14	8.06 ± 0.06 ^{Ab}	7.72 ± 0.13 ^{Bb}	7.92 ± 0.04 ^{Bb}
	21	8.07 ± 0.05 ^{Bb}	8.06 ± 0.07 ^{Ba}	8.33 ± 0.09 ^{Aa}
	28	8.55 ± 0.06 ^{Aa}	7.54 ± 0.08 ^{Cc}	8.39 ± 0.07 ^{Ba}
	Average values	8.17 ± 0.04	7.76 ± 0.08	8.09 ± 0.08

Two-way ANOVA analysis was applied to estimate the effect of time of cold storage and starter strain on cell density values. The least significant difference comparison values (LSD, 95% confidence interval) were calculated for each factor. Lupin beverage: time, 0.12 log cfu g^{-1}; strain, 0.10 log cfu g^{-1}. Pea beverage: time, 0.14 log cfu g^{-1}; strain, 0.11 log cfu g^{-1}. Superscript letters indicate significant differences within rows, while lowercase letters indicate significant differences within columns.

Here, we underline the ability of *L. acidophilus* ATCC 4356, *L. fermentum* DSM 20052 and *L. paracasei* DSM 20312 to survive and grow in legume-based beverages during cold storage, at viable loads higher than 7 log cfu mL^{-1}, respecting the minimum value recommended for the daily intake of lactic acid bacteria throughout these products [42].

In particular, *L. acidophilus* ATCC 4356 and *Lc. paracasei* DSM 20312 showed a significant increase in cell density in both pea and lupin beverages throughout storage, whereas *Lm. fermentum* DSM 20052 showed an increase in viable load followed by a decrease in the late stages of cold storage in both beverages.

Results of this work are in agreement with those of Liao et al. [43], who reported lactic acid bacteria loads higher than 8 log cfu mL^{-1} in adzuki bean beverages fermented with *Lactococcus lactis* or *Lacticaseibacillus rhamnosus* GG, and stored for 28 days at 4 °C.

Even though the preparation of legume-based water extracts can be carried out following different protocols, we can conclude that lactic acid bacteria are able to ferment legume extracts and survive in these beverages during cold storage, as also demonstrated for the strains *Lacticaseibacillus rhamnosus* GR-1 and *Streptococcus thermophilus* found in cowpea beverages after 28 days [44].

The two-way ANOVA of both lupin and pea grain beverages showed that the cell density was significantly affected by the strain and incubation period, as well as their interaction ($p \leq 0.05$). The interaction between the strain and incubation period factors significantly influenced the microbial cell densities of *Lactobacillus* strains.

The high viable cell loads of the three LAB strains determined the decrease in pH values for both the fermented legume-based beverages throughout cold storage. In particular, the pH decreased from 4.81 to 4.00 in lupin-based beverages fermented with *L. acidophilus* and 4.08 to 3.30 in pea-based beverages fermented with *Lc. paracasei.* Despite the high viable concentration found for all strains in both fermented beverages during cold storage, no post-acidification phenomenon was found. This condition is considered useful to preserve both the microbial viability of probiotic strains in foods and the sensory characteristics of the fermented product [45–48]. During cold storage, the viable cell load increased by one log, on average. The growth of these strains during cold storage, probably resulting from the availability of different types of nutrients [49], was already reported for other lactic acid bacteria strains in different legume-based products [49,50].

The lactobacilli strains selected in this work, even though isolated from different matrices, were found to be able to grow in the legume-based water extracts and to survive for one month of cold storage. These results are in agreement with the ability of autochthonous and allochthonous lactic acid bacteria as starters of vegetable matrices [35,51,52] and their survival during cold storage.

3.3.1. Proteins and Free Peptides/Amino Acids throughout Cold Storage

Table 7 shows the concentration of total protein in fermented legume-based beverages throughout 28 days of cold storage.

The amount of total free peptides and amino acids remained almost the same during cold storage, as shown in Table 8. Thus, it could be assumed that the water-soluble proteins extracted during the soaking and blending process are a preferential source of organic nitrogen and are metabolized by these strains for their survival. A slight decrease in total protein content during cold storage was found only in beverages inoculated with *L. acidophilus* ATCC 4356 and *L. paracasei* DSM 20312 (Table 7). A limited degree of proteolysis, depending on the strain and fermentation duration, was also reported by Arteaga et al. [53] in lacto-fermented pea protein isolate. On the contrary, Schlegel et al. [54] found that *Limosilactobacillus reuteri* and *Lentilactobacillus parabuchneri* hydrolyzed medium- and low-molecular-weight polypeptides from lupin protein isolate. Based on the extraction and fermentation process here displayed, we speculate that the proteolysis degree of fermented legume grains is the result of the process of protein extraction, the protein profile of the matrix, the LAB strain, and the fermentation conditions.

Table 7. Concentration of total protein (mg mL^{-1}) of fermented beverages and in unfermented water extracts during 28 days of refrigerated (4 °C) storage.

Beverage	Days of Cold Storage	Unfermented Extract	Fermented Beverages		
			L. acidophilus ATCC 4356	*Lm. fermentum* DSM 20052	*Lc. paracasei* DSM 20312
Lupins	0	1.22 ± 0.09 Ab	0.78 ± 0.17 Ba	0.97 ± 0.09 Ba	1.05 ± 0.06 Aa
	7	1.58 ± 0.05 Aa	0.60 ± 0.26 Ba	0.78 ± 0.17 Ba	0.66 ± 0.08 Bb
	14	1.79 ± 0.05 Aa	0.66 ± 0.37 Ca	0.96 ± 0.23 Ba	0.76 ± 0.10 Bb
	21	1.66 ± 0.06 Aa	0.54 ± 0.26 Ca	0.82 ± 0.10 Ba	0.74 ± 0.07 Bb
	28	1.27 ± 0.05 Ab	0.34 ± 0.37 Cb	0.73 ± 0.18 Bb	0.51 ± 0.05 Bc
Peas	0	0.26 ± 0.04 Ab	0.05 ± 0.04 Ca	0.06 ± 0.04 Bc	0.12 ± 0.03 Ba
	7	0.24 ± 0.04 Ac	0.06 ± 0.04 Ba	0.05 ± 0.05 Bc	0.08 ± 0.02 Ba
	14	0.24 ± 0.07 Ac	0.04 ± 0.03 Ba	0.05 ± 0.04 Bc	0.07 ± 0.03 Ba
	21	0.35 ± 0.05 Ab	nd	0.14 ± 0.05 Bb	0.07 ± 0.04 Ca
	28	0.50 ± 0.07 Aa	nd	0.25 ± 0.06 Ba	0.07 ± 0.05 Ca

Two-way ANOVA analysis was applied to estimate the effect of time of cold storage and starter strain on total protein values. The least significant difference comparison values (LSD, 95% confidence interval) were calculated for each factor. Lupin beverage: time, 0.25 mg mL^{-1}; strain, 0.22 mg mL^{-1}. Pea beverage: time, 0.07 mg mL^{-1}; strain, 0.06 mg mL^{-1}. Superscript letters indicate significant differences within rows, while lowercase letters indicate significant differences within columns. nd = not detected.

Table 8. Concentration of free peptides and amino acids (mg mL^{-1}) in fermented beverages and in unfermented water extracts during 28 days of refrigerated (4 °C) storage.

Beverage	Days of Cold Storage	Unfermented Extract	Fermented Beverages		
			L. acidophilus ATCC 4356	*Lm. fermentum* DSM 20052	*Lc. paracasei* DSM 20312
Lupin	0	3.94 ± 0.05 Ba	3.56 ± 0.09 Ba	4.72 ± 0.04 Aa	3.43 ± 0.03 Ba
	7	4.31 ± 0.07 Aa	2.82 ± 0.48 Cb	3.65 ± 0.05 Bb	3.00 ± 0.04 Ca
	14	4.16 ± 0.14 Aa	2.75 ± 0.65 Bb	3.86 ± 0.07 Ab	2.91 ± 0.07 Ba
	21	3.63 ± 0.06 Ab	2.80 ± 0.55 Bb	3.75 ± 0.06 Ab	2.63 ± 0.01 Bb
	28	3.39 ± 0.15 Ab	2.52 ± 1.17 Bb	4.51 ± 0.02 Aa	2.85 ± 0.02 Bb
Pea	0	3.26 ± 0.05 Ab	2.26 ± 0.06 Cd	3.22 ± 0.05 Aa	2.76 ± 0.03 Bc
	7	3.34 ± 0.04 Ab	2.63 ± 0.04 Bb	2.58 ± 0.09 Bc	2.55 ± 0.08 Bd
	14	3.47 ± 0.02 Aa	3.09 ± 0.10 Ba	2.77 ± 0.11 Db	2.95 ± 0.04 Cb
	21	3.53 ± 0.06 Aa	2.55 ± 0.05 Dc	2.72 ± 0.05 Cb	3.43 ± 0.03 Ba
	28	3.05 ± 0.05 Ac	2.72 ± 0.08 Cb	2.84 ± 0.07 Bb	3.00 ± 0.04 Ab

Two-way ANOVA analysis was applied to estimate the effect of time of cold storage and starter strain on total free peptide and amino acid values. The least significant difference comparison values (LSD, 95% confidence interval) were calculated for each factor. Lupin beverage: time, 0.57 mg mL^{-1}; strain, 0.51 mg mL^{-1}. Pea beverage: time, 0.10 mg mL^{-1}; strain, 0.09 mg mL^{-1}. Superscript letters indicate significant differences within rows, while lowercase letters indicate significant differences within columns.

As already found for total protein content, the concentrations of free peptides and amino acids were significantly affected by the strain but not by the storage period.

Fermented plant-based foods are often designed as dairy alternatives. However, often, these foods are claimed to contain lower concentrations of nutrients, mainly proteins and peptides, than their milk-based counterparts. It is noteworthy that the concentrations of free peptides and amino acids here reported, and measured by the OPA method, were always higher that those reported by Bhattacharya et al. [55] for 10 commercial dairy products, including milk, ranging from 60 to 130 mg L^{-1}, as measured by ion-exchange chromatography.

3.3.2. Free Amino Acids

The concentrations of amino acids released after the fermentation, as well as those still detectable at the end of cold storage, were quantified by HPLC. Table 9 summarizes the concentrations of different groups of amino acids, which are individually detailed in Tables S1 and S2. As already observed for the total concentrations of free peptides and amino acids, the total amount of free amino acids decreased in all fermented legume extracts. The highest reduction in the concentration of total free amino acids was observed in legume grain water extracts fermented with *Lm. fermentum* DSM 20052 (Table 9). As far as the concentration of each amino acid group is concerned, no relevant changes were observed in essential amino acids and 7-aminobutyric acid. However, pea water extract fermented with *Lm. fermentum* DSM 20052 showed a great reduction in essential amino acids, branched-chain amino acids, and other amino acids (Table 9).

Table 9. Concentrations of free amino acids (expressed in mg L^{-1}) in fermented beverages and in unfermented water extracts at the beginning (T0) and at the end (T28) of refrigerated (4 °C) storage period.

Beverage	Groups of Amino Acids	Unfermented Control		Fermented Beverages					
				L. acidophilus ATCC 4356		*Lm. Fermentum* DSM 20052		*Lc. paracasei* DSM 20312	
		T0	T28	T0	T28	T0	T28	T0	T28
Lupin	Essential amino acids [1]	53.93	48.72	23.80	20.21	16.54	13.05	19.56	17.35
	Br. chain amino acids [2]	72.48	69.79	16.17	12.84	8.15	6.59	2.11	1.51
	Other amino acids	528.43	494.60	404.68	331.77	265.70	203.55	378.34	326.65
	ɣ-aminobutyric acid	106.64	105.93	105.27	96.95	101.55	97.55	104.54	102.92
	Total amount	761.48	719.04	549.92	461.77	391.94	320.74	504.55	448.43
Pea	Essential amino acids [1]	107.83	105.53	108.65	103.06	64.25	50.62	112.22	99.01
	Br. chain amino acids [2]	106.21	106.49	120.92	101.97	91.66	77.29	121.14	107.48
	Other amino acids	753.18	760.11	738.47	710.34	629.57	513.56	756.34	691.66
	ɣ-aminobutyric acid	40.37	38.06	45.48	41.15	45.35	39.57	45.90	38.61
	Total amount	1007.59	1010.19	1013.52	956.52	830.83	681.04	1035.6	936.76

[1], Thr + Met + Phe + Trp; [2], Val + Leu + Ile.

As reported in Tables S1 and S2, cysteine and tyrosine were largely consumed by all strains, and aspartic acid was reduced preferentially by the *Lm. fermentum* DSM 20052, whereas ɣ-aminobutyric acid remained almost stable in all fermented extracts. In some cases, the reduction in the total content of free amino acids after fermentation was accompanied by an increase in the concentration of certain amino acids. We can speculate that the increased concentration of some amino acids could be related to the reduction in the total protein concentration recorded during cold storage, as reported above (Table 7).

Free amino acids in lupin-based beverages decreased on average by 42% after fermentation. Similar results were also reported for yogurt by Germani et al. [56], who observed a reduction after fermentation of more than 30% in the total amount of free amino acids.

In the case of pea-based beverages, a sharp reduction in the free amino acid concentration was found only after fermentation with *Lm. fermentum* DSM 20052. It is interesting to note that the free amino acid content of fermented pea beverages is in line with that of cow milk (ca. 450 µmol/L) [57]. It is possible to conclude that, as reported for the total concentration of free peptides and amino acids (Table 8), the concentration of total free amino acids decreased after fermentation but remained almost stable during cold storage (Tables S1 and S2).

At the end of the refrigerated period, the total free amino acid content ranged from 223.19 mg L^{-1} to 915.37 mg L^{-1} in lupin-based beverages fermented with *Lm. fermentum* DSM 20052 and pea-based ones fermented with *L. acidophilus* ATCC 4356, respectively (Tables S1 and S2). The consumption of fermented legume extracts characterized by a high cell density of probiotic strains and high free amino acid content could improve the concentration of post-prandial blood amino acids. Indeed, Jägeret et al. [58] found that pea extract fermented with the probiotic strains *Lc. paracasei* DSM 20312 and *L. acidophilus* ATCC 4356 increased amino acid absorption after pea protein ingestion.

In addition, it is interesting to note that a glass (150 mL) of fermented lupin-based beverage contains approximately the same amount of ɣ-aminobutyric acid potentially able to lower systolic blood pressure as demonstrated in humans consuming 50 g per day of GABA-enriched cheese [59].

3.3.3. Sensory Properties

The sensory characteristics of the legume-based beverages were evaluated every seven days during the cold storage period using a simplified check-all-that-apply (CATA) method. Due to the innovative characteristics of these lab-scale fermented beverages and the absence of any similar sensory experience in the untrained panelists, the sensory acceptance of these beverage was considered the main result to be achieved. Thus, questionnaires were compiled describing only a single, the main, trait for each macro-descriptor. The simplified CATA method here applied considered only the acceptability (good/not good) of each descriptor, resulting in a binary response score.

These scores were then organized in a contingency table, combined, and normalized. In some cases, the same descriptor (e.g., acid taste) was considered acceptable (score 1) or not according to the personal preferences of panelists. Here, we considered that scores higher than 0.7 are representative of a sufficient level of acceptability.

Since there was no information about the sensory descriptors of fermented legume-based beverages in the literature, the descriptors of the CATA method, belonging to macro-descriptors "Appearance", "Odor", and "Taste", were freely defined by each panelist and then compared to each other. Due to the water-like consistency of these beverages, the "Texture" macro-descriptor, necessary to describe yogurt-like fermented beverages, was not included.

All descriptors provided by panelists for each fermented beverage are reported in Supplementary Tables S3 and S4 and are herein summarized in Table 10.

Table 10. Comparison between acceptability scores of macro-sensory descriptors of fermented legume grain beverages and unfermented water extracts.

Beverage	Macro-Descriptors	Water Extract	Fermented Beverages		
			L. acidophilus ATCC 4356	*Lm. fermentum* DSM 20052	*Lc. paracasei* DSM 20312
Lupin	Appearance	0.7	0.6	0.6	0.6
	Odor	0.8	0.7	0.5	0.7
	Taste	0.0	0.6	0.4	0.6
Pea	Appearance	0.4	0.5	0.3	0.5
	Odor	0.8	0.8	0.4	0.9
	Taste	0.7	0.7	0.7	0.9

The sensory characteristics of the legume-based beverages were largely and specifically affected by the lactic acid fermentation. In comparison with the unfermented control, lactic acid fermentation of the lupin water extract moderately affected the three sensory macro-descriptors, whereas fermentation of the pea water extract was positively affected by the inoculation of *L. acidophilus* ATCC 4356 and *Lc. paracasei* DSM 20312.

Among the three lactobacilli assayed, *Lm. fermentum* DSM 20052 was the strain that produced limited or no improvements in acceptability scores for all legume-based beverages. None of the three lactobacilli were able to produce a significant improvement in lupin acceptability scores. In this case, the taste was the worst macro-descriptor, since it was characterized by an unpleasant, bitter, and persistent taste in the innermost area of the tongue.

This unacceptable sensory characteristic was reduced thanks to lactic acid fermentation, increasing the taste acceptability from 0 to 0.6. In particular, a milk flavor in samples fermented with *L. acidophilus* ATCC 4356, and vegetable notes in samples fermented with *Lm. fermentum* DSM 20052 and *Lc. paracasei* DSM 20312, partially masked the bitterness (Table S3). Reduced bitterness was recently found in lupin protein isolates fermented with *Latilactobacillus sakei* subsp. *carnosus* [54].

As far as the appearance of legume-based beverages is concerned, lactic acid fermentation did not negatively affect the appearance of legume samples, with lupin samples characterized by a transparent straw yellow color and with a negligible amount of sediment, and pea samples characterized by a greenish-yellow color (Tables S3 and S4).

All fermented lupin-based beverages were characterized by a cooked ham odor (Table S3). Similarly, Schlegel et al. [54] found notes of cooked products (cooked potato, roasty, and oatmeal) in lupin fermented with lactic acid bacteria. The best result for the macro-descriptor "Odor" was assigned to pea extracts fermented with *Lc. paracasei* DSM20132, which, as a result of fermentation, produced pleasant notes of "green peas", and "fruity", "floral", and "fresh-cut grass" notes, resulting in a score of 0.9 (Table S4). These results agree with El Youssef et al. [60], which found "green flavor/vegetal" and "leguminous plant" as the main descriptors of pea protein fermented with lactic acid bacteria. The sensory characteristics of the legume-based beverages showed moderate changes under cold storage, leading to only an increase in odor and taste of acidity (data not shown).

Our results partially agree with other works in which both extracts of lupin and pea grains were fermented, even though differences in sensory notes described cannot be correctly compared due to differences in fermenting strains, extraction processes, and fermentation steps [61,62]. Even though the method applied is less informative than others based on hedonic scales, it was able to define the level of acceptability of each beverage and underline which sensory trait needs to be improved in order to increase the average level of acceptability.

4. Conclusions

This work demonstrates the ability of some strains of lactobacilli to ferment legume water extracts and to survive during cold storage. This result appears particularly interesting in the case of the probiotic strain *L. acidophilus* ATCC 4356. The organoleptic profile of lupin- and pea-based beverages was positively affected by the starter, with the best results obtained with *Lc. paracasei* DSM 20312. Independently of the fermenting strain, high protein and amino acid content was found in lupin-based beverages. In conclusion, the appropriate combination of fermenting strain and legume grains could lead to the production of a legume-based milk substitute containing high concentrations of free peptides and amino acids. Since highly viable lactic acid bacteria were found after up to 28 days of cold storage, these beverages could also be a potential carrier of probiotic lactic acid bacteria. Experiments set up to demonstrate the survival of probiotic microbial cells in the human gastrointestinal tract could offer information about the potential health benefits of these beverages. Moreover, further studies will require a more in-depth beverage sensory characterization in order to improve the preliminary results obtained by the application of the simplified CATA method here employed.

Supplementary Materials: The following supporting information can be downloaded at: https://www.mdpi.com/article/10.3390/foods11213346/s1, Table S1: Concentration of free aminoacids (expressed in mg L^{-1}) in lupin grain-based beverages at the beginning (T0) and at the end (T28) of refrigerate (4 °C) storage period. Table S2: Concentration of free aminoacids (expressed in mg L^{-1}) in pea grain-based beverages at the beginning (T0) and at the end (T28) of refrigerate (4 °C) storage period. Table S3: Results of sensory analysis of lupin-based beverage before and after fermentation, allocated to macrodescriptors appearance, odour and taste. Table S4: Results of sensory analysis of pea-based beverage before and after fermentation, allocated to macrodescriptors appearance, odour and taste.

Author Contributions: F.B. was responsible for funding acquisition; F.B. and C.D. conceived the work and designed the experimental plan; C.D. was responsible for microbiological analyses and sensory evaluation; L.P. was involved in sensory evaluation and microbiological assays and carried out statistical analysis; M.V. was mainly involved in chemical analyses; C.G.R. was involved in manuscript review and editing; C.G.R., F.B. and C.D. wrote the first draft of the manuscript. All authors have read and agreed to the published version of the manuscript.

Funding: This research was funded by the project "Innovation and enhancement of productivity, sustainability and profitability of the Apulian typical legumes supply chain" (acronym PSR_Leg), practical code (barcode dds 94250033001), admitted to the Call "Submeasure 16.2—Support for pilot projects and the development of new products, practices, processes and technologies" funded by the Apulia Region under the 2014–2020 Rural Development Programme.

Institutional Review Board Statement: Not applicable.

Informed Consent Statement: Not applicable.

Data Availability Statement: The datasets generated for this study are available on request to the corresponding author.

Acknowledgments: The authors are grateful to Marinella Cavallo, Paolo Valerio and Simonetta Martena (CNR-ISPA) for their skilled administrative support provided during the realization of this work.

Conflicts of Interest: The authors declare no conflict of interest.

References

1. Silva, A.R.; Silva, M.M.; Ribeiro, B.D. Health issues and technological aspects of plant-based alternative milk. *Food Res. Int.* **2020**, *131*, 108972. [CrossRef] [PubMed]
2. Alonso, M.E.; González-Montaña, J.R.; Lomillos, J.M. Consumers' concerns and perceptions of farm animal welfare. *Animals.* **2020**, *10*, 385. [CrossRef]
3. Karwacka, M.; Ciurzyńska, A.; Lenart, A.; Janowicz, M. Sustainable development in the agri-food sector in terms of the carbon footprint: A Review. *Sustainability* **2020**, *12*, 6463. [CrossRef]
4. Fortune Business Insights: Dairy Alternatives Market Size, Share and Industry Analysis, by Source (Soy, Almond, Coconut, Rice, Oats and others), Product Type (Non-Dairy Milk, Butter, Cheeses, Yogurts, Ice Cream, and others), Distribution Channel (Supermarkets/Hypermarkets, Convenience Stores, Online Retail, and others), and Regional Forecasts, 2021–2028. Fortune Business Insights. 2021. Available online: https://www.fortunebusinessinsights.com/industry-reports/dairy-alternatives-market-100221 (accessed on 20 July 2022).
5. Chaturvedi, S.; Chakraborty, S. Review on potential non-dairy synbiotic beverages: A preliminary approach using legumes. *Int. J. Food Sci. Technol.* **2021**, *56*, 2068–2077. [CrossRef]
6. De Pasquale, I.; Pontonio, E.; Gobbetti, M.; Rizzello, C.G. Nutritional and functional effects of the lactic acid bacteria fermentation on gelatinized legume flours. *Int. J. Food Microbiol.* **2020**, *316*, 108426. [CrossRef] [PubMed]
7. Kårlund, A.; Gómez-Gallego, C.; Korhonen, J.; Palo-oja, O.M.; El-Nezami, H.; Kolehmainen, M. Harnessing microbes for sustainable development: Food fermentation as a tool for improving the nutritional quality of alternative protein sources. *Nutrients* **2020**, *12*, 1020. [CrossRef] [PubMed]
8. Mital, B.K.; Steinkraus, K.H. Fermentation of soy milk by lactic acid bacteria. A review. *J. Food Prot.* **1979**, *42*, 895–899. [CrossRef]
9. Schindler, S.; Wittig, M.; Zelena, K.; Krings, U.; Bez, J.; Eisner, P.; Berger, R.G. Lactic fermentation to improve the aroma of protein extracts of sweet lupin (*Lupinus angustifolius*). *Food Chem.* **2011**, *128*, 330–337. [CrossRef]
10. Song, H.Y.; Yu, R.C. Optimization of culture conditions for gamma-aminobutyric acid production in fermented adzuki bean milk. *J. Food Drug Anal.* **2018**, *26*, 74–81. [CrossRef]
11. Gan, R.Y.; Shah, N.P.; Wang, M.F.; Lui, W.Y.; Corke, H. *Lactobacillus plantarum* WCFS1 fermentation differentially affects antioxidant capacity and polyphenol content in mung bean (*Vigna radiata*) and soya bean (*Glycine max*) milks. *J. Food Process. Preserv.* **2017**, *41*, e12944. [CrossRef]
12. Saraphanchotiwitthaya, A.; Sripalakit, P. Production of γ-aminobutyric acid from red kidney bean and barley grain fermentation by *Lactobacillus brevis* TISTR 860. *Biocatal. Agric. Biotechnol.* **2018**, *16*, 49–53. [CrossRef]
13. Wang, S.; Chelikani, V.; Serventi, L. Evaluation of chickpea as alternative to soy in plant-based beverages, fresh and fermented. *LWT* **2018**, *97*, 570–572. [CrossRef]
14. Tangyu, M.; Fritz, M.; Aragao-Börner, R.; Ye, L.; Bogicevic, B.; Bolten, C.J.; Wittmann, C. Genome-based selection and application of food-grade microbes for chickpea milk fermentation towards increased L-lysine content, elimination of indigestible sugars, and improved flavour. *Microb. Cell Factories* **2021**, *20*, 109. [CrossRef] [PubMed]
15. Skrzypczak, K.; Jabłońska-Ryś, E.; Gustaw, K.; Sławińska, A.; Waśko, A.; Radzki, W.; Michalak-Majewska, M.; Gustaw, W. Reinforcement of the antioxidative properties of chickpea beverages through fermentation carried out by probiotic strain *Lactobacillus plantarum* 299v. *J. Pure Appl. Microbiol.* **2019**, *13*, 1–12. [CrossRef]
16. Verni, M.; De Mastro, G.; De Cillis, F.; Gobbetti, M.; Rizzello, C.G. Lactic acid bacteria fermentation to exploit the nutritional potential of Mediterranean faba bean local biotypes. *Food Res. Int.* **2019**, *125*, 108571. [CrossRef]
17. Verni, M.; Demarinis, C.; Rizzello, C.G.; Baruzzi, F. Design and characterization of a novel fermented beverage from lentil grains. *Foods* **2020**, *9*, 893. [CrossRef]
18. Pontonio, E.; Raho, S.; Dingeo, C.; Centrone, D.; Carofiglio, V.E.; Rizzello, C.G. Nutritional, functional, and technological characterization of a novel gluten-and lactose-free yogurt-style snack produced with selected lactic acid bacteria and *Leguminosae* flours. *Front. Microbiol.* **2020**, *11*, 1664. [CrossRef]
19. Lopes, M.; Pierrepont, C.; Duarte, C.M.; Filipe, A.; Medronho, B.; Sousa, I. Legume beverages from chickpea and lupin, as new milk alternatives. *Foods* **2020**, *9*, 1458. [CrossRef]
20. Tangyu, M.; Muller, J.; Bolten, C.J.; Wittmann, C. Fermentation of plant-based milk alternatives for improved flavour and nutritional value. *Appl. Microbiol. Biotechnol.* **2019**, *103*, 9263–9275. [CrossRef]
21. Drewnowski, A.; Gomez-Carneros, C. Bitter taste, phytonutrients, and the consumer: A review. *Am. J. Clin. Nutr.* **2000**, *72*, 1424–1435. [CrossRef]
22. Wang, Y.C.; Yu, R.C.; Yang, H.Y.; Chou, C.C. Sugar and acid contents in soymilk fermented with lactic acid bacteria alone or simultaneously with bifidobacteria. *Food Microbiol.* **2003**, *20*, 333–338. [CrossRef]

23. Horáčková, Š.; Muehlhansova, A.; Slukova, M.; Schulzová, V.; Plockova, M. Fermentation of soymilk by yoghurt and bifidobacteria strains. *Czech J. Food Sci.* **2015**, *33*, 313–319. [CrossRef]
24. Schindler, S.; Zelena, K.; Krings, U.; Bez, J.; Eisner, P.; Berger, R.G. Improvement of the aroma of pea (*Pisum sativum*) protein extracts by lactic acid fermentation. *Food Biotechnol.* **2012**, *26*, 58–74. [CrossRef]
25. Siebert, K.J. Modeling the flavor thresholds of organic acids in beer as a function of their molecular properties. *Food Qual Prefer.* **1999**, *10*, 129–137. [CrossRef]
26. *ISO 4833-1:2013/Amd 1:2022*; Microbiology of the Food Chain—Horizontal Method for the Enumeration of Microorganisms—Part 1: Colony Count at 30 °C by the Pour Plate Technique. International Organization for Standardization: Geneva, Switzerland, 2013.
27. Makki, G.M.; Kozak, S.M.; Jencarelli, K.G.; Alcaine, S.D. Evaluation of the efficacy of commercial protective cultures against mold and yeast in queso fresco. *J. Dairy Sci.* **2020**, *103*, 9946–9957. [CrossRef] [PubMed]
28. Matarante, A.; Baruzzi, F.; Cocconcelli, P.S.; Morea, M. Genotyping and toxigenic potential of *Bacillus subtilis* and *Bacillus pumilus* strains occurring in industrial and artisanal cured sausages. *Appl. Environ. Microbiol.* **2004**, *70*, 5168–5176. [CrossRef]
29. Bradford, M.M. A rapid and sensitive method for the quantitation of microgram quantities of protein utilizing the principle of protein-dye binding. *Anal. Biochem.* **1976**, *72*, 248–254. [CrossRef]
30. Church, F.C.; Swaisgood, H.E.; Porter, D.H.; Catignani, G.L. Spectrophotometric assay using o-phthaldialdehyde for determination of proteolysis in milk and isolated milk proteins. *J. Dairy Sci.* **1983**, *66*, 1219–1227. [CrossRef]
31. De Pasquale, I.; Verni, M.; Verardo, V.; Gómez-Caravaca, A.M.; Rizzello, C.G. Nutritional and functional advantages of the use of fermented black chickpea flour for semolina-pasta fortification. *Foods* **2021**, *10*, 182. [CrossRef]
32. Lazo, O.; Claret, A.; Guerrero, L. A comparison of two methods for generating descriptive attributes with trained assessors: Check-all-that-apply (CATA) vs. free choice profiling (FCP). *J. Sens. Stud.* **2016**, *31*, 163–176. [CrossRef]
33. Santana, R.V.; dos Santos, D.C.; Santana, A.C.A.; de Oliveira Filho, J.G.; de Almeida, A.B.; de Lima, T.M.; Guimarães Silva, F.; Egea, M.B. Quality parameters and sensorial profile of clarified "Cerrado" cashew juice supplemented with *Saccharomyces boulardii* and different sweeteners. *LWT* **2020**, *128*, 109319. [CrossRef]
34. He, W.; Liu, S.; Heponiemi, P.; Heinonen, M.; Marsol-Vall, A.; Ma, X.; Yang, B.; Laaksonen, O. Effect of *Saccharomyces cerevisiae* and *Schizosaccharomyces pombe* strains on chemical composition and sensory quality of ciders made from Finnish apple cultivars. *Food Chem.* **2021**, *345*, 128833. [CrossRef] [PubMed]
35. Di Cagno, R.; Surico, R.F.; Minervini, G.; Rizzello, C.G.; Lovino, R.; Servili, M.; Taticchi, A.; Urbani, S.; Gobbetti, M. Exploitation of sweet cherry (*Prunus avium* L.) puree added of stem infusion through fermentation by selected autochthonous lactic acid bacteria. *Food Microbiol.* **2011**, *28*, 900–909. [CrossRef]
36. Al-Sahlany, S.T.; Niamah, A.K. Bacterial viability, antioxidant stability, antimutagenicity and sensory properties of onion types fermentation by using probiotic starter during storage. *Nutr. Food Sci.* **2022**, *52*, 901–916. [CrossRef]
37. Sanchez, I.; Palop, L.; Ballesteros, C. Biochemical characterization of lactic acid bacteria isolated from spontaneous fermentation of 'Almagro' eggplants. *Int. J. Food Microbiol.* **2000**, *59*, 9–17. [CrossRef]
38. Seseña, S.; Palop, M.L.I. An ecological study of lactic acid bacteria from Almagro eggplant fermentation brines. *J. Appl. Microbiol.* **2007**, *103*, 1553–1561. [CrossRef] [PubMed]
39. Cissé, H.; Kagambèga, B.; Sawadogo, A.; Tankoano, A.; Sangaré, G.; Traoré, Y.; Ouoba, I.I.L.; Savadogo, A. Molecular characterization of *Bacillus*, lactic acid bacteria and yeast as potential probiotic isolated from fermented food. *Sci. Afr.* **2019**, *6*, e00175. [CrossRef]
40. Sukmawati, S.; Rosalina, F.; Sipriyadi, S.; Dewi, N.K.; Yunita, M.; Sarhan, A.R.T.; Rahayu, Y.; Kusumawati, E. Bacterial diversity of mangrove ecosystem in Klawalu Sorong, West Papua, Indonesia. *Biodiversitas J. Biol. Divers.* **2022**, *23*, 1427–1432.
41. Lee, M.; Hong, G.E.; Zhang, H.; Yang, C.Y.; Han, K.H.; Mandal, P.K.; Lee, C.H. Production of the isoflavone aglycone and antioxidant activities in black soymilk using fermentation with *Streptococcus thermophilus* S10. *Food Sci. Biotechnol.* **2015**, *24*, 537–544. [CrossRef]
42. Ziarno, M.; Bryś, J.; Parzyszek, M.; Veber, A. Effect of Lactic Acid Bacteria on the Lipid Profile of Bean-Based Plant Substitute of Fermented Milk. *Microorganisms* **2020**, *8*, 1348. [CrossRef]
43. Liao, W.; Wang, C.; Shyu, Y.; Yu, R.; Ho, K. Influence of preprocessing methods and fermentation of adzuki beans on γ-aminobutyric acid (GABA) accumulation by lactic acid bacteria. *J. Funct. Foods* **2013**, *5*, 1108–1115. [CrossRef]
44. Aduol, K.O.; Onyango, A.N.; Imathiu, S.M. Proximate, Microbial and Sensory Characteristics of Cowpea Milk Fermented with Probiotic Starter Cultures. *EJFOOD* **2020**, *2*, 1–7. [CrossRef]
45. Donkor, O.N.; Henriksson, A.; Vasiljevic, T.; Shah, N.P. Effect of acidification on the activity of probiotics in yoghurt during cold storage. *Int. Dairy J.* **2006**, *16*, 1181–1189. [CrossRef]
46. Shah, N.P.; Lankaputhra, W.E.; Britz, M.L.; Kyle, W.S. Survival of *Lactobacillus acidophilus* and *Bifidobacterium bifidum* in commercial yoghurt during refrigerated storage. *Int. Dairy J.* **1995**, *5*, 515–521. [CrossRef]
47. Baruzzi, F.; Quintieri, L.; Caputo, L.; Cocconcelli, P.; Borcakli, M.; Owczarek, L.; Jasinska, U.T.; Skapska, S.; Morea, M. Improvement of Ayran quality by the selection of autochthonous microbial cultures. *Food Microbiol.* **2016**, *60*, 92–103. [CrossRef] [PubMed]
48. Baruzzi, F.; de Candia, S.; Quintieri, L.; Caputo, L.; De Leo, F. Development of a synbiotic beverage enriched with bifidobacteria strains and fortified with whey proteins. *Front. Microbiol.* **2017**, *8*, 640. [CrossRef]
49. Parra, K.; Ferrer, M.; Piñero, M.; Barboza, Y.; Medina, L.M. Use of *Lactobacillus acidophilus* and *Lactobacillus casei* for a potential probiotic legume-based fermented product using pigeon pea (*Cajanus cajan*). *J. Food Prot.* **2013**, *76*, 265–271. [CrossRef]

50. Saito, V.S.T.; Dos Santos, T.F.; Vinderola, C.G.; Romano, C.; Nicoli, J.R.; Araújo, L.S.; Costa, M.M.; Andrioli, J.L. Uetanabaro, A.P.T. Viability and resistance of lactobacilli isolated from cocoa fermentation to simulated gastrointestinal digestive steps in soy yogurt. *J. Food Sci.* **2014**, *79*, M208–M213. [CrossRef]
51. Di Cagno, R.; Surico, R.F.; Siragusa, S.; De Angelis, M.; Paradiso, A.; Minervini, F.; De Gara, L.; Gobbetti, M. Selection and use of autochthonous mixed starter for lactic acid fermentation of carrots, French beans or marrows. *Int. J. Food Microbiol.* **2008**, *127*, 220–228. [CrossRef]
52. Cabello-Olmo, M.; Oneca, M.; Torre, P.; Díaz, J.V.; Encio, I.J.; Barajas, M.; Araña, M. Influence of storage temperature and packaging on bacteria and yeast viability in a plant-based fermented food. *Foods* **2020**, *9*, 302. [CrossRef]
53. Arteaga, V.G.; Leffler, S.; Muranyi, I.; Eisner, P.; Schweiggert-Weisz, U. Sensory profile, functional properties and molecular weight distribution of fermented pea protein isolate. *Curr. Res. Food Sci.* **2021**, *4*, 1–10. [CrossRef] [PubMed]
54. Schlegel, K.; Leidigkeit, A.; Eisner, P.; Schweiggert-Weisz, U. Technofunctional and sensory properties of fermented lupin protein isolates. *Foods* **2019**, *8*, 678. [CrossRef] [PubMed]
55. Bhattacharya, M.; Salcedo, J.; Robinson, R.C.; Henrick, B.M.; Barile, D. Peptidomic and glycomic profiling of commercial dairy products: Identification, quantification and potential bioactivities. *NPJ Sci. Food* **2019**, *3*, 1–13. [CrossRef] [PubMed]
56. Germani, A.; Luneia, R.; Nigro, F.; Vitiello, V.; Donini, L.M.; Del Balzo, V. The yogurt amino acid profile's variation during the shelf-life. *Ann. Ig.* **2014**, *26*, 205–212. [PubMed]
57. Schad, G.J.; Bollig, B. Fast and Simple Determination of Free Amino Acids in Milk. *LC GC Eur.* **2015**, *28*, 418–419.
58. Jäger, R.; Zaragoza, J.; Purpura, M.; Iametti, S.; Marengo, M.; Tinsley, G.M.; Anzalone, A.J.; Oliver, J.M.; Fiore, W.; Biffi, A.; et al. Probiotic administration increases amino acid absorption from plant protein: A placebo-controlled, randomized, double-blind, multicenter, crossover study. *Probiotics Antimicrob. Proteins* **2020**, *12*, 1330–1339. [CrossRef] [PubMed]
59. Pouliot-Mathieu, K.; Gardner-Fortier, C.; Lemieux, S.; St-Gelais, D.; Champagne, C.P.; Vuillemard, J.C. Effect of cheese containing gamma-aminobutyric acid-producing lactic acid bacteria on blood pressure in men. *Pharma Nutr.* **2013**, *1*, 141–148. [CrossRef]
60. El Youssef, C.; Bonnarme, P.; Fraud, S.; Péron, A.C.; Helinck, S.; Landaud, S. Sensory improvement of a pea protein-based product using microbial co-cultures of lactic acid bacteria and yeasts. *Foods* **2020**, *9*, 349. [CrossRef]
61. Chaturvedi, S.; Chakraborty, S. Evaluation of prebiotic properties of legume-based synbiotic beverages. *J. Food Process. Preserv.* **2022**, *46*, e16685. [CrossRef]
62. Jiménez-Martínez, C.; Hernández-Sánchez, H.; Dávila-Ortiz, G. Production of a yogurt-like product from *Lupinus campestris* seeds. *J. Sci. Food Agric.* **2003**, *83*, 515–522. [CrossRef]

foods

MDPI

Article

Nutritional Function and Flavor Evaluation of a New Soybean Beverage Based on *Naematelia aurantialba* Fermentation

Tao Sun [1], Hao Jiang [1], Kai Yang [1], Xingkai Li [1], Shiyu Wang [1], Haoyu Yao [1], Rui Wang [1], Sha Li [1], Yian Gu [1], Peng Lei [1,*], Hong Xu [1,*] and Dafeng Sun [2]

1 State Key Laboratory of Materials-Oriented Chemical Engineering, College of Food Science and Light Industry, Nanjing Tech University, Nanjing 211816, China; sun-tao@njtech.edu.cn (T.S.); jianghao@njtech.edu.cn (H.J.); 201961119005@njtech.edu.cn (K.Y.); 201921019042@njtech.edu.cn (X.L.); 201921019081@njtech.edu.cn (S.W.); 201921019097@njtech.edu.cn (H.Y.); ruiwang2013@njtech.edu.cn (R.W.); lisha@njtech.edu.cn (S.L.); yian.gu@hotmail.com (Y.G.)

2 Kunming Edible Fungi Institute of All China Federation of Supply and Marketing Cooperatives, Kunming 650032, China; sdafeng@163.com

* Correspondence: lei-peng@njtech.edu.cn (P.L.); xuh@njtech.edu.cn (H.X.); Tel.: +86-187-6168-1790 (P.L.); +86-25-58139433 (H.X.)

Abstract: The soy beverage is a healthy product rich in plant protein; however, its unpleasant flavor affects consumer acceptance. The aim of this study was to determine the feasibility of using *Naematelia aurantialba* as a strain for the preparation of fermented soybean beverages (FSB). Increases in Zeta potential, particle size, and viscosity make soy beverages more stable. We found that nutrient composition was increased by fermenting *N. aurantialba*, and the antioxidant activity of soybean beverages significantly increased after 5 days of fermentation. By reducing the content of beany substances such as hexanal and increasing the content of 1-octen-3-ol, the aroma of soybean beverages fermented by *N. aurantialba* changed from "beany, green, and fatty" to "mushroom and aromatic". The resulting FSB had reduced bitterness but considerably increased sourness while maintaining the fresh and sweet taste of unfermented soybean beverages (UFSB). This study not only provides a theoretical basis for the market promotion of FSB but also provides a reference for basidiomycetes-fermented beverages.

Keywords: sensory evaluation; hierarchical cluster analyses; volatile organic compounds; GC-MS; GC-IMS; principal component

Citation: Sun, T.; Jiang, H.; Yang, K.; Li, X.; Wang, S.; Yao, H.; Wang, R.; Li, S.; Gu, Y.; Lei, P.; et al. Nutritional Function and Flavor Evaluation of a New Soybean Beverage Based on *Naematelia aurantialba* Fermentation. *Foods* **2022**, *11*, 272. https://doi.org/10.3390/foods11030272

Academic Editors: Carlo Giuseppe Rizzello and Palmira De Bellis

Received: 22 December 2021
Accepted: 18 January 2022
Published: 20 January 2022

1. Introduction

With the recent enhancement of nutritional, healthy, and green dietary ideas, plant protein drinks are gaining increasing attention, and their development remains of ongoing interest in the beverage industry [1]. Soybeans are one of the most important sources of plant proteins due to their high protein and essential amino acid content [2]. The soy beverages are considered alternatives to milk, especially for individuals allergic to milk proteins, intolerant of lactose, or on a vegetarian diet [2]. Commercialized beverages should not only be rich in nutrients but also have a good flavor [3]. However, the beany flavor of soy beverages is the main factor limiting their acceptance by consumers [1,2,4]. The compounds causing this beany flavor are mainly decomposition products of soybean lipids catalyzed by lipoxygenases (LOX) and hydroperoxide lyases [5]. Generally, two methods are used to reduce the beany flavor of soy beverages [6]. First, LOX knockout soybean cultivars are used [7]. However, previous research has revealed that removing complete LOX isozymes from soybeans can have a negative impact on the plant's defense mechanisms [8]. Second, methods such as temperature control, acid treatment, enzyme treatment, and supercritical carbon dioxide extraction have been employed to remove flavors from processing conditions, but the high energy consumption and cost of such methods make the preparation cost of soybean beverages too high to benefit commercialization [9].

Fermentation using probiotics or edible fungi is considered an inexpensive and safe method of improving the nutritional composition and flavor compounds of plant-based foods and increasing their shelf stability [10–13]. At present, Lactobacillus species, such as *L. plantarum*, *L. bulgaricus*, and *L. paracasei*, are used to improve the sensory properties of soymilk and reduce or mask the beany flavor of soybeans [14]. However, the development of plant-based soybean beverage varieties, particularly fermentation type microbial species diversity, will be hindered if only strains of the lactic acid bacteria genus are used as fermented soybean beverages. As a result, researchers all around the world are seeking strategies to overcome this constraint by employing alternative probiotics. The products being developed involve the use of various types of bacteria and fungi, particularly edible fungi, in the same medicines and foods. Previous studies have found that fermenting soybean beverages with basidiomycetes has great application potential. *Phellinus igniarius* and *Agrocybe cylindracea* not only showed good growth conditions in soy beverages, but also improved antioxidant properties and inhibited epidermal tumor proliferation in mice [15]. *Ganoderma lucidum* was used to ferment soy beverages to improve their health properties, and *Lycoperdon pyriforme* was used to adjust their aroma and eliminate the beany flavor [11–16]. However, there are still some problems facing basidiomycete-fermented soybean beverages; specifically, most of the currently used strains are mycelia, which degenerate easily and result in poor quality of the final fermented products.

Naematelia aurantialba, also known as *Tremella aurantialba*, is an indigenous Chinese edible and medicinal fungus [17]. Previous research has found that polysaccharides, saponin, phenolic, and flavonoid compounds in *N. aurantialba* are responsible for antioxidation, antiinflammatory, antitumor, and immunomodulatory effects [18–21]. In the absence of an associate, *N. aurantialba* acts as a basidiomycete whose spores multiply in an outgrowth manner and do not develop into mycelium [22,23]. Our previous study showed that the most favored nitrogen source for exopolysaccharide production by *N. aurantialba* NX-20 was soybean protein, which means that it could not only utilize soybean beverage for growth but also secrete its most active substance, *N. aurantialba* polysaccharide (NAPS) [24,25]. While ultra-high molecular weight NAPS perform multiple biological activities, they also act as a water-soluble pseudoplastic fluid to increase the stability and viscosity of soybean beverages. Thus, soybean beverages fermented with NAPS do not require the incorporation of complex and expensive stabilizers, resulting in reduced production costs. In addition, the flavor of fermented liquor from *N. aurantialba* is pleasant and has a high potential for application in the removal or masking of the beany flavor of fermented soybean beverages [26].

Therefore, in this study, an unfermented soybean beverage (UFSB) was fermented by *N. aurantialba* to produce a fermented soybean beverage (FSB). The physicochemical properties and nutritional components of the soybean beverages were examined, and their antioxidant capacity was investigated. In addition, HS-GC-IMS and HS-SPME-GC-MS were utilized to explore the changes in volatile odor in soybean beverages. Finally, the aroma and taste characteristics of the soybean beverage were evaluated by sensory evaluation and electronic tongue technology.

2. Materials and Methods

2.1. Preparation of Soy Beverages

To prepare the UFSB from fresh soybeans, the soybeans were washed with distilled water and soaked for 12 h at 25 °C. The beverage was then prepared using a soybean milk machine at a solid–liquid ratio of 1:5 (mass ratio). It was then filtered through four layers of cotton gauze and autoclaved at 105 °C for 20 min.

To prepare the fermented soy beverage (FSB) from fresh soybeans, the isolated strain, *N. aurantialba* NX-20, which had been preserved in the China General Microbiological Culture Collection Center (CGMCC 18588), was inoculated into the seed medium (potato dextrose broth) and cultured at 25 °C for 3 d. The seed liquid was then transferred to 100 mL of fermentation medium at 10% inoculation and cultured at 25 °C for 5 d. The fermentation medium consisted of 4 g of glucose and 100 mL of UFSB. After fermentation, the broth was boiled for 30 min to inactivate the *N. aurantialba* spore cells, and then the samples were frozen at −80 °C until ready for further analysis. In addition, the spread plate method was used to detect the quality of microorganisms in unsterilized soy fermented beverages. The phosphate-buffered saline (PBS) was used to make gradient dilutions of the fermentation broth. The number of cells was then counted after potato dextrose agar (PDA) culture (4 d, 5 °C). The amount of *N. aurantialba* in unsterilized soy fermented beverages was approximately $10^9 \pm 5 \times 10^7$ CFU/mL.

2.2. Changes in Physicochemical Properties of Soy Beverages

The particle size and zeta potential of the samples were analyzed using a nanoparticle potentiometer (NICOMP Z3000, Port Richey, FL, USA) [27]. The test sample was diluted with de-ionized water (1:12.5) and stirred at 300 rpm for approximately 10 min before analysis. The average size of the samples was determined based on polarization intensity difference scattering [27].

The viscosity of the soybean beverage was measured using an Ubbelohde viscometer (NDJ-8S, Nanjing, China) in a 25 °C water bath [28].

A HunterLab UltraScan PRO (Hunter Associate Laboratory Inc., Reston, VA, USA) colorimeter was used to examine the samples [29]. The samples were placed in optical glass cells with a fixed route length of 50 mm and a 25 mm aperture for reflectance measurement (Large Area View—LAV). The average of five readings was used to arrive at the findings. The CIELab L* (lightness), a* (redness), and b* (yellowness) values are used by the UltraScan PRO to measure the reflected color of food products.

A digital pH meter (PHS-3, Shanghai INESA Scientific Instrument Co., Ltd., Shanghai, China) was used to measure the pH.

2.3. Analysis of Nutrients

The total soluble protein in a sample was measured using the Bradford method according to the manufacturer's instructions [30].

The total fat in a sample was measured using a Soxhlet apparatus, which is based on the Soxhlet extractor method [11].

The total sugar content in the sample was determined using the phenol sulfuric acid method. The reducing sugars were determined using the DNS method. Determination of glucose concentration was conducted using the SBA-40c biosensor (Biological Institute of Shandong Academy of Science, Jinan, China) [24].

The amount of total dietary fiber in the sample was determined using the enzymatic gravimetric method [30].

The total acid content in a sample acid-base titration was determined based on the enzymatic gravimetric method [30].

Free amino acids in the samples were analyzed using an automated amino acid analyzer (S-433D, Sykam, Munich, Germany) [31].

The Folin–Ciocalteu method and aluminum chloride were used to determine the total flavonoid and phenolic contents, respectively; each method was performed according to the manufacturer's instructions on the kits (Nanjing JianCheng Bioengineering Institute, Nanjing, China) [32].

The NAPS content in the samples was determined by HPLC-GPC. Based on the detection of NAPS by HPLC (Shimadzu 20A, Tokyo, Japan), two OHpak SB-806M HQ columns (8.0 mm × 300 mm, Shodex, Tokyo, Japan) were used in succession [24]. The mobile phase consisted of ultrapure water at a flow rate of 1.0 mL/min. Detection was performed using an evaporative light scattering detector (ELSD, ELSD-16) with a drift tube temperature of 60 °C and ventilation rate of 3.0 L/min. The NAPS content was calculated according to the standard linear equation formed by different concentrations of NAPS standards [24].

2.4. Analysis of Antioxidant Capacity

ABTS (2, 2′-azino-bis (3-ethylbenzothiazoline-6-sulfonic acid), DPPH (2,2-Diphenyl-1-picrylhydrazyl), Fenton, and pyrogallol autoxidation methods were used to analyze ABTS radical scavenging activity, DPPH radical scavenging activity, scavenging rate of hydroxyl radical activity, and scavenging rate of superoxide anion radical activity. Each method was performed according to the manufacturer's instructions on the kits (Nanjing JianCheng Bioengineering Institute, Nanjing, China) [33,34].

2.5. Electronic Tongue Measurement

The flavor profiles of the soy beverages were analyzed using a potential electro-tongue (SA402B; Insent, Tokyo, Japan) [35,36]. The sensor of the e-tongue was first pre-equilibrated using 0.01 moL·L^{-1} of sodium chloride, 0.01 moL·L^{-1} of sodium glutamate, and 0.01 moL·L^{-1} hydrochloric acid, prior to detection. The e-tongue had an acquisition time of 120 s for each part of the sample, one data collection per second, nine acquisition times for each part of the sample, then 30 s of washing, and the results were given as the average of three data points.

2.6. Analysis of Volatile Compounds

2.6.1. Headspace Gas Chromatography-Ion Mobility Spectrometry (HS-GC-IMS) Analysis

The volatile organic compounds (VOCs) of the UFSB and FSB were analyzed by HS-GC-IMS analysis based on FlavourSpec® Flavor analyzers (Gesellschaft für Analytische Sensorsysteme mbH, Dortmund, Germany) and an Agilent 490 gas chromatograph (Agilent Technologies, Palo Alto, CA, USA) [37,38]. Approximately 1 g of sample was weighed and placed in a 20 mL headspace bottle and analyzed by an autosampler after incubation at 60 °C for 20 min (injection volume—500 μL, injection needle temperature—85 °C). The GC analytical column was a FS-SE-54-CB-1 capillary column (15 m × 0.53 mm, RESTEK, Bellefonte, PA, USA). The carrier gas, which was nitrogen, followed a programmed flow rate: initially the flow rate was 2 mL/min for 2 min, and then it was increased to 10 mL/min for 8 min, 100 mL/min for 10 min, and finally 150 mL/min for 10 min. The total GC run time was 30 min. The samples were separated in a column at 60 °C and then ionized in an IMS ionization chamber using a 3H ionization source (300 MBq activity). A 9.8 cm drift tube with a constant voltage (5 kV at 45 °C) and a nitrogen flow of 150 mL/min was used. Each spectrum was reported as an average of 12 scans. Qualitative and quantitative analyses of VOCs were performed using the National Institute of Standards and Technology (NIST) database built into the software, the IMS database, and reference materials [38].

2.6.2. Headspace Solid-Phase Microextraction Gas Chromatography–Mass Spectrometry (HS-SPME-GC-MS) Analysis

The volatile compounds in the FSB and UFSB samples were determined by HS-SPME-GC-MS [39]. One gram of the sample was transferred to a 20 mL headspace bottle (Agilent, Palo Alto, CA, USA) containing a saturated solution of NaCl to inhibit any enzyme reactions. The vials were sealed using crimp-top caps with a TFE-silicone headspace septa (Agilent). During the SPME analysis, each vial was placed at 60 °C for 10 min, and then a 65 μm divinylbenzene/carboxen/polydimethylsiloxane fiber (Supelco, Bellefonte, PA, USA) was exposed to the headspace of the sample for 20 min at 60 °C. After sampling, desorption of the VOCs from the fiber coating was carried out in the injection port of the GC apparatus (Model 7890 B; Agilent) at 250 °C for 5 min in splitless mode. The identification and quantification of VOCs was carried out using an Agilent Model 7890B GC and a 7000D mass spectrometer (Agilent), equipped with a 30 m × 0.25 mm × 1.0 μm DB-5MS (5% phenyl-polymethylsiloxane) capillary column. Helium was used as the carrier gas at a linear velocity of 1.0 mL/min. The temperatures of the injector and detector were kept at 250 and 280 °C, respectively. The oven temperature was set to 40 °C (5 min) and increased at 6 °C/min to 280 °C, and maintained for 5 min. Mass spectra were recorded in the electron impact ionization mode at 70 eV. The temperatures of the quadrupole mass detector, ion source, and transfer line were set to 150, 230, and 280 °C, respectively. Mass measurements were taken at 1 s intervals. Spectra were scanned in the range of m/z 30–350 amu [40].

The identification of volatile compounds was achieved by comparing the mass spectra with the data system library and linear retention index. The flavor description of volatile compounds was determined using the volatile compounds in food (VCF) database (https://www.vcf-online.nl/VcfHome.cfm, accessed on 10 July 2021), a good subject company information system (http://www.thegoodscentscompany.com/index.html, accessed on 10 July 2021), and references [11,41].

2.7. Analysis of Relative Odor Activity Value

The relative odor activity value (*ROAV*) was calculated using the following equation

$$ROAV \approx 100 \times \frac{C_{sample}}{C_{standard}} \times \frac{T_{standard}}{T_{sample}} \quad (1)$$

where C_{sample} and T_{sample} represented each volatile component's relative content and odor threshold, respectively. $C_{standard}$ and $T_{standard}$ represented the relative contents and odor thresholds, respectively, of the constituents that contributed most to the total odor in the sample [41].

2.8. Sensory Evaluation

The odors of the unfermented and fermented samples, including sour, mushroom, aromatic, fatty, green, and beany flavors, were evaluated by 20 experienced, trained assessors (10 males and 10 females). The intensity of each category was recorded based on a 7-point system, with 1 being low intensity and 7 being high intensity [41].

2.9. Statistical Analysis

All the experiments were performed using at least three independent samples. The SPSS software (version 19.0; SPSS Inc., Chicago, IL, USA) was used for statistical analysis and the results are expressed as the mean ± SD. The metabolites were characterized by analyzing mass spectra using the software Qualitative Analysis Workflow B.08.00. One-way analysis of variance and the least significant difference test were used, and the significance level is 0.05 ($\alpha = 0.05$).

In addition, for GC-IMS analysis, the laboratory analytical viewer was used to visualize the analyzed spectra, where each point referred to a VOCs that could be quantitatively analyzed by establishing a standard curve and reporter plugin for direct comparison of spectral differences between samples (two-dimensional top view and three-dimensional spectra). The gallery plot plugin was used to perform the comparison of fingerprinting patterns between samples.

Principal component (PCA) and hierarchical cluster analyses were performed using the statistical function prcomp within the R package (version 4.2, www.r-project.org, accessed on 1 July 2021) [42].

The sensory analysis was carried out on three parallel samples for each variable, and the data was expressed as mean value ± standard error.

3. Results and Discussion

3.1. Physicochemical Properties and Nutrient Composition of the Soybean Beverage

3.1.1. Analysis of Physicochemical Properties of the Soybean Beverage

The physical and chemical properties of UFSB and FSB are shown in Table 1. The mean particle sizes of UFSB and FSB were 198 nm and 926 nm, respectively. The increased particle sizes of FSB might be because fermentation produced NAPS with molecular weights of up to 2924.6 kDa, or because some proteins produced by fermentation were denatured during sterilization [43]. Furthermore, the Zeta potentials of UFSB and FSB were −12.41 and −32.54 mv, respectively. The term "Zeta potential" is widely used to indicate the degree of electrostatic repulsion between adjacent, similarly charged, colloidal particles in suspension, with higher surface charges enhancing electrostatic repulsion between droplets and overcoming van der Waals' forces and hydrophobic attractions [44]. The absolute Zeta potential value increased in FSB, indicating that after *N. aurantialba* fermentation, the stability of the soy beverage increased [44]. Pectin, chitosan, carboxymethyl cellulose, and other polysaccharides are commonly employed as stabilizers and thickeners to alter the zeta potential, average particle size, and viscosity of drinks to improve their stability, and NAPS may have a similar function [45,46]. The NAPS is an anionic polysaccharide and the polysaccharide protein complex generated by it binds with proteins, increasing the electrostatic and steric repulsion forces between the soybean beverage droplets and, thereby, increasing the viscosity [24]. To test the latter hypothesis, we compared the viscosities of both soybean beverages and found that the viscosities of FSB and UFSB were 804 mPa·s and 3.4 mPa·s, respectively. This indicated that NAPS indeed increased the viscosity of the soybean beverages following fermentation. Such an increase in viscosity might improve beverage stability but may be distasteful to consumers [47]. Taking this into consideration, FSB in the market can be diluted using purified water. The color values and appearances of both beverages are shown in Table 1. The a*-value (red/green) and b*-value (blue/yellow) of the color measurement were statistically significant, but not the lightness of L*. This meant that the UFSB had gotten redder and bluer than the FSB. However, the hue shift was hardly noticeable to the naked eye, and all beverages appeared "milk white, faint yellow." The pH of FSB and UFSB was 4 and 6, respectively. This indicates that the soybean beverage fermented with *N. aurantialba* generated more acidic chemicals, resulting in an acidic taste due to increased glucuronic acid synthesis. Lactic acid bacteria-fermented milk with a pH of around 4 is available on the market [48].

Table 1. Physicochemical parameters and nutritional ingredients of the unfermented and fermented soy beverages with *N. aurantialba*.

Parameter		FSB	UFSB
particle size distribution (nm)	D_{50}	926 ± 32^{A}	198 ± 20^{B}
Zeta potential (mv)		-32.5 ± 2.10^{A}	-12.4 ± 2.40^{B}
pH		4.00 ± 0.03^{A}	6.50 ± 0.02^{B}
appearance		milk white, faint yellow	milk white, faint yellow
viscosity (mPa·s)		804 ± 5.6^{A}	3.40 ± 0.5^{B}
L*a*b color space	L*	78.5 ± 0.58^{A}	77.6 ± 1.05^{A}
	a*	-1.41 ± 0.02^{A}	-1.21 ± 0.13^{B}
	b*	8.32 ± 0.09^{A}	7.13 ± 0.09^{B}
protein (g/kg)		17.0 ± 1.3^{A}	28.0 ± 1.5^{B}
fatty acid (g/kg)		20.0 ± 0.43^{A}	21.2 ± 6.4^{A}
total sugar (g/L)		13.1 ± 0.21^{A}	2.94 ± 0.18^{B}
reducing sugar (g/L)		4.28 ± 0.41^{A}	1.03 ± 0.28^{B}
glucose (g/L)		3.12 ± 0.12^{A}	0.17 ± 0.05^{B}
dietary fiber (%)		4.20 ± 0.50^{A}	4.60 ± 0.80^{A}
total acid (g/kg)		5.40 ± 0.06^{A}	3.40 ± 0.04^{B}
total amino acids (μg/mL)		32.8 ± 0.008^{A}	32.0 ± 0.011^{B}
total polyphenol (mg/L)		2313 ± 6.48^{A}	1285 ± 8.24^{B}
total flavonoid (mg/L)		1080 ± 3.59^{A}	603 ± 2.65^{B}
NAPS (g/L)		8.68 ± 0.12^{A}	0^{B}

Each value is expressed as mean ± SD ($n = 3$); means with different capital letters within a row indicate significant differences ($p < 0.05$); L* value represents the lightness of the sample, and a* value and b* value indicate the redness and yellowness, respectively.

3.1.2. Analysis of Soy Beverage Nutrient Composition

The levels of protein, fatty acid, total sugar, reducing sugar, dietary fiber, total acid, moisture, ash content, total amino acids, total polyphenols, total flavonoids, and NAPS in the FSB and UFSB are shown in Table 1. The protein content of FSB (17.0 ± 1.3 g/kg) was lower than that of UFSB (28.0 ± 1.5 g/kg), respectively. It may be that the proteins in the soybean beverage were used by *N. aurantialba* as a nitrogen source for the fungi's growth, which might explain the higher levels of total free amino acids in FSB than in UFSB. No significant differences in the levels of fatty acids or dietary fiber were observed before or after fermentation, which is consistent with previous studies using *L. pyriforme* to develop fermented soybean beverages [11]. The total free amino acid content and composition of the soybean beverage after fermentation by *N. aurantialba* increased. The essential amino acid levels of FSB increased by 47.4% compared to that of UFSB, along with the production of cysteine and proline, which were absent from UFSB, implying that fermentation of soybean beverages by *N. aurantialba* increased the amino acid levels (Table S1). According to Yang et al., the higher content of free amino acids produced in fermented soymilk is due to the high activity of proteolytic enzymes in *Grifola frondosa*, which could be used to digest soybean proteins, suggesting that *N. aurantialba* may also contain high proteolytic enzyme activity that could be used to digest soybean proteins [31]. In addition, the increased levels of free amino acids meant that soy beverages were more easily absorbed after fermentation [1].

Reducing sugar and total sugar concentrations were compared between the two fermented beverages, and the levels of total and reducing sugar concentrations in the FSB were significantly higher than in the UFSB. The changes in reducing sugars were mainly due to the exogenous addition of glucose at 40 g/L for fungal growth. The increase in total sugar levels was mainly due to *N. aurantialba* metabolism, which utilized a large amount of glucose as a matrix to generate 8.68 g/L of NAPS (Table 1 and Figure S1). Because of their unique health properties, polysaccharides are frequently used to improve the health value of beverages [49]. In *Caenorhabditis elegans*, a beverage rich in polysaccharides of *Cyclocarya paliurus* can protect against oxidative stress and reduce fat deposition without affecting critical physiological activities [50]. Furthermore, frequent ingestion of yogurt containing hallabong peel polysaccharide boosts natural killer NK cell capacity while decreasing proinflammatory cytokine levels [51]. As a result, after *N. aurantialba* fermentation, the bioactivity of soy drinks may increase.

The content of total acid is also one of the important factors in evaluating the composition of beverages. The total acid level in the soy beverage was 0.54 g/kg after fermentation, which showed an increase of 37% compared with its unfermented counterparts; this indicated the decreased pH of the FSB. This means that the FSB had the characteristics of a "sour" lactic acid bacteria beverage. Lactic acid bacteria beverages tend to all have a lower pH due to the large amount of lactic acid produced by fermentation [40]. Furthermore, *Ganoderma lucidum* fermentation reduced the pH of pumpkin juice from 7 to 4, giving the drink a sour flavor [52].

In addition, the reason that *N. aurantialba* presented acidity might be because of the production of uronic acid as well as other organic acids by bacteriophage metabolism.

Phenolic compounds and flavonoids, such as rutin, catechin, and naringin, are widely found in raw materials from plants, microbial, and other sources, and are known to have important antioxidant activities, so we examined the contents of total polyphenols and flavonoids in UFSB and FSB [32]. In FSB and UFSB, total polyphenol content was 2313.3 mg/L and 1285.5 mg/L, respectively. Additionally, the total flavonoid content in FSB and UFSB was 1040.2 mg/L and 603.44 mg/L, respectively. Thus, the total antioxidant content of FSB was higher than that of UFSB, which means that FSB has potential as a functional food. Because phenolics are the major antioxidant components in mushrooms, and their contents can be used to evaluate the antioxidant capacity of beverages to some extent, it is important to detect the presence of phenolics in FSB. Islam et al. found that the content of polyphenols and flavonoids in *N. aurantialba* was 0.80 mg GAE/g and 0.13 mg GAE/g, respectively [53]. The kinds of phenolics might include homogenous acid, protocatechuic acid, and p-hydroxybenzoic acid, for example, and the findings could be evaluated using LC-MS due to the non-specificity of the Folin–Ciocalteu reagent [21].

3.2. Antioxidant Capacity of Soy Beverages

Antioxidant capacity is an important indicator often used to assess whether fermented beverages have health care value [54]. Figure 1 displays the antioxidant capacity of FSB. As no single method can be used to describe the overall antioxidant capacity of a sample, four antioxidant detection techniques were used to examine the antioxidant capacities of FSB and UFSB: ABTS radical scavenging, DPPH radical scavenging, OH radical scavenging, and superoxide anion radical scavenging. The ABTS radical scavenging capacity of FSB was 81.4%, while that of UFSB was 42.8%; it increased in FSB by 90.1%. The DPPH radical scavenging capacity of FSB was 94.75%, while that of UFSB was 50.88%. It increased in FSB by 86.22%. The hydroxyl radical scavenging capacity of FSB was 70.35%, while that of UFSB was 20.35%; after fermentation, this value increased by 71.07%. FSB had a 90.22% superoxide anion radical scavenging capacity, whereas for UFSB it was 30.28%; thus, fermentation increased it 2.31-fold. The free radical scavenging capacity and total antioxidant capacity of FSB were significantly higher than those of UFSB, possibly because the total polyphenol and flavonoid levels, which exhibited antioxidant activity, of FSB were 1.8 and 1.7 times higher than those of UFSB, respectively. In addition, NAPS, a natural macromolecular polysaccharide, also has a certain antioxidant capacity [17,55]. Based on the antioxidant analysis, it was not surprising that the antioxidant activity of soybean beverage fermented by *N. aurantialba* increased significantly. These differences may be attributed to the reaction rate of free radical sources to polysaccharides, total phenols and flavonoids, and other active substances in FSB. Therefore, FSB with effective antioxidant capacity may be an effective health drink for humans.

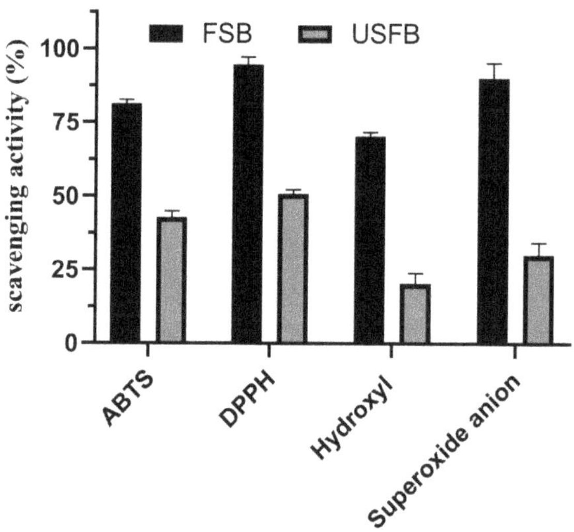

Figure 1. Antioxidant capacity of soy beverage with *N. aurantialba*.

3.3. Volatile Fingerprints of Soybean Beverage by HS-GC-IMS

HS-GC-IMS was used to investigate flavor profiles in the soybean beverages before and after fermentation, and the data is presented as 3D visualized topography (Figure 2A) and 2D top view plots (Figure 2B). The results showed that there were large differences in signal properties and intensities between FSB and UFSB. The first two principal components accounted for 96% and 2% of the total variance, respectively (Figure 2C). In addition, the PCA results showed that FSB and UFSB occupied relatively independent spaces in the profiles, implying that their flavor profiles were quite different. To gain more insight into the specific types of compounds identified, the volatile fingerprints of FSB were successfully established (Figure 2D). From the volatile fingerprint, we obtained a color, and a darker color indicated a higher intensity. In addition, some single compounds with high concentrations or proton affinities produced more than one signal, corresponding to dimers and trimers, which were retained at similar times but migrated at different times [56]. The volatile fingerprint information indicated that 69 typical aromas were tentatively identified in the FSB, with some components exhibiting two peaks corresponding to monomeric and dimeric forms, and the dimers being eluted after the monomers. The compounds in region a′ of Figure 2D, including 2-pentylfuran, 1-hexanol, (E)-2-hexenal, hexanal, pentanal, and ethyl acetate, were abundant in the UFSB sample. In contrast, the signal intensities of volatile compounds in region b′ consisting of 1-octen-3-ol, 3-octanone, 1-octen-3-one (M), and 1-octen-3-one (D), significantly increased after fermentation.

The peak intensities and relative contents of VOCs in the soy beverages are shown in Table 2 and Table S2. These compounds were classified into six categories, including 12 alcohols, 26 aldehydes, 3 esters, 11 ketones, 1 furan, and 15 unidentified compounds. Hexanal, regarded as one of the most important contributors to the beany flavor of soybean, decreased significantly in both peak intensity and relative content after fermentation [9]. In addition, other substances that were also regarded as the main sources of the beany flavor, such as (E)-2-hexenal, pentanal, 2-pentylfuran, 1-hexanol, and other compounds, also showed a significant decrease in signal intensity. To date, little data is available on the mechanism by which basidiomycetes degrade or adsorb beany flavor compounds such as hexanal, hexanol, and (E)-2-hexenal [57]. Based on studies on using Bifidobacterium to ferment soymilk, it can be speculated that the key enzymes needed to oxidize aldehydes to acids seem to be aldehyde dehydrogenase and aldehyde dehydrogenase [58].

In addition, the generation of more volatile compounds during fermentation may enhance some positive aroma properties in soy beverages, thereby eliminating or masking the beany flavor of soybeans [6,12,59]. Because of their low odor threshold and strong odor properties, aldehydes were the most detected compounds and important contributors to the aroma of FSB [60]. Substances such as pentanal (fruit, berry), benzeneacetaldehyde (green, sweet, cocoa), and (E)-2-pentenal (green, fruity) showed a greater increase in signal intensities in soy beverages after fermentation. Two compounds, pentanal and benzeneacetaldehyde, which are considered to be two pleasant odors of the fruiting body of *Tricholoma matsutake*, are mainly produced by the oxidation of polyunsaturated fatty acid double bonds [37,61]. Ketones are generated by polyunsaturated fatty acid oxidation, the Maillard reaction, amino acid degradation, or microbial oxidation [62]. The signal intensities of 1-octen-3-one (D), 1-octen-3-one (M), and 3-octanone in FSB increased by 8.67-, 3.71-, and 1.26-fold, respectively, compared with those of UFSB. Moreover, in addition to the large increases in signal intensities of 1-octen-3-one and 3-octanone with the mushroom odor, significant increases in the signal intensities of 2-pentanone, 2-butanone, 2-heptanone (M), 2-heptanone (D), 2,3-butanedione, and acetone with a fruity odor were also observed. Compared with corresponding ketones and aldehydes, the threshold of alcohols is higher and can be divided into saturated and unsaturated alcohols. Unsaturated alcohols have low thresholds and distinctive odors [37]. Among the alcohols that should be noted is 1-octen-3-ol, which is known as the mushroom alcohol and T. matsutake alcohol and is considered to be one of the major substances responsible for the flavor of mushrooms [37,61]. Fermented soy beverages had a 2.78-fold increase in 1-octen-3-ol compared to UFSB. In addition, the signal intensities of some substances with an alcoholic flavor, such as 2-methyl-1-propanol, 1-pentanol, 3-methyl-1-butanol (M), and 3-methyl-1-butanol (D), were also significantly increased, but it was speculated that they had limited influence on the overall flavor because of their higher odor thresholds.

It must be emphasized that a series of eight carbon compounds are key contributors to the mushroom flavor, including (E)-2-octenal, 1-octen-3-one, 1-octen-3-ol, and 3-octanone (Table 2 and Table S2). These substances are considered typical volatile compounds in edible mushrooms and are major contributors to the unique flavor of mushrooms, probably produced by the metabolism of fatty acids by edible mushrooms [63]. This means that FSB is highly likely to have a pleasant mushroom aroma.

3.4. Volatile Fingerprints of Soybean Beverage by HS-SPME-GC-MS

To further evaluate the effects of *N. aurantialba* fermentation on the flavor of soybean beverages, HS-SPME-GC-MS was used to analyze the VOCs in the samples. Similar to the HS-GC-IMS results, the number and area of VOCs peaks changed with fermentation, indicating that fermentation altered VOCs formation and, consequently, the flavor of the soy beverage (Figure 3A). As shown in Figure 3B, the results of PCA showed that the first two principal components accounted for 48.49% and 22.8% of the total variance, and FSB and UFSB occupied relatively independent spaces in the profiles, which was similar to the results of GC-IMS, further illustrating that the flavor substances in the samples were different.

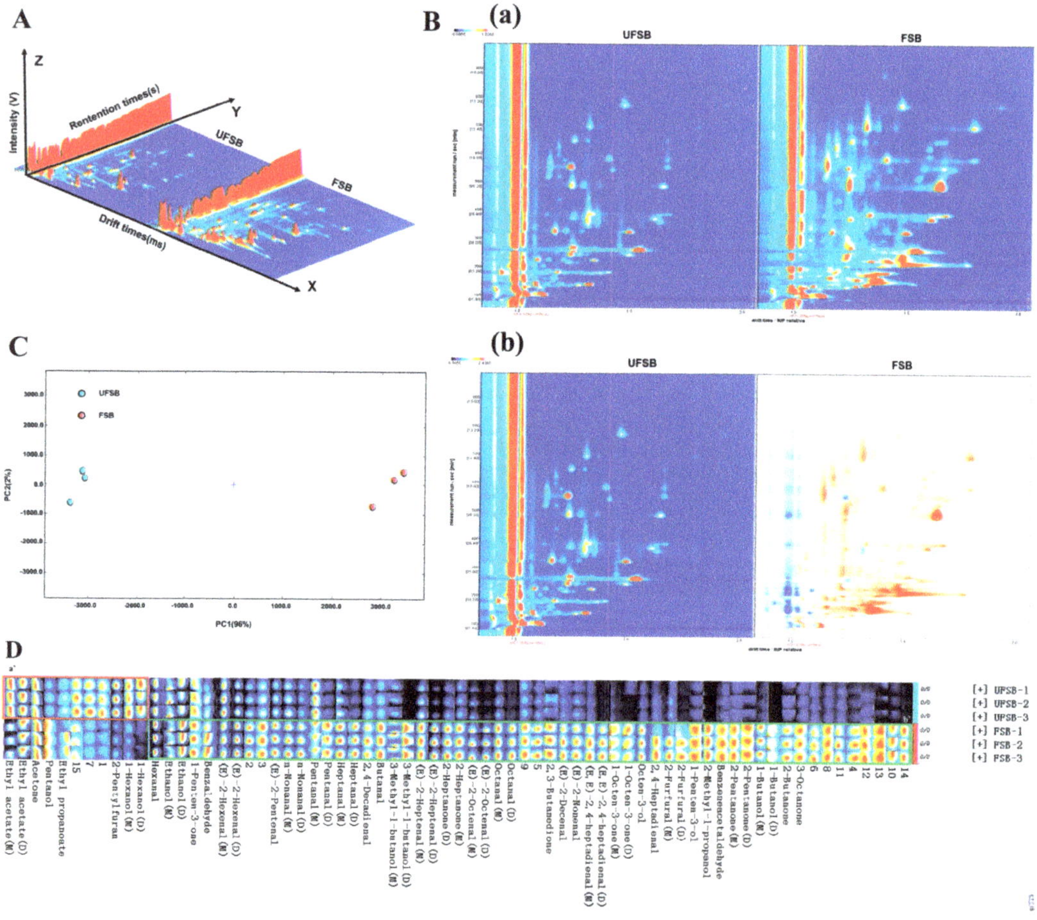

Figure 2. The volatile organic compounds analysis in soy beverage by HS-GC-IMS. (**A**) 3D-topographic features of samples, (**B**) 2D-topographic top view plot of samples (**a**): normal top view plot, (**b**): 2D-topographic top view plot deduction), (**C**) PCA score of samples, and (**D**) fingerprint spectra of samples.

Table 2. Identification of the volatile compounds in fermented soy beverage by HS-GC-IMS.

Count	Compounds	RI	Rt (s)	Dt (a.u.)	Odor Description	Relative Amount (%)	
						UFSB	FSB
					Aldehydes		
1	2,4-Decadienal	1285	1352	1.42	fatty green	1.20 ± 0.011 [A]	1.03 ± 0.054 [B]
2	(E)-2-Decenal (M)	1233	1148	1.48	fatty, fish, hay	0.520 ± 0.052 [A]	0.530 ± 0.025 [A]
3	(E)-2-Nonenal (D)	1147	879	1.41	fatty green	0.280 ± 0.052 [A]	0.330 ± 0.030 [A]
4	n-Nonanal (M)	1106	772	1.48	rose fresh fruity	2.33 ± 0.080 [A]	1.76 ± 0.057 [B]
5	n-Nonanal (D)	1106	772	1.94		0.310 ± 0.035 [A]	0.280 ± 0.016 [A]
6	(E)-2-Octenal (M)	1072	695	1.34	fatty green	2.85 ± 0.430 [A]	4.44 ± 0.239 [B]
7	(E)-2-Octenal (D)	1072	694	1.81		0.480 ± 0.023 [A]	1.93 ± 0.205 [B]
8	(E, E)-2,4-heptadienal (M)	1036	621	1.19	fatty green	0.620 ± 0.036 [A]	1.06 ± 0.095 [B]
9	(E, E)-2,4-heptadienal (D)	1035	619	1.61		0.090 ± 0.002 [A]	0.370 ± 0.032 [B]
10	2,4-Heptadienal	1019	590	1.20	green fruity	0.330 ± 0.092 [A]	0.380 ± 0.124 [A]
11	Octanal (M)	1015	581	1.42	green fatty	1.20 ± 0.044 [A]	1.77 ± 0.029 [B]
12	Octanal (D)	1014	580	1.82		0.210 ± 0.021 [A]	0.650 ± 0.037 [B]
18	(E)-2-Heptenal (M)	964	485	1.26	green fatty	3.59 ± 0.245 [A]	4.37 ± 0.068 [B]
19	(E)-2-Heptenal (D)	964	484	1.66		2.34 ± 0.088 [A]	9.72 ± 0.031 [B]
20	Benzaldehyde	981	518	1.15	bitter almond, sweet cherry	0.900 ± 0.159 [A]	0.650 ± 0.128 [A]
21	Heptanal (M)	905	387	1.36	fatty green	1.45 ± 0.118 [A]	1.33 ± 0.020 [B]
22	Heptanal (D)	904	385	1.69		0.580 ± 0.023 [A]	1.27 ± 0.035 [B]
27	(E)-2-hexenal (M)	858	325	1.18	green fatty	3.55 ± 0.057 [A]	1.59 ± 0.037 [B]
28	(E)-2-hexenal (D)	855	322	1.51		3.58 ± 0.217 [A]	2.88 ± 0.062 [B]
29	2-Furfural (M)	833	298	1.09	sweet woody almond	0.260 ± 0.028 [A]	0.260 ± 0.100 [A]
30	2-Furfural (D)	831	295	1.33		0.040 ± 0.0003 [A]	0.130 ± 0.035 [B]
31	hexanal	796	260	1.55	green fatty	17.3 ± 0.708 [A]	7.38 ± 0.039 [B]
32	(E)-2-pentenal	756	224	1.36	green fruity	1.27 ± 0.095 [A]	1.05 ± 0.002 [B]
33	pentanal (M)	699	180	1.19	fruity berry	1.49 ± 0.074 [A]	0.620 ± 0.030 [B]
34	pentanal (D)	701	182	1.43		0.730 ± 0.02 [A]	1.70 ± 0.109 [B]
38	butanal	605	135	1.29	pungent cocoa green	0.890 ± 0.039 [A]	0.900 ± 0.021 [A]
54	benzeneacetaldehyde	1071	693	1.26	green sweet cocoa	0.230 ± 0.004 [A]	0.260 ± 0.018 [A]
13	1-Octen-3-ol	998	552	1.16	mushroom	2.06 ± 0.072 [A]	3.48 ± 0.082 [B]
25	1-hexanol (M)	886	359	1.32	fruity alcoholic sweet green	2.55 ± 0.032 [A]	0.500 ± 0.065 [B]
26	1-hexanol (D)	884	357	1.64		0.600 ± 0.094 [A]	0.080 ± 0.016 [B]
35	2-Methyl-1-propanol	632	146	1.37	ethereal winey cortex	0.250 ± 0.012 [A]	5.49 ± 0.124 [B]
40	ethanol (M)	483	94.5	1.05	strong alcoholic	10.5 ± 0.367 [A]	5.51 ± 0.062 [B]
41	ethanol (D)	486	95.2	1.14		1.99 ± 0.251 [A]	1.74 ± 0.080 [A]
50	1-butanol (M)	668	163	1.17	sweet balsam whiskey	0.430 ± 0.042 [A]	1.10 ± 0.023 [B]
51	1-butanol (D)	667	162	1.39		0.040 ± 0.004 [A]	1.53 ± 0.017 [B]
52	3-Methyl-1-butanol (M)	744	214	1.24	alcoholic fruity	1.35 ± 0.059 [A]	0.990 ± 0.030 [B]
53	3-Methyl-1-butanol (D)	745	215	1.50		0.410 ± 0.023 [A]	7.70 ± 0.062 [B]
43	1-penten-3-ol	692	175	1.34	fruity green	0.450 ± 0.007 [A]	0.900 ± 0.004 [B]
44	1-Pentanol	777	243	1.51	sweet balsam	0.650 ± 0.044 [A]	0.660 ± 0.011 [A]
14	2-Pentylfuran	999	553	1.25	green beany vegetable	5.28 ± 0.040 [A]	1.09 ± 0.132 [B]
15	3-Octanone	993	543	1.31	sweet mushroom	0.170 ± 0.020 [A]	0.200 ± 0.002 [A]
16	1-Octen-3-one (M)	987	531	1.27	mushroom	0.150 ± 0.007 [A]	0.370 ± 0.040 [B]
17	1-Octen-3-one (D)	986	528	1.68		0.120 ± 0.076 [A]	0.590 ± 0.050 [B]
23	2-heptanone (M)	895	372	1.26	fruity sweet	0.750 ± 0.017 [A]	0.910 ± 0.008 [B]
24	2-heptanone (D)	896	373	1.63		1.14 ± 0.106 [A]	1.79 ± 0.037 [B]
39	2-Butanone	592	130	1.24	fruity	0.200 ± 0.042 [A]	0.330 ± 0.014 [B]
42	Acetone	512	103	1.13	apple pear	7.75 ± 0.893 [A]	5.01 ± 0.168 [B]
43	1-penten-3-one	689	173	1.08	pungent peppery onion	0.250 ± 0.039 [A]	0.100 ± 0.003 [B]
47	2,3-butanedione	591	130	1.17	sweet creamy	0.210 ± 0.049 [A]	0.270 ± 0.008 [A]
48	2-pentanone (M)	687	172	1.12	sweet fruity	0.080 ± 0.006 [A]	0.110 ± 0.001 [B]
49	2-pentanone (D)	691	175	1.36		0.480 ± 0.002 [A]	0.820 ± 0.011 [B]
36	ethyl acetate (M)	606	136	1.10	fruity sweet green	3.20 ± 0.154 [A]	0.770 ± 0.022 [B]
37	ethyl acetate (D)	612	138	1.33		5.60 ± 0.213 [A]	3.02 ± 0.054 [B]
45	ethyl propanoate	710	188	1.15	sweet fruity	0.120 ± 0.014 [A]	0.050 ± 0.021 [B]

Note: Rt: Represents the retention time in the capillary GC column; RI: represents the retention index calculated using n-ketones C4–C9 as external standard on FS-SE-54-CB-1 column; each value is expressed as mean ± SD ($n = 3$); "(M)", Monomer; "(D)", Dime; for Relative amount: means with different capital letters within a row indicate significant differences ($p < 0.05$); relative amount, the percentage of each compound peak intensity to total peak intensity of all compounds.

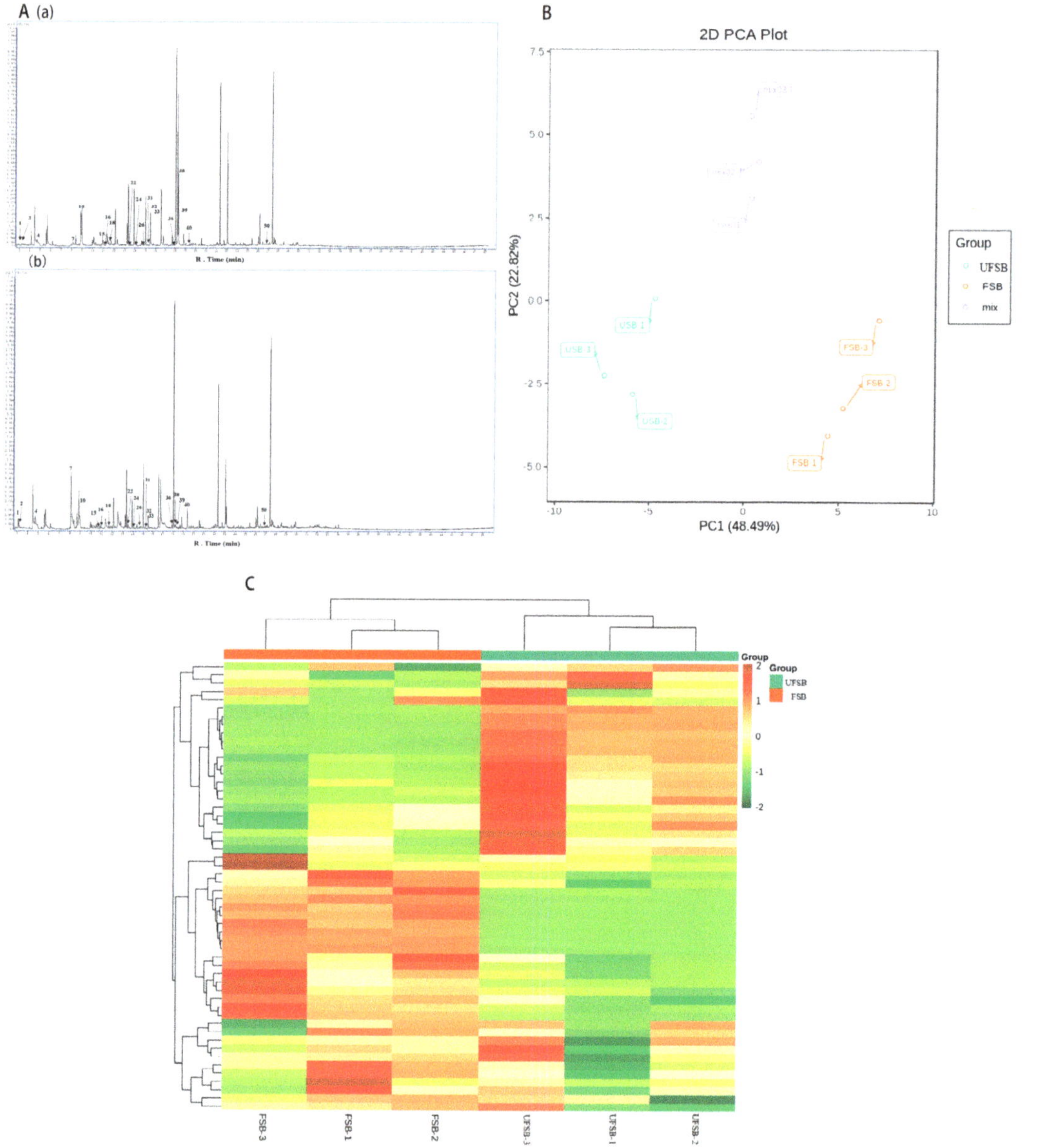

Figure 3. The volatile organic compounds analysis in soy beverage by HS-SPME-GC-MS. (**A**) GC-MS total ion current chromatograms of (**a**) the unfermented soy beverage and (**b**) the fermented soy beverage with *N. aurantialba*, (**B**) PCA score of samples, (**C**) heat map visualization of samples.

In UFSB and FSB, ten classes 49 and ten classes 54 recognized volatile chemicals were found, including heterocyclic compounds, aromatics, alkanes, alkenes, alcohols, aldehydes, ester ketones, and terpenoids (Table 3 and Table S3, Figure 3C). The most detected compounds were heterocycles, and the peak areas of the four heterocycles, including oxetane, 3-(1-methylethyl)-, thiophene, 2-pentyl-, thiophene, 2-hexyl-, 1-pentanone, and 1-(2-furanyl)-, increased significantly. All compounds, except for oxetane, 3-(1-methylethyl), which had no odor description, were Maillard reaction-derived compounds with an odor description of "fruity, sweet." Researchers have found that these substances are characteristic aromas in meats such as pork, beef, and lamb, and may also be responsible for the meat flavor of some mushrooms [64]. Interestingly, thiophene, 2-pentyl-, was a derivative of cysteine, which was produced from a soy beverage after fermentation by *N. aurantialba* (Table S1).

Table 3. Identification of the volatile compounds in fermented soy beverage by HS-SPME-GC-MS.

No.	RT	Compounds	RI	NIST_RI	Odor Description	Peak Intensity	
						UFSB	FSB
				Heterocyclic compound			
1	3.06	Oxetane, 3-(1-methylethyl)-	718.12	664	-	ND	1,621,445 ± 196,936
10	8.96	2-pentylfuran	987.89	993	green earthy beany	6,498,125 ± 534,065 [a]	712,224 ± 128,211 [b]
14	10.8	(E)-2-(1-pentenyl)-furan	1051.11	1048	roasted	186,144 ± 31,660 [a]	125,155 ± 7559 [b]
22	13.5	Thiophene, 2-pentyl-	1143.43	1090	fruit, sweet	ND	628,882 ± 20,192
26	14.78	Thiophene, 2-hexyl-	1187.4	1277	floral fruity gassy	ND	251,010 ± 32,886
36	17.91	1-Pentanone, 1-(2-furanyl)-	1299.53	1176	sweet caramel	184,692 ± 27,820 [a]	557,517 ± 87,545 [b]
				Alcohol			
8	8.67	1-Octen-3-ol	977.65	982	mushroom	984,538 ± 121,842 [a]	3,477,799 ± 1,627,475 [b]
				Aldehyde			
7	8.11	Benzaldehyde	957.34	960	bitter almond, sweet cherry	1,346,850 ± 149,896 [a]	16,021,725 ± 803,440 [b]
11	9.58	(E, E)-2,4-heptadienal	1009.64	1011	fatty green	458,123 ± 83,423 [a]	292,071 ± 65,426 [b]
15	10.95	2-octenal	1056.04	1060	fatty green herbal	510,065 ± 53,147 [a]	150,411 ± 12,114 [b]
24	14.01	Benzaldehyde, 4-ethyl-	1160.8	1171	bitter almond	316,945 ± 43,024 [a]	72,065 ± 1910 [b]
39	18.34	2,4-Decadienal, (E,E)-	1315.79	1317	fatty green	1,186,788 ± 64,214 [a]	57,711 ± 9133 [b]
				Ester			
27	14.85	Methyl salicylate	1189.63	1234	wintergreen mint	134,718 ± 47,046 [a]	80,900 ± 8110 [b]
28	15	Octanoic acid, ethyl ester	1194.8	1198	sweet fruity	4,720,782 ± 406,796 [a]	7,761,488 ± 1,037,991 [b]
40	19.39	2(3H)-Furanone, dihydro-5-pentyl-	1355.43	1366	coconut creamy sweet buttery	120,699 ± 23,131 [a]	5,154,713 ± 83,079 [b]

Note: Each value is expressed as mean ± SD (n = 3); for peak intensity: means with different lowercase letters within a row indicate significant differences ($p < 0.05$); ND, not detected; "-", not described.

The peak areas of the two furan heterocycles, 2-pentylfuran and (E)-2-(1-pentenyl)-furan, decreased significantly, and they were one of the browning reaction's volatiles (Tables 3 and S3). However, their flavor was described as "green and fatty." A decrease in their levels may improve the beany flavor of the soybean beverage. Under the action of microorganisms, benzaldehyde, a compound produced by the degradation of phenylalanine under the action of microorganisms, showed increased levels following fermentation, which was consistent with the results of the GC-IMS [65]. In addition, the levels of various VOCs, such as (E, E)-2,4-heptadienal, 2-octenal, and 4-ethyl-benzaldehyde, significantly decreased, possibly contributing to the reduction in the herbaceous and green aroma of soy beverages caused by excessive amounts of such compounds.

In mushroom-fermented beverages, 1-octen-3-ol increased 2.5-fold as the primary representative of alcoholic aroma substances, which is compatible with the GC-IMS findings. The signal intensities of these esters (2 (3H)-furanone, dihydro-5-pentyl, octanoic acid, ethyl ester, and methyl salicylate) were substantially greater in FSB than in UFSB, showing that *N. aurantialba* contributes to the improved fragrance intensity and complexity of the soybean beverages.

3.5. Analysis of the Relative Odor Activity Values

To evaluate the contribution of different compounds to the odor of soy beverages, the ROAV was determined and calculated to assess the overall odor levels (Table 4).

From GC-IMS analysis, the main volatile flavor compounds in the control group were octanal, 1-octen-3-one, (E)-2-nonenal, 2,4-decadienal, hexanal, 1-octen-3-ol, 2-pentylfuran, and a (E)-2-octenal), whereas those in the FSB group were 1-octen-3-one (mutagen), and others were octanal, (E)-2-nonenal, in that order, compounds such as 2,4-decadienal, and 1-octen-3-ol. After fermentation, the ROAV values of the beany flavor substances, mainly aldehydes, such as hexanal, octanal, 2,4-decadiena, and (E)-2-nonenal, decreased by 80%, 20%, 60%, and 40%, respectively.

From the GC-MS analysis, the main volatile flavor components of the control group were in the order of compounds such as (E, E)-2,4-decadienal (fatty, green), 2-pentylfuran (green), and 2-octenal (fatty, green). After 5 d of fermentation, the FSB group had a total of eight VOCs with high ROAV (ROAV > 1), including 1-octen-3-ol, (E, E)-2,4-decadienal, 2-octenal, 2 (3H)-furanone, dihydro-5-pentyl-, octanoic acid, ethyl ester, and 2-pentylfuran. The study by Aschemann Witzel et al. similarly demonstrated the degradation of green VOCs, such as 2,4-decadienal, in soybean beverages fermented using basidiomycetes to improve the flavor of soybean beverages. However, the mechanism of (E, E)-2,4-decadienal degradation by basidiomycetes is yet to be fully understood [11]. In addition, the strongest contributor to the typical aroma of FSB fermented by *B. piriformis* was 1-octen-3-one, which is similar to our findings that 1-octen-3-one provided a pleasant characteristic odor following fermentation by basidiomycetes [11].

Thus, although the detection sensitivity and VOCs were distinguishable between GC-IMS and GC-MS, the results of these two methods collectively indicated, to some extent, that the beany flavor substances of soybean beverages decreased after fermentation, and that the mushroom, fruity, and sweet aromas dominated.

Table 4. Relative odor activity values (ROAV) and odor description of volatile compounds of soy beverage.

CAS.	Compounds	Odor Thresholds (mg/kg)	ROAV				Odor Description
			GC-IMS		GC-MS		
			UFSB	FSB	UFSB	FSB	
100-52-7	Benzaldehyde	0.3	<1	<1	-	-	bitter almond, sweet cherry
110-62-3	Pentanal	0.012	1.0	<1	-	-	fruity berry
111-71-7	Heptanal	0.01	1.2	<1	-	-	fatty green
122-78-1	Benzeneacetaldehyde	0.004	<1	<1	-	-	green sweet cocoa
123-72-8	Butanal	0.00526	1.4	<1	-	-	pungent cocoa green
124-13-0	Octanal	0.0001	100.0	75.4	-	-	green fatty
124-19-6	n-Nonanal	0.0035	5.5	2.2	-	-	rose fresh fruity
1576-87-0	(E)-2-Pentenal	0.98	<1	<1	-	-	green fruity
18829-55-5	(E)-2-Heptenal	0.013	2.3	1.4	-	-	green fatty
18829-56-6	(E)-2-Nonenal	0.000065	35.9	21.9	-	-	fatty green

Table 4. *Cont.*

CAS.	Compounds	Odor Thresholds (mg/kg)	ROAV				Odor Description
			GC-IMS		GC-MS		
			UFSB	FSB	UFSB	FSB	
2363-88-4	2,4-Decadienal	0.0003	33.3	14.7	-	-	fatty green
2363-89-5	2-octenal	0.0002	-	-	6.4	32.6	fatty green herbal
4313-03-5	(E, E)-2,4-heptadienal	0.0154	<1	<1	<1	<1	fatty green
4748-78-1	Benzaldehyde, 4-ethyl-	0.12323	-	-	<1	<1	bitter almond
5910-85-0	2,4-Heptadienal	0.15	<1	<1	-	-	green fruity
66-25-1	Hexanal	0.0045	31.9	7.0	-	-	green fatty
6728-26-3	(E)-2-Hexenal	0.04	<1	<1	-	-	green fatty
3913-81-3	(E)-2-Decenal	0.25	<1	<1	-	-	fatty, fish, hay
25152-84-5	2,4-Decadienal, (E,E)-	0.000027	-	-	100.0	84.9	fatty green
2548-87-0	(E)-2-Octenal	0.004	5.9	4.7	-	-	fatty green
1998/1/1	2-Furfural	9.562	<1	<1	-	-	sweet woody almond
104-61-0	2(3H)-Furanone, dihydro-5-pentyl-	0.0097	-	-	<1	23.4	coconut creamy sweet buttery
105-37-3	Ethyl propanoate	0.01	<1	<1	-	-	sweet fruity
106-32-1	Octanoic acid, ethyl ester	0.0193	-	-	<1	17.6	sweet fruity
119-36-8	Methyl salicylate	0.04	-	-	<1	<1	wintergreen mint
104-62-1	Formic acid, 2-phenylethyl ester	0.27	-	-	<1	<1	rose green hyacinth watercress herbal
141-78-6	Ethyl acetate	7.5	<1	<1	-	-	fruity sweet green
106-68-3	3-Octanone	0.0214	<1	<1	-	-	sweet mushroom
107-87-9	2-Pentanone	153	<1	<1	-	-	sweet fruity
110-43-0	2-Heptanone	0.14	<1	<1	-	-	fruity sweet
1629-58-9	1-Penten-3-one	0.0013	1.6	<1	-	-	pungent peppery onion
431-03-8	2,3-Butanedione	0.001	1.7	1.1	-	-	sweet creamy
4312-99-6	1-Octen-3-one(M)	0.000016	78.1	100	-	-	mushroom
67-64-1	Acetone	100	<1	<1	-	-	apple pear
78-93-3	2-Butanone	35.4002	<1	<1	-	-	fruity
111-27-3	1-Hexanol	0.2	<1	<1	<1	<1	fruity alcoholic sweet green
123-51-3	3-Methyl-1-butanol	0.22	<1	<1	-	-	alcoholic fruity
3391-86-4	1-Octen-3-ol	0.0015	11.4	9.9	<1	100	mushroom
515-00-4	Bicyclo[3.1.1]hept-2-ene-2-methanol, 6,6-dimethyl-	0.007	-	-	<1	<1	woody green
616-25-1	1-Penten-3-ol	0.3581	<1	<1	-	-	fruity green
626-93-7	2-Hexanol	6.7	-	-	<1	<1	fruity sweet green
64-17-5	Ethanol	2900	<1	<1	-	-	strong alcoholic
71-36-3	1-Butanol	0.5	<1	<1	-	-	sweet balsam whiskey
71-41-0	1-Pentanol	0.3581	<1	<1	-	-	sweet balsam
78-83-1	2-Methyl-1-propanol	8	<1	<1	-	-	ethereal winey cortex
3777-69-3	2-Pentylfuran	0.0058	7.6	<1	2.8	5.5	green beany vegetable

"-", not detected.

3.6. Analysis of the Relative Odor Activity Values

To evaluate the contribution of different compounds to the odor of soy beverages, the ROAV was determined and calculated to assess the overall odor levels.

From GC-IMS analysis, the main volatile flavor compounds in the control group were octanal, 1-octen-3-one, (E)-2-nonenal, 2,4-decadienal, hexanal, 1-octen-3-ol, 2-pentylfuran, a(E)-2-octenal, whereas those in the FSB group were 1-octen-3-one (mutagen), and others were octanal, (E)-2-nonenal, in that order, and compounds such as 2,4-decadienal and 1-octen-3-ol. After fermentation, the ROAV values of the beany flavor substances, mainly aldehydes, such as hexanal, octanal, 2,4-decadiena, and (E)-2-nonenal, decreased by 80%, 20%, 60%, and 40%, respectively.

From the GC-MS analysis, the main volatile flavor components of the control group were in the order of compounds such as (E, E)-2,4-decadienal (fatty, green), 2-pentylfuran (green), and 2-octenal (fatty, green). After 5 d of fermentation, the FSB group had a total of eight VOCs with high ROAV (ROAV > 1), including 1-octen-3-ol, (E, E)-2,4-decadienal, 2-octenal, 2 (3H)-furanone, dihydro-5-pentyl -, octanoic acid, ethyl ester, and 2-pentylfuran. The study by Aschemann Witzel et al. similarly demonstrated the degradation of green VOCs, such as 2,4-decadienal, in soybean beverages fermented using basidiomycetes to improve the flavor of soybean beverages [11]. However, the mechanism of (E, E)-2,4-decadienal degradation by basidiomycetes is yet to be fully understood. In addition, the strongest contributor to the typical aroma of FSB fermented by *B. piriformis* was 1-octen-3-one, which is similar to our findings that 1-octen-3-one provided a pleasant characteristic odor following fermentation by basidiomycetes [11].

Thus, although the detection sensitivity and VOCs were distinguishable between GC-IMS and GC-MS, the results of these two methods collectively indicated, to some extent, that the beany flavor substances of soybean beverages decreased after fermentation, and that the mushroom, fruity, and sweet aromas dominated.

3.7. Sensory Property of Soy Beverages

To evaluate the flavor of FSB, we ran sensory analyses. The results of the two samples are shown in Figure 4A. Sensory evaluation showed that the characteristic odor of UFSB was beany, green, and fatty. The characteristic flavors of FSB were mushroom and aromatic. This is consistent with the results of the ROAV analysis. In addition, the FSB similarly exhibited an acidic note, which consisted of the low pH of the FSB. Overall, FSB had a softer, consumer-satisfying odor.

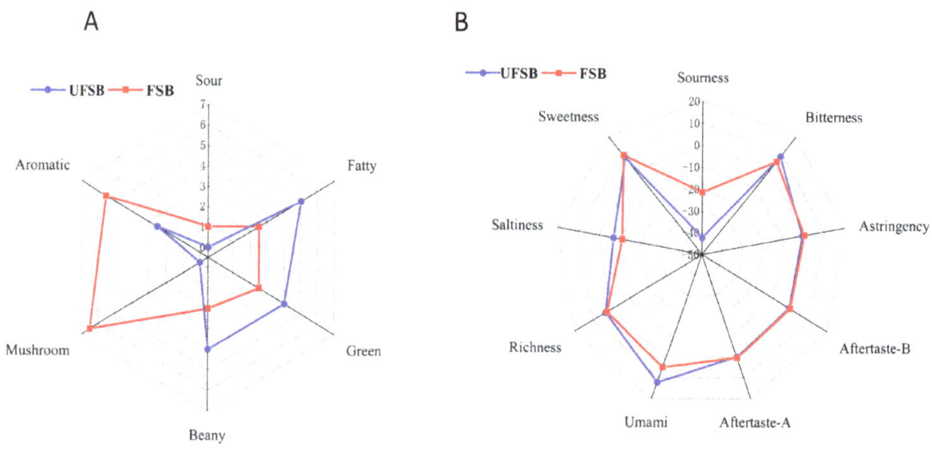

Figure 4. Radar chart of FSB and UFSB. (**A**) Sensory property of soy beverages, (**B**) electronic tongue measurement of soy beverages.

3.8. Electronic Tongue Measurement

A radar plot is shown in Figure 4B, showing the results of our investigations into the compounds contributing to the umami, bitter, astringent, tangy, sour, and salty notes in soy beverages. It is evident from the figure that there were significant differences between the two groups of samples, with a large rise in sourness, a small decrease in both bitter and bitter echoes, a smaller rise in sweet and astringency, and a slight decrease in umami and richness in FSB compared to UFSB. It is known that fermentation decreases the bitter taste of soy beverages but increases sourness while maintaining the fresh and sweet taste of natural soybean. Thus, FSB has potential in terms of its success in the fermented soybean beverage industry. The higher levels of glucose in the FSB could have masked the bitterness [66]. In future research, we would like to investigate whether we can optimize the glucose supply for initial fermentation such that the strain utilizes all of the glucose, thus resolving the issue where the difference in glucose levels in the sample influences sensory evaluation. Microbial fermentation has been reported to alter the taste of beverages to some extent [12,59]. For example, lactic acid bacteria are used to ferment ginkgo juice; acetic acid bacteria are used to ferment kombucha tea; yeasts are used to ferment juices such as pomegranate, apple, and strawberry juices, changing their flavor profiles [67–69]. Despite the fact that basidiomycete-fermented beverages have not been substantially explored, Zhao et al. found that following *G. lucidum* fermentation, the flavor of pumpkin juice transforms from a disagreeable "stewing" to a pleasant "fruity and flowery" note [52]. In general, changes in the flavor profiles of microbial-fermented beverages are closely associated with the metabolites of the strains themselves. For example, *L. plantarum* was used to ferment *Z. jujuba* juice, and the resulting juice was mainly characterized by acidic notes with decreased bitter and astringent notes due to the production of large amounts of lactic acid, aromatic compounds, and sulfur organics when metabolized [70].

4. Conclusions

In this study, we investigated the feasibility of fermented soybean beverages to enrich the diversity of plant-based products. FSB was prepared from soybean as a raw material, using *N. aurantialba* as a species. Soy beverages become more stable because of the increase in Zeta potential, particle size, and viscosity. The nutrient composition was increased by the fermentation of *N. aurantialba*. The antioxidant activity of soybean beverages significantly increased after 5 d of fermentation. By reducing the content of beany substances such as hexanal and increasing the content of 1-octen-3-ol, the aroma of soybean beverage fermented by *N. aurantialba* changed from "beany, green, fatty" to "mushroom and aromatic". FSB not only reduces bitterness but also greatly increases sourness while maintaining the fresh and sweet taste of UFSB. The results of this study indicate that the soy beverage fermented by *N. aurantialba* could not only improve the health characteristics of soy beverages but also reduce the beany taste of soy beverages, which means that it has the potential to be marketed. However, the degradation pathways of *N. aurantialba* for these beany-flavored substances should be investigated in future studies using isotope labeling experiments. In addition, multi-omics technologies such as transcriptomics, metabolomics, and proteomics can be used to uncover the underlying causes of how *N. aurantialba* fermentation impacts volatile chemicals in soymilk and to better understand the VOCs changing mechanism. More research into these elements will help the development of basidiomycete fermentation technologies in the food industry.

Supplementary Materials: The following supporting information can be downloaded at: https://www.mdpi.com/article/10.3390//foods11030272/s1, Figure S1: The HPLC-GPC spectrum of soybean beverage, Table S1: Free amino acid of the unfermented and fermented soy beverages with *N. aurantialba*, Table S2: Volatile compounds identified in samples by HS-GC-IMS, Table S3: Volatile compounds identified in samples by HS-SPME-GC-MS.

Author Contributions: Conceptualization, P.L. and H.X.; methodology, P.L.; software, R.W. and Y.G.; validation, D.S.; formal analysis, T.S. and S.W.; investigation, H.J. and H.Y.; resources, P.L.; data curation, T.S. and X.L.; writing—original draft preparation, T.S.; writing—review and editing, T.S. and K.Y.; visualization, S.L.; supervision, P.L.; project administration, P.L. and H.X. All authors have read and agreed to the published version of the manuscript.

Funding: The research was financially supported by National Key Research and Development Program of China (2021YFC2101703), National Students' platform for innovation and entrepreneurship training program (202110291042Z), the National Natural Science Foundation of China (31800101), Jiangsu Agricultural Science and Technology Innovation Fund (CX(21)3158), State Key Laboratory Open Foundation (ZK201905), and the Key Research and Development Project of Jiangsu Province (BE2019390).

Institutional Review Board Statement: Not applicable.

Informed Consent Statement: Not applicable.

Data Availability Statement: Not applicable.

Acknowledgments: We thank the volatile compounds in food (VCF) database (https://www.vcf-online.nl/VcfHome.cfm, accessed on 10 July 2021) and a good subject company information system (http://www.thegoodscentscompany.com/index.html, accessed on 10 July 2021) for providing most of flavor description of volatile compounds in this study.

Conflicts of Interest: The authors declare no conflict of interest.

References

1. Sethi, S.; Tyagi, S.K.; Anurag, R.K. Plant-based milk alternatives an emerging segment of functional beverages: A review. *J. Food Sci. Technol.* **2016**, *53*, 3408–3423. [CrossRef] [PubMed]
2. Aschemann-Witzel, J.; Gantriis, R.F.; Fraga, P.; Perez-Cueto, F.J. Plant-based food and protein trend from a business perspective: Markets, consumers, and the challenges and opportunities in the future. *Crit. Rev. Food Sci. Nutr.* **2020**, *61*, 3119–3128. [CrossRef]
3. Freeland-Graves, J.H.; Nitzke, S. Position of the academy of nutrition and dietetics: Total diet approach to healthy eating. *J. Acad. Nutr. Diet.* **2013**, *113*, 307–317. [CrossRef] [PubMed]
4. Paul, A.A.; Kumar, S.; Kumar, V.; Sharma, R. Milk Analog: Plant based alternatives to conventional milk, production, potential and health concerns. *Crit. Rev. Food Sci. Nutr.* **2020**, *60*, 3005–3023. [CrossRef]
5. Yuan, S.; Chang, S.K.-C. Selected odor compounds in soymilk as affected by chemical composition and lipoxygenases in five soybean materials. *J. Agric. Food Chem.* **2007**, *55*, 426–431. [CrossRef] [PubMed]
6. Wang, B.; Zhang, Q.; Zhang, N.; Bak, K.H.; Soladoye, O.P.; Aluko, R.E.; Fu, Y.; Zhang, Y. Insights into formation, detection and removal of the beany flavor in soybean protein. *Trends Food Sci. Technol.* **2021**, *112*, 336–347. [CrossRef]
7. Wang, J.; Kuang, H.; Zhang, Z.; Yang, Y.; Yan, L.; Zhang, M.; Song, S.; Guan, Y. Generation of seed lipoxygenase-free soybean using CRISPR-Cas9. *Crop J.* **2020**, *8*, 432–439. [CrossRef]
8. Ma, L.; Li, B.; Han, F.; Yan, S.; Wang, L.; Sun, J. Evaluation of the chemical quality traits of soybean seeds, as related to sensory attributes of soymilk. *Food Chem.* **2015**, *173*, 694–701. [CrossRef]
9. Yu, H.; Liu, R.; Hu, Y.; Xu, B. Flavor profiles of soymilk processed with four different processing technologies and 26 soybean cultivars grown in China. *Int. J. Food Prop.* **2017**, *20*, S2887–S2898. [CrossRef]
10. Ghosh, K.; Ray, M.; Adak, A.; Halder, S.K.; Das, A.; Jana, A.; Parua, S.; Vágvölgyi, C.; Mohapatra, P.K.D.; Pati, B.R. Role of probiotic Lactobacillus fermentum KKL1 in the preparation of a rice based fermented beverage. *Bioresour. Technol.* **2015**, *188*, 161–168. [CrossRef]
11. Nedele, A.-K.; Gross, S.; Rigling, M.; Zhang, Y. Reduction of green off-flavor compounds: Comparison of key odorants during fermentation of soy drink with Lycoperdon pyriforme. *Food Chem.* **2020**, *334*, 127591. [CrossRef]
12. Mäkinen, O.E.; Wanhalinna, V.; Zannini, E.; Arendt, E.K. Foods for special dietary needs: Non-dairy plant-based milk substitutes and fermented dairy-type products. *Crit. Rev. Food Sci. Nutr.* **2016**, *56*, 339–349. [CrossRef]
13. Cao, Z.-H.; Green-Johnson, J.M.; Buckley, N.D.; Lin, Q.-Y. Bioactivity of soy-based fermented foods: A review. *Biotechnol. Adv.* **2019**, *37*, 223–238. [CrossRef] [PubMed]
14. Zhu, Y.-Y.; Thakur, K.; Feng, J.-Y.; Cai, J.-S.; Zhang, J.-G.; Hu, F.; Wei, Z.-J. B-vitamin enriched fermented soymilk: A novel strategy for soy-based functional foods development. *Trends Food Sci. Technol.* **2020**, *105*, 339–349. [CrossRef]
15. Shon, Y.-H.; Nam, K.-S. Inhibition of cytochrome P450 isozymes and ornithine decarboxylase activities by polysaccharides from soybeans fermented with Phellinus igniarius or Agrocybe cylindracea. *Biotechnol. Lett.* **2004**, *26*, 43–55. [CrossRef]
16. Yang, H.; Zhang, L. Changes in some components of soymilk during fermentation with the basidiomycete Ganoderma lucidum. *Food Chem.* **2009**, *112*, 1–5. [CrossRef]
17. Du, X.; Zhang, Y.; Mu, H.; Lv, Z.; Yang, Y.; Zhang, J. Structural elucidation and antioxidant activity of a novel polysaccharide (TAPB1) from Tremella aurantialba. *Food Hydrocoll.* **2015**, *43*, 459–464. [CrossRef]

18. Yuan, Q.; Zhang, X.; Ma, M.; Long, T.; Xiao, C.; Zhang, J.; Liu, J.; Zhao, L. Immunoenhancing glucuronoxylomannan from Tremella aurantialba Bandoni et Zang and its low-molecular-weight fractions by radical depolymerization: Properties, structures and effects on macrophages. *Carbohydr. Polym.* **2020**, *238*, 116184. [CrossRef] [PubMed]
19. Deng, C.; Sun, Y.; Fu, H.; Zhang, S.; Chen, J.; Xu, X. Antioxidant and immunostimulatory activities of polysaccharides extracted from Tremella aurantialba mycelia. *Mol. Med. Rep.* **2016**, *14*, 4857–4864. [CrossRef]
20. Zhang, Z.; Lian, B.; Huang, D.; Cui, F. Compare activities on regulating lipid-metabolism and reducing oxidative stress of diabetic rats of Tremella aurantialba broth's extract (TBE) with its mycelia polysaccharides (TMP). *J. Food Sci.* **2009**, *74*, H15–H21. [CrossRef] [PubMed]
21. Guo, Y.-J.; Deng, G.-F.; Xu, X.-R.; Wu, S.; Li, S.; Xia, E.-Q.; Li, F.; Chen, F.; Ling, W.-H.; Li, H.-B. Antioxidant capacities, phenolic compounds and polysaccharide contents of 49 edible macro-fungi. *Food Funct.* **2012**, *3*, 1195–1205. [CrossRef]
22. Bandoni, R.J. Conjugation in Tremella mesenterica. *Can. J. Bot.* **1963**, *41*, 467–474. [CrossRef]
23. Ingold, C. Basidiospore germination in Tremella foliacea. *Trans. Br. Mycol. Soc.* **1982**, *79*, 561–563. [CrossRef]
24. Sun, T.; Wang, R.; Sun, D.; Li, S.; Xu, H.; Qiu, Y.; Lei, P.; Sun, L.; Xu, X.; Zhu, Y. High-efficiency production of Tremella aurantialba polysaccharide through basidiospore fermentation. *Bioresour. Technol.* **2020**, *318*, 124268. [CrossRef]
25. Sun, T.; Zhang, Y.; Jiang, H.; Yang, K.; Wang, S.; Wang, R.; Li, S.; Lei, P.; Xu, H.; Qiu, Y.; et al. Whole genome sequencing and annotation of naematelia aurantialba (Basidiomycota, Edible-Medicinal Fungi). *J. Fungi* **2022**, *8*, 6. [CrossRef]
26. Dai, C.; Huang, X.; Lv, R.; Zhang, Z.; Sun, J.; Aheto, J.H. Analysis of volatile compounds of Tremella aurantialba fermentation via electronic nose and HS-SPME-GC-MS. *J. Food Saf.* **2018**, *38*, e12555. [CrossRef]
27. Hassan, K.A.; Mujtaba, M.A. Antibacterial efficacy of garlic oil nano-emulsion. *AIMS Agric. Food* **2019**, *4*, 194–205. [CrossRef]
28. Li, H.; Xu, H.; Li, S.; Feng, X.; Xu, H.; Ouyang, P. Effects of dissolved oxygen and shear stress on the synthesis and molecular weight of welan gum produced from *Alcaligenes* sp. CGMCC2428. *Process Biochem.* **2011**, *46*, 1172–1178. [CrossRef]
29. Oliveira, L.C.; Schmiele, M.; Steel, C.J. Development of whole grain wheat flour extruded cereal and process impacts on color, expansion, and dry and bowl-life texture. *LWT* **2017**, *75*, 261–270. [CrossRef]
30. Jiang, K.; Tang, B.; Wang, Q.; Xu, Z.; Sun, L.; Ma, J.; Li, S.; Xu, H.; Lei, P. The bio-processing of soybean dregs by solid state fermentation using a poly γ-glutamic acid producing strain and its effect as feed additive. *Bioresour. Technol.* **2019**, *291*, 121841. [CrossRef]
31. Yang, H.; Zhang, L.; Xiao, G.; Feng, J.; Zhou, H.; Huang, F. Changes in some nutritional components of soymilk during fermentation by the culinary and medicinal mushroom Grifola frondosa. *LWT* **2015**, *62*, 468–473. [CrossRef]
32. Jaganath, I.B.; Crozier, A. Dietary flavonoids and phenolic compounds. *Plant Phenolics Hum. Health Biochem. Nutr. Pharmacol.* **2010**, *1*, 1–50.
33. Gu, J.; Zhang, H.; Zhang, J.; Wen, C.; Ma, H.; Duan, Y.; He, Y. Preparation, characterization and bioactivity of polysaccharide fractions from *Sagittaria sagittifolia* L. *Carbohydr. Polym.* **2020**, *229*, 115355. [CrossRef]
34. Su, Y.; Li, L. Structural characterization and antioxidant activity of polysaccharide from four auriculariales. *Carbohydr. Polym.* **2020**, *229*, 115407. [CrossRef]
35. Phuhongsung, P.; Zhang, M.; Bhandari, B. 4D printing of products based on soy protein isolate via microwave heating for flavor development. *Food Res. Int.* **2020**, *137*, 109605. [CrossRef]
36. Menis-Henrique, M.E.C. Methodologies to advance the understanding of flavor chemistry. *Curr. Opin. Food Sci.* **2020**, *33*, 131–135. [CrossRef]
37. Li, M.; Du, H.; Lin, S. Flavor changes of tricholoma matsutake singer under different processing conditions by using HS-GC-IMS. *Foods* **2021**, *10*, 531. [CrossRef]
38. Chen, Y.; Li, P.; Liao, L.; Qin, Y.; Jiang, L.; Liu, Y. Characteristic fingerprints and volatile flavor compound variations in Liuyang Douchi during fermentation via HS-GC-IMS and HS-SPME-GC-MS. *Food Chem.* **2021**, *361*, 130055. [CrossRef]
39. Yan, Y.; Wang, M.; Jin, B.; Yang, J.; Li, S. Performance evaluation and microbial community analysis of the biofilter for removing grease and volatile organic compounds in the kitchen exhaust fume. *Bioresour. Technol.* **2021**, *319*, 124132. [CrossRef] [PubMed]
40. Tieman, D.; Zhu, G.; Resende, M.F.; Lin, T.; Nguyen, C.; Bies, D.; Rambla, J.L.; Beltran, K.S.O.; Taylor, M.; Zhang, B. A chemical genetic roadmap to improved tomato flavor. *Science* **2017**, *355*, 391–394. [CrossRef] [PubMed]
41. Yi, C.; Li, Y.; Zhu, H.; Liu, Y.; Quan, K. Effect of Lactobacillus plantarum fermentation on the volatile flavors of mung beans. *LWT* **2021**, *146*, 111434. [CrossRef]
42. Hothorn, T.; Zeileis, A. Partykit: A modular toolkit for recursive partytioning in R. *J. Mach. Learn. Res.* **2015**, *16*, 3905–3909.
43. Shi, F.; Tian, X.; McClements, D.J.; Chang, Y.; Shen, J.; Xue, C. Influence of molecular weight of an anionic marine polysaccharide (sulfated fucan) on the stability and digestibility of multilayer emulsions: Establishment of structure-function relationships. *Food Hydrocoll.* **2021**, *113*, 106418. [CrossRef]
44. Wei, Y.; Cai, Z.; Wu, M.; Guo, Y.; Tao, R.; Li, R.; Wang, P.; Ma, A.; Zhang, H. Comparative studies on the stabilization of pea protein dispersions by using various polysaccharides. *Food Hydrocoll.* **2020**, *98*, 105233. [CrossRef]
45. Sun, W.; Zheng, Y.; Chen, S.; Chen, J.; Zhang, H.; Fang, H.; Ye, X.; Tian, J. Applications of polysaccharides as stabilizers in acidified milks. *Food Rev. Int.* 2021; 1–17, *ahead-of-print*. [CrossRef]
46. Shao, P.; Feng, J.; Sun, P.; Xiang, N.; Lu, B.; Qiu, D. Recent advances in improving stability of food emulsion by plant polysaccharides. *Food Res. Int.* **2020**, *137*, 109376. [CrossRef] [PubMed]

47. Martínez-Monteagudo, S.I.; Kamat, S.; Patel, N.; Konuklar, G.; Rangavajla, N.; Balasubramaniam, V. Improvements in emulsion stability of dairy beverages treated by high pressure homogenization: A pilot-scale feasibility study. *J. Food Eng.* **2017**, *193*, 42–52. [CrossRef]
48. Adediran, A.; Aforijiku, S. Carbohydrate fermentation profile and physiological studies of lactic acid bacteria from native raw cow milk. *J. Adv. Microbiol. Res.* **2020**, *20*, 83–93. [CrossRef]
49. Niyigaba, T.; Liu, D.; Habimana, J.D.D. The extraction, functionalities and applications of plant polysaccharides in fermented foods: A review. *Foods* **2021**, *10*, 3004. [CrossRef]
50. Lin, C.; Lin, Y.; Xiao, J.; Lan, Y.; Cao, Y.; Chen, Y. Effect of momordica saponin-and cyclocarya paliurus polysaccharide-enriched beverages on oxidative stress and fat accumulation in caenorhabditis elegans. *J. Sci. Food Agric.* **2021**, *101*, 3366–3375. [CrossRef]
51. Lee, M.-H.; Kim, M.; Kim, M.; Kwak, J.H.; Chang, D.H.; Yu, W.K.; Lee, S.-H.; Lee, J.H. Consumption of dairy yogurt with the polysaccharide rhamnogalacturonan from the peel of the Korean citrus hallabong enhances immune function and attenuates the inflammatory response. *Food Funct.* **2016**, *7*, 2833–2839. [CrossRef]
52. Zhao, J.; Liu, W.; Chen, D.; Zhou, C.; Song, Y.; Zhang, Y.; Ni, Y.; Li, Q. Microbiological and physicochemical analysis of pumpkin juice fermentation by the basidiomycetous fungus Ganoderma lucidum. *J. Food Sci.* **2015**, *80*, C241–C251. [CrossRef]
53. Islam, T.; Yu, X.; Xu, B. Phenolic profiles, antioxidant capacities and metal chelating ability of edible mushrooms commonly consumed in China. *LWT* **2016**, *72*, 423–431. [CrossRef]
54. Puchau, B.; Zulet, M.A.; de Echávarri, A.G.; Hermsdorff, H.H.M.; Martínez, J.A. Dietary total antioxidant capacity: A novel indicator of diet quality in healthy young adults. *J. Am. Coll. Nutr.* **2009**, *28*, 648–656. [CrossRef]
55. Du, X.; Zhang, J.; Lv, Z.; Ye, L.; Yang, Y.; Tang, Q. Chemical modification of an acidic polysaccharide (TAPA1) from Tremella aurantialba and potential biological activities. *Food Chem.* **2014**, *143*, 336–340. [CrossRef] [PubMed]
56. Rodríguez-Maecker, R.; Vyhmeister, E.; Meisen, S.; Martinez, A.R.; Kuklya, A.; Telgheder, U. Identification of terpenes and essential oils by means of static headspace gas chromatography-ion mobility spectrometry. *Anal. Bioanal. Chem.* **2017**, *409*, 6595–6603. [CrossRef] [PubMed]
57. Chen, Y., Jr. Effects of Micronization, Ethanol Washing, and Enzymatic Hydrolysis Processing Alone or in Combination on Trypsin Inhibitors, Lipoxygenase Activities and Selected "Beany" Flavour Related Compounds in Soybean Flour. Master's Thesis, University of Manitoba, Winnipeg, MB, Canada, 2015.
58. Vermeulen, N.; Czerny, M.; Gänzle, M.G.; Schieberle, P.; Vogel, R.F. Reduction of (E)-2-nonenal and (E, E)-2, 4-decadienal during sourdough fermentation. *J. Cereal Sci.* **2007**, *45*, 78–87. [CrossRef]
59. El Youssef, C.; Bonnarme, P.; Fraud, S.; Péron, A.-C.; Helinck, S.; Landaud, S. Sensory improvement of a pea protein-based product using microbial co-cultures of lactic acid bacteria and yeasts. *Foods* **2020**, *9*, 349. [CrossRef] [PubMed]
60. Brunton, N.P.; Cronin, D.A.; Monahan, F.J.; Durcan, R. A comparison of solid-phase microextraction (SPME) fibres for measurement of hexanal and pentanal in cooked turkey. *Food Chem.* **2000**, *68*, 339–345. [CrossRef]
61. Guo, Y.; Chen, D.; Dong, Y.; Ju, H.; Wu, C.; Lin, S. Characteristic volatiles fingerprints and changes of volatile compounds in fresh and dried Tricholoma matsutake Singer by HS-GC-IMS and HS-SPME-GC–MS. *J. Chromatogr. B* **2018**, *1099*, 46–55. [CrossRef]
62. Ames, J.M. Control of the Maillard reaction in food systems. *Trends Food Sci. Technol.* **1990**, *1*, 150–154. [CrossRef]
63. Hiraide, M.; Miyazaki, Y.; Shibata, Y. The smell and odorous components of dried shiitake mushroom, Lentinula edodes I: Relationship between sensory evaluations and amounts of odorous components. *J. Wood Sci.* **2004**, *50*, 358–364. [CrossRef]
64. Tian, Y.; Zhao, Y.; Huang, J.; Zeng, H.; Zheng, B. Effects of different drying methods on the product quality and volatile compounds of whole shiitake mushrooms. *Food Chem.* **2016**, *197*, 714–722. [CrossRef]
65. Hidalgo, F.J.; Zamora, R. Formation of phenylacetic acid and benzaldehyde by degradation of phenylalanine in the presence of lipid hydroperoxides: New routes in the amino acid degradation pathways initiated by lipid oxidation products. *Food Chem. X* **2019**, *2*, 100037. [CrossRef]
66. Mennella, J.A.; Bobowski, N.K.; Reed, D.R. The development of sweet taste: From biology to hedonics. *Rev. Endocr. Metab. Disord.* **2016**, *17*, 171–178. [CrossRef]
67. Wang, Y.; Tao, Y.; Zhang, X.; Shao, S.; Han, Y.; Chu, D.-T.; Xie, G.; Ye, X. Metabolic profile of ginkgo kernel juice fermented with lactic aicd bacteria: A potential way to degrade ginkgolic acids and enrich terpene lactones and phenolics. *Process Biochem.* **2019**, *76*, 25–33. [CrossRef]
68. De Filippis, F.; Troise, A.D.; Vitaglione, P.; Ercolini, D. Different temperatures select distinctive acetic acid bacteria species and promotes organic acids production during Kombucha tea fermentation. *Food Microbiol.* **2018**, *73*, 11–16. [CrossRef] [PubMed]
69. Berenguer, M.; Vegara, S.; Barrajón, E.; Saura, D.; Valero, M.; Martí, N. Physicochemical characterization of pomegranate wines fermented with three different Saccharomyces cerevisiae yeast strains. *Food Chem.* **2016**, *190*, 848–855. [CrossRef] [PubMed]
70. Cai, W.; Tang, F.; Zhao, X.; Guo, Z.; Zhang, Z.; Dong, Y.; Shan, C. Different lactic acid bacteria strains affecting the flavor profile of fermented jujube juice. *J. Food Process. Preserv.* **2019**, *43*, e14095. [CrossRef]

foods

MDPI

Article

Lentil Fortification and Non-Conventional Yeasts as Strategy to Enhance Functionality and Aroma Profile of Craft Beer

Laura Canonico [1], Alice Agarbati [1], Emanuele Zannini [2], Maurizio Ciani [1,*] and Francesca Comitini [1,*]

[1] Dipartimento Scienze Della Vita e Dell'Ambiente, Università Politecnica Delle Marche, Via Brecce Bianche, 60131 Ancona, Italy
[2] School of Food and Nutritional Sciences, University College Cork, T12 K8AF Cork, Ireland
* Correspondence: m.ciani@univpm.it (M.C.); f.comitini@univpm.it (F.C.)

Abstract: During the last few years, consumer demand has been increasingly oriented to fermented foods with functional properties. This work proposed to use selected non-conventional yeasts (NCY) *Lachancea thermotolerans* and *Kazachstania unispora* in pure and mixed fermentation to produce craft beer fortified with hydrolyzed red lentils (HRL). For this, fermentation trials using pils wort (PW) and pils wort added with HRL (PWL) were carried out. HRL in pils wort improved the fermentation kinetics both in mixed and pure fermentations without negatively affecting the main analytical characters. The addition of HRL determined a generalized increase in amino acids concentration in PW. *L. thermotolerans* and *K. unispora* affected the amino acid profile of beers (with and without adding HRL). The analysis of by-products and volatile compounds in PW trials revealed a significant increase of some higher alcohols with *L. thermotolerans* and ethyl butyrate with *K. unispora*. In PWL, the two NCY showed a different behavior: an increment of ethyl acetate (*K. unispora*) and β-phenyl ethanol (*L. thermotolerans*). Sensory analysis showed that the presence of HRL characterized all beers, increasing the perception of the fruity aroma in both pure and mixed fermentation.

Keywords: beer; non-conventional yeasts; lentil; functional beer

Citation: Canonico, L.; Agarbati, A.; Zannini, E.; Ciani, M.; Comitini, F. Lentil Fortification and Non-Conventional Yeasts as Strategy to Enhance Functionality and Aroma Profile of Craft Beer. *Foods* **2022**, *11*, 2787. https://doi.org/10.3390/foods11182787

Academic Editors: Carlo Giuseppe Rizzello and Palmira De Bellis

Received: 29 July 2022
Accepted: 6 September 2022
Published: 9 September 2022

1. Introduction

Nowadays, the growing interest in the craft beer market leads to a constant search for advanced processes, newly selected microorganisms, and raw materials leading to alternative products. Indeed, craft beers, often including "specialty beers", have experienced exponential growth over the last two decades, primarily driven by premiumization and consumers' willingness to seek new, intimate, and unique drinking experiences [1,2].

Although there are various options for obtaining different sensory profiles in craft beers (using special malts or adjuncts, hop varieties, water quality, etc.), the choice of yeast strains for wort fermentation and beer conditioning is crucial [3]. Indeed, most aroma-active compound production is strictly linked to the yeast strain that characterizes the beer in its style and final taste [4,5].

Although several yeast strains are commercially accessible, the availability of new starter strains remains an essential differentiating factor among craft beers produced in different microbreweries. Recently, as well as for enology, the brewing sector is growing the attention towards selected non-*Saccharomyces* yeasts for their possibility to confer unique aromas to the final product. This aspect has relevance considering craft beer as an unpasteurized, unfiltered, and re-fermented in-bottle beverage [6]. Specialty beers are products obtained following the classic style process with the addition of fruits, herbs and spices, various flavorings (e.g., liquorice, smoke, hot pepper), and alternative fermentable substrates (e.g., honey, maple syrup, molasses) [7]. Generally, specialty beers are all for beer styles which do not fit elsewhere. Low-calorie, low-alcohol or non-alcohol, novel-flavored, gluten-free, and functional beers are specialty beers of particular interest [8].

In addition, beer contains some health-promoting substances with positive impacts on the body, including minerals, vitamins, polyphenols, fiber, and relatively low levels of ethanol. Thus, beer can serve as a promising basis for developing a wide variety of functional beverages.

Functional beers are products obtained by adding beneficial health value, intended either as functional ingredients or functional fermenting yeasts [9]. An absolute novelty is represented by probiotic beer among the functional beers, obtained by incorporating probiotic microorganisms. Craft beer, an unpasteurized and unfiltered product, is potentially a vehicle for delivering probiotics. Chemical or biological acidification via *Lactobacilli* and bifidobacterial fermentation secure the microbial stability of the final beer. However, excessive levels of acetic acid could be produced if *Bifidobacteria* are incorporated.

Moreover, because viability is crucial for the efficacy of probiotics, attention must be paid to the sensitivity of probiotics to hop bitter acids, which can inhibit the survival of Gram-positive lactic acid bacteria [10,11]. Probiotics are not only bacteria; indeed, *Saccharomyces cerevisiae* var. *boulardii* is a probiotic yeast strain. Detailed information on the properties of probiotic yeast strains has been previously reported [12,13]. It will be of interest to explore these strains for specialty beer brewing. A novel unfiltered and unpasteurized probiotic beer could be produced by fermenting the wort with a probiotic strain of *S. cerevisiae*.

Studies have shown that foods and drinks with live probiotics are more effective in providing health effects than products containing inactive probiotics. Craft beer with live yeasts can be considered a new tool for beneficial health effects [14–16].

Another strategy to obtain functional beer could be to combine functional or probiotic yeasts with functional ingredients. Between them, a promising ingredient could be legumes such as chickpeas, lentils, and soy as an important source of protein for human nutrition. Legume products are essential in our daily diet to lead a healthy life [17,18]. From a nutritional point of view, legumes are of particular interest for the human diet as they are rich in fiber [19] and proteins: albumin and globulins are the dominant proteins present in legume seeds, with about 70% of legume proteins is produced by globulins [20,21]. Legumes also contain significant amounts of vitamins and other micronutrients. In this regard, the lentil (*Lens culinaris*) is a grain legume that represents an important protein source (25–30%). In Italy, lentils of Colfiorito and Castelluccio are an excellent product [22]. The goal of functional craft beer could be reached using wort enriched with hydrolyzed protein and fiber sources and by using probiotic yeasts [23].

In this scenario, the purpose of this study is to use non-conventional wild yeasts (NCY) to produce functional craft beer with reduced alcohol content. The selected NCY *Lachancea thermotolerans* and *Kazachstania unispora* were previously tested for their probiotic properties and subjected to safety assessment studies [24]. Fermentation trials were carried out using wort added with hydrolyzed red lentils (HRL) as a source of additional proteins and selected NCY in pure co-culture processes. The growth evolution, wort affinity, and viability during and after storage of selected NCY and the analytical, aromatic, and sensorial profile of functional beers were evaluated.

2. Materials and Methods

2.1. Yeast Strains

Three strains used in this study belong to the species *Lthermotolerans* (B13), *K.unispora* (M3-B3) and *S. cerevisiae* (2PV) coming from different un-anthropized environments and spontaneously processed foods. These strains were isolated, identified and characterized as probiotic and/or functional strains and tested in wort fermentation trials [23,24]. All of the yeast strains were maintained at 4 °C for short-term storage in YPD agar (yeast extract 10 g/L, peptone 20 g/L, dextrose 20 g/L, (agar 18 g/L) (Oxoid, Basingstoke, UK) and in YPD broth supplemented with 80% (*w*/*v*) glycerol at −80 °C, for long-term storage.

2.2. Fermentation Trials

L. thermotolerans and *K. unispora* were selected and used in pure and mixed fermentations with *S. cerevisiae*. In mixed fermentations *S. cerevisiae*/NYC yeast ratios were: 1:20 (*L. thermotolerans*), and 1:50 (*K. unispora*). The fermentation trials were carried out in 500-mL flasks containing 500 mL of wort at 20 ± 1 °C locked with a Müller hydraulic valve. The flasks were inoculated with 72 h pre-cultures grown in 10% malt extract at 20 ± 1 °C. The fermentation kinetics were monitored by measuring the weight loss of the flasks due to the CO_2 evolved until the end of the fermentation (i.e., constant weight for 3 consecutive days). The fermentations were carried out in triplicate trials under static conditions.

2.3. Pils Wort and Lentil Wort Preparation

The trials were conducted in two worts: pils wort (PW) and pils wort added with HRL (PWL). PW comes from a batch of 1500 L of malted barley wort to produce Pilsner beer with the following main analytical characters: pH 5.4, specific gravity 12.2 °Plato, and 20 IBU. Two set of fermentation trials were conducted to evaluate the fermentation potential of yeast strains. First, PWL was prepared using pils wort and HRL. HRL was prepared using a mixture of lentil flour (70%) and water with the addition of α-amylase Hitempase STXL Kerry Group (Tralee, Ireland); and proteolytic enzyme Bioprotease P1 from Kerry Group (Tralee, Ireland). The washing procedures were conducted following the protocol reported by Canonico et al. [23] (1 h, from 45 °C to 75 °C) After that the substrate was boiled and centrifuged, obtaining the resulting wort that was added at 20% to PW.

2.4. By-Products and Volatile Compounds

(Glucose, sucrose, maltose were determined using specific enzymatic kits (kit k-masug) Megazyme, Wicklow, Ireland), while the protein content in final beers was measured using Lowry method. [25]. Direct injection of final beers prepared following Canonico et al. [26] procedure into a gas chromatography system (GC-2014; Shimadzu, Kjoto, Japan) was used to quantify acetaldehyde ethyl acetate, n-propanol, isobutanol, amyl and isoamyl alcohols. The main volatile compounds were determined by solid-phase microextraction (HS-SPME) method using a fiber Divinylbenzene/Carboxen/Polydimethylsiloxane (DVB/CAR/PDMS) (Sigma-Aldrich, St. Louis, MO, USA). The compounds were desorbed by inserting the fiber into gas chromatograph GC (GC-2014; Shimadzu, Kjoto, Japan) identified and quantified by comparisons with external calibration curves for each compound [26].

2.5. Sensory Analysis

The secondary fermentation was carried out in 330 mL bottles with the addition of 5.5 g/L of sucrose at 18–20 °C for 7–10 days. After bottle re-fermentation, the beers were stored at 4 °C and underwent sensory analysis using a scale from 1 to 10 [27] based on a list of descriptors related to both the aromatic notes (e.g., floral, fruity, toasty) and the main structural features (e.g., sweet, acidity, flavor, astringency, bitterness, olfactory persistence). A group of 10 trained testers carried this out. The data were elaborated with statistical analyses to obtained information about the contribution on each descriptor on the organoleptic quality of beer. Informed consent was obtained from all subjects involved in the study of sensory analysis.

2.6. Yeast Vitality Assay after 3 Months of Bottling

The vitality of the strains after 3 months of bottling was carried out using viable cell counts on WL Nutrient Agar (Oxoid, Hampshire, UK) and Lysine Agar (Oxoid, Hampshire, UK) for the differentiation of NCY yeast from *S. cerevisiae* strain.

2.7. Nutritional Values Amino Acid Composition of Final Beers

Determination of moisture and dry matter in food for human use by gravimetry was determined by Method/Document acronym: ISTISAN 1996/34 Met B; Method/Document title: Calculation of Carbohydrates and Energy Value in food for human use was determined by Method/Document acronym: MP 0297 rev 6 2018; Method/Document acronym: Determination of amino acids in food, fertilizers and soil improvers by ion chromatography was determined by Method/Document title: MP 1471 rev 6 202; Method/Document title: Determination of ash/crude ash in food for human and zootechnical use by gravimetry was determined by Method/Document acronym: MP 2271 rev 0 2018; Method/Document title: Determination of total fat substances in food for human use by gravimetry (method with acid hydrolysis) was determined by Method/Document acronym: ISTISAN 1996/34 Met A.

2.8. Statistical Analysis

Analysis of variance (ANOVA) was applied to the experimental data for the main analytical character of the beers. The significant differences were determined using Duncan tests, and the results were considered significant if the associated p values were <0.05. Principal component analysis (PCA) was applied to discriminate between the means of the contents of volatile compounds. The statistical software package JMP 11 ® was used for statistical analysis.

3. Results

3.1. Fermentation Kinetics

Figure 1 shows the fermentation kinetics of *S. cerevisiae*, *K. unispora*, and *L. thermotolerans* strains in pure and in mixed fermentation (*S. cerevisiae*/*K. unispora* and *S. cerevisiae*/*L. thermotolerans*), in both PW (Figure 1a) and PWL (Figure 1b).

K. unispora and *L. thermotolerans* in pure fermentation showed slower fermentation kinetics than *S. cerevisiae* pure culture, both in their respective mixed fermentations and in both worts tested together. As expected, *S. cerevisiae* pure culture showed the highest fermentation kinetics compared to the other fermentation trials, with the maximum loss of CO_2 within the first 3–4 days of fermentation in both PW and PWL worts. However, mixed fermentation (in both worts) and the presence of lentil (PWL)-enhanced fermentation kinetics of *K. unispora* and *L. thermotolerans* pure fermentations.

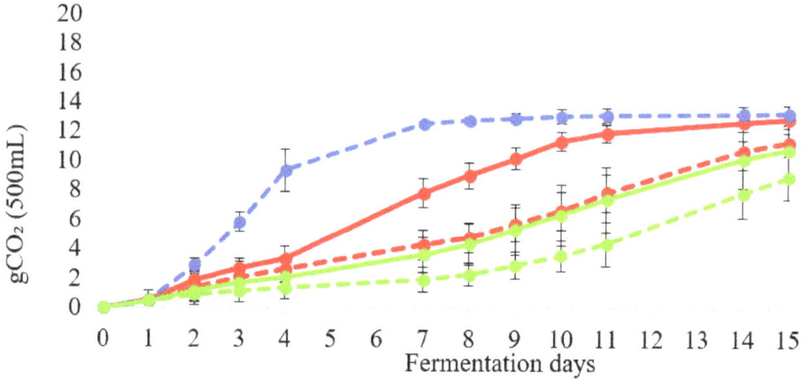

(a)

Figure 1. *Cont.*

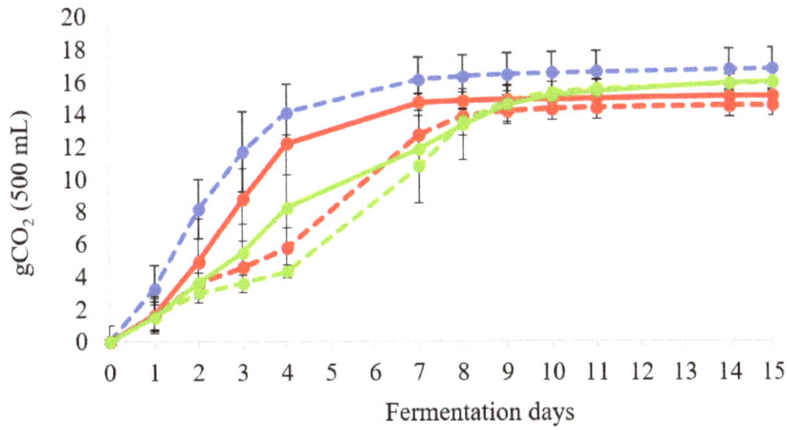

(**b**)

Figure 1. Fermentation kinetics of the pure and mixed fermentations in (**a**) PW and (**b**) PWL. Pure culture of *S. cerevisiae* (- - -); *L. thermotolerans* (- - -), and *K. unispora* (- - -), and mixed fermentation of *S. cerevisiae*/*L. thermotolerans* (——) and *S. cerevisiae*/*K. unispora* (——).

3.2. Main Analytical Characteristics

The data of the main analytical characters of the beers obtained at the end of the primary fermentation on PW and PWL are reported Table 1.

Glucose and sucrose were completely consumed in all fermentation trials, while only *S. cerevisiae* trials did not show residual maltose in both PW and PWL. Pure and mixed NCY fermentation trials showed residual maltose. This residue was similar in all PW fermentation trials (12–13 g/L), while *K. unispora* pure fermentation showed higher residual maltose (30 g/L), indicating a lower fermentation activity. In PWL a general enhancement of maltose consumption in NCY fermentation trials was shown. The ethanol content was generally lower with NCY in pure and mixed fermentations, even if this reduction is significant only in a few cases. No significant difference between the final beer was shown regarding the other analytical character.

Regarding the amino acid profile (Table 2), the final beer on PW exhibited a significant increase in aspartic acid and phenylalanine content with *K. unispora,* in both pure and mixed fermentations.

Still, in PW *L. thermotolerans* pure fermentation exhibited a significant increase in glutamic acid, lysine, and asparagine content. The other amino acids did not show significant differences among the trials tested. As expected, the final beers with lentil (PWL) exhibited a higher content in all amino acids tested compared to PW. Among the trials carried out in PWL, *L. thermotolerans* pure and mixed fermentations exhibited a general significant reduction in amino acid content (except for phenylalanine, methionine, and ornithine), while *K. unispora* (pure and mixed fermentations) exhibited a similar amino acid content with the *S. cerevisiae* pure culture.

To assess the influence of *L. thermotolerans* and *K. unispora* on both wort trials, the data were elaborated by principal component analysis (PCA) (Figure 2).

Table 1. The main analytical characters of final beer on PW (pils wort) and PWL (pils wort added with 20% HRL). Data are the means ± standard deviations. Data with different superscript letters ($^{a, b, c}$) within each row between the fermentation trials tested on same wort (Duncan tests; $p < 0.05$). The initial composition of the sugars in pils wort were: Glucose 11.92 g/L; Sucrose 24.2 g/L; Maltose 74.73 g/L. Protein content: 12.02 g/L. The initial composition of the sugars in pils wort added with lentil were: Glucose 6.74 g/L; Sucrose 37.72 g/L; Maltose 84.70 g/L. Protein content: 26.39 g/L. %). ND: not detected. LoQ: limit of quantification (1 g/L).

	PW					PWL				
The Main Fermentation Parameters	*S. cerevisiae*	*L. thermotolerans*	*K. unispora*	*S. cerevisiae/L. thermotolerans*	*S. cerevisiae/K. unispora*	*S. cerevisiae*	*L. thermotolerans*	*K. unispora*	*S. cerevisiae/L. thermotolerans*	*S. cerevisiae/K. unispora*
Residual Glucose g/L	ND	ND	ND	ND	ND	ND	ND	ND	ND	ND
Residual Sucrose g/L	ND	ND	ND	ND	ND	ND	ND	ND	ND	ND
Residual Maltose g/L	ND	13.52 ± 1.50 ^{b}	30.14 ± 0.65 ^{a}	12.91 ± 0.15 ^{b}	11.95 ± 1.61 ^{b}	0.11 ± 1.10 ^{c}	4.77 ± 0.1 ^{b}	4.44 ± 2.164 ^{b}	5.97 ± 0.704 ^{b}	11.95 ± 1.61 ^{a}
Protein g/L	12.53 ± 3.87 ^{a}	33.29 ± 3.71 ^{a}	25.51 ± 16.45 ^{a}	18.49 ± 11.32 ^{a}	17.43 ± 13.18 ^{a}	26.27 ± 12.99 ^{a}	30.12 ± 3.89 ^{a}	25.51 ± 11.68 ^{a}	22.05 ± 8.43 ^{a}	17.43 ± 13.18 ^{a}
Ethanol %*v/v*	3.38 ± 0.14 ^{a}	3.04 ± 0.1 $^{a, b}$	3.03 ± 0.12 $^{a, b}$	2.99 ± 0.10 ^{b}	3.07 ± 0.18 $^{a, b}$	3.76 ± 0.17 ^{a}	3.3 ± 0.17 $^{a, b}$	3.78 ± 0.10 ^{a}	3.44 ± 0.06 $^{a, b}$	3.07 ± 0.18 ^{b}
Moisture g/100 g	95.90 ± 0.38 ^{a}	95.73 ± 0.38 ^{a}	95.85 ± 0.38 ^{a}	96.18 ± 0.38 ^{a}	95.89 ± 0.38 ^{a}	94.19 ± 0.38 $^{a, b}$	93.82 ± 0.38 ^{b}	94.34 ± 0.38 $^{a, b}$	93.95 ± 0.38 $^{a, b}$	95.89 ± 0.38 ^{a}
Fatty g/100 g	0.050 ± 0.033 ^{a}	<LoQ ^{a}	<LoQ ^{a}	<LoQ ^{a}	<LoQ ^{a}	<LoQ ^{a}	<LoQ ^{a}	<LoQ ^{a}	<LoQ ^{a}	<LoQ ^{a}
Ashes g/100 g	0.15 ± 0.03 ^{a}	0.17 ± 0.03 ^{a}	0.19 ± 0.04 ^{a}	0.19 ± 0.04 ^{a}	0.20 ± 0.04 ^{a}	03.0 ± 00.4 ^{a}	0.29 ± 0.04 ^{a}	0.34 ± 0.04 ^{a}	0.30 ± 0.04 ^{a}	0.20 ± 0.04 ^{a}
Carbohydrates g/100 g	3.43 ± 0.39 ^{a}	3.63 ± 0.39 ^{a}	3.50 ± 0.39 ^{a}	3.14 ± 0.39 ^{a}	3.47 ± 0.39 ^{a}	4.30 ± 0.39 ^{a}	4.74 ± 03.9 ^{a}	4.15 ± 0.39 ^{a}	4.63 ± 0.39 ^{a}	3.47 ± 0.39 ^{a}
Energy value kcal/100 g	16 ± 2 ^{a}	16 ± 2 ^{a}	16 ± 2 ^{a}	15 ± 2 ^{a}	16 ± 2 ^{a}	22 ± 2 $^{a, b}$	24 ± 2 ^{a}	21 ± 2 $^{a, b}$	23 ± 2 $^{a, b}$	16 ± 2 ^{b}
Dry substance g/100 g	4.10 ± 0.38 ^{b}	4.27 ± 0.38 ^{b}	6.7 ± 0.38 ^{a}	3.82 ± 0.38 ^{b}	4.11 ± 0.38 ^{b}	5.81 ± 0.38 ^{a}	6.18 ± 0.38 ^{a}	5.66 ± 0.38 ^{a}	6.05 ± 0.38 ^{a}	4.11 ± 0.38 ^{b}

Table 2. Amino acid composition of final beers produced in pure and mixed fermentations on PW (pils wort) and (PWL) pils wort added with lentil. Data are means ± standard deviations from three independent experiments. Data with different superscript letters ($^{a, b, c}$) within each row between the fermentation trials tested on the same wort (Duncan tests (0.05%). LoQ: limit of quantification (1 g/L).

	PW (Pils Wort)					PWL (Pils + Lentil Wort)				
Amino Acid Composition (mg/L)	*S. cerevisiae*	*L. thermotolerans*	*K. unispora*	*S. cerevisie/L. thermotolerans*	*S. cerevisiae/K. unispora*	*S. cerevisiae*	*L. thermotolerans*	*K. unispora*	*S. cerevisie/L. thermotoler-ans*	*S. cerevisiae/K. unispora*
Aspartic acid	$45 \pm 8^{a,b}$	$31 \pm 7^{a,b}$	49 ± 8^{a}	18 ± 7^{b}	54 ± 9^{a}	257 ± 26^{a}	110 ± 13^{b}	226 ± 23^{a}	105 ± 12^{b}	224 ± 23^{a}
Glutamic acid	$53 \pm 8^{a,b}$	56 ± 9^{a}	26 ± 8^{b}	$36 \pm 8^{a,b}$	$28 \pm 7^{a,b}$	495 ± 48^{a}	246 ± 25^{b}	220 ± 23^{b}	234 ± 24^{b}	277 ± 28^{b}
Alanine	80 ± 11^{a}	81 ± 10^{a}	81 ± 10^{a}	66 ± 9^{a}	84 ± 11^{a}	318 ± 31^{a}	$218 \pm 22^{a,b}$	$309 \pm 31^{a,b}$	209 ± 21^{b}	313 ± 31^{a}
Arginine	72 ± 10^{a}	69 ± 10^{a}	79 ± 11^{a}	48 ± 8^{a}	80 ± 10^{a}	$277 \pm 27^{a,b}$	191 ± 20^{b}	303 ± 30^{a}	180 ± 19^{b}	303 ± 30^{a}
Asparagine	$18 \pm 7^{a,b}$	31 ± 7^{a}	$16 \pm 7^{a,b}$	<LoQ ^{b}	$21 \pm 7^{a,b}$	$135 \pm 15^{a,b}$	$105 \pm 13^{a,b}$	$134 \pm 15^{a,b}$	84 ± 11^{b}	151 ± 16^{a}
Proline	319 ± 32^{a}	326 ± 3^{a}	349 ± 35^{a}	327 ± 32^{a}	341 ± 34^{a}	369 ± 37^{a}	330 ± 33^{a}	392 ± 39^{a}	323 ± 32^{a}	368 ± 36^{a}
Phenyl alanine	$32 \pm 8^{a,b}$	$44 \pm 8^{a,b}$	56 ± 9^{a}	17 ± 7^{b}	54 ± 9^{a}	219 ± 22^{a}	$154 \pm 16^{a,b}$	215 ± 22^{a}	142 ± 16^{b}	220 ± 22^{a}
Glycine	29 ± 7^{a}	26 ± 7^{a}	30 ± 7^{a}	28 ± 7^{a}	30 ± 7^{a}	138 ± 15^{a}	71 ± 10^{b}	118 ± 14^{a}	72 ± 10^{b}	$116 \pm 13^{a,b}$
Glutamine	15 ± 7^{a}	10 ± 7^{a}	12 ± 7^{a}	<LoQ ^{a}	14 ± 7^{a}	65 ± 9^{a}	13 ± 7^{b}	51 ± 8^{a}	16 ± 7^{b}	52 ± 8^{a}
Isoleucine	14 ± 7^{a}	19 ± 7^{a}	20 ± 7^{a}	<LoQ ^{a}	20 ± 7^{a}	144 ± 16^{a}	89 ± 11^{b}	140 ± 15^{a}	75 ± 10^{b}	151 ± 16^{a}
Histidine	12 ± 7^{a}	30 ± 8^{a}	26 ± 7^{a}	18 ± 7^{a}	35 ± 8^{a}	78 ± 10^{a}	44 ± 8^{b}	$63 \pm 9^{a,b}$	43 ± 8^{b}	$68 \pm 9^{a,b}$
Leucine	30 ± 7^{a}	38 ± 8^{a}	42 ± 8^{a}	17 ± 7^{b}	42 ± 8^{a}	260 ± 26^{a}	166 ± 18^{b}	251 ± 25^{a}	151 ± 16^{b}	264 ± 27^{a}
Lysine	<LoQ ^{b}	20 ± 7^{a}	$12 \pm 7^{a,b}$	<LoQ ^{b}	$12 \pm 7^{a,b}$	206 ± 21^{a}	112 ± 13^{b}	208 ± 21^{a}	101 ± 12^{b}	218 ± 22^{a}
Methionine	<LoQ	<LoQ	<LoQ	<LoQ	<LoQ	59 ± 9^{a}	39 ± 8^{a}	54 ± 9^{a}	34 ± 8^{a}	58 ± 9^{a}
Ornithine	<LoQ	<LoQ	<LoQ	<LoQ	<LoQ	20 ± 7^{a}	14 ± 7^{a}	21 ± 7^{a}	14 ± 7^{a}	21 ± 7^{a}
Serine	14 ± 7^{a}	18 ± 7^{a}	14 ± 7^{a}	<LoQ	17 ± 7^{a}	192 ± 20^{a}	103 ± 12^{b}	183 ± 19^{a}	87 ± 11^{b}	193 ± 20^{a}
Tyrosine	49 ± 8^{a}	61 ± 9^{a}	66 ± 10^{a}	36 ± 8^{a}	64 ± 9^{a}	209 ± 21^{a}	141 ± 16^{b}	$193 \pm 20^{a,b}$	135 ± 15^{b}	$192 \pm 20^{a,b}$
Threonine	<LoQ	<LoQ	<LoQ	<LoQ	<LoQ	98 ± 12^{a}	48 ± 8^{b}	95 ± 11^{a}	37 ± 8^{b}	106 ± 12^{a}
Valine	39 ± 8^{a}	56 ± 9^{a}	52 ± 8^{a}	28 ± 7^{a}	52 ± 9^{a}	304 ± 30^{a}	$200 \pm 20^{b,c}$	$280 \pm 28^{a,b}$	183 ± 19^{c}	$292 \pm 29^{a,b}$
Gamma aminobutyric acid	71 ± 10^{a}	64 ± 9^{a}	82 ± 10^{a}	70 ± 10^{a}	82 ± 10^{a}	136 ± 15^{b}	104 ± 12^{b}	283 ± 28^{a}	104 ± 12^{b}	242 ± 24^{a}
Total free amino acid	892 ± 45^{a}	980 ± 46^{a}	1012 ± 48^{a}	709 ± 41^{b}	1030 ± 48^{a}	3979 ± 104^{a}	2498 ± 72^{c}	$3739 \pm 97^{a,b}$	$2\,329 \pm 68^{c}$	3829 ± 97^{b}

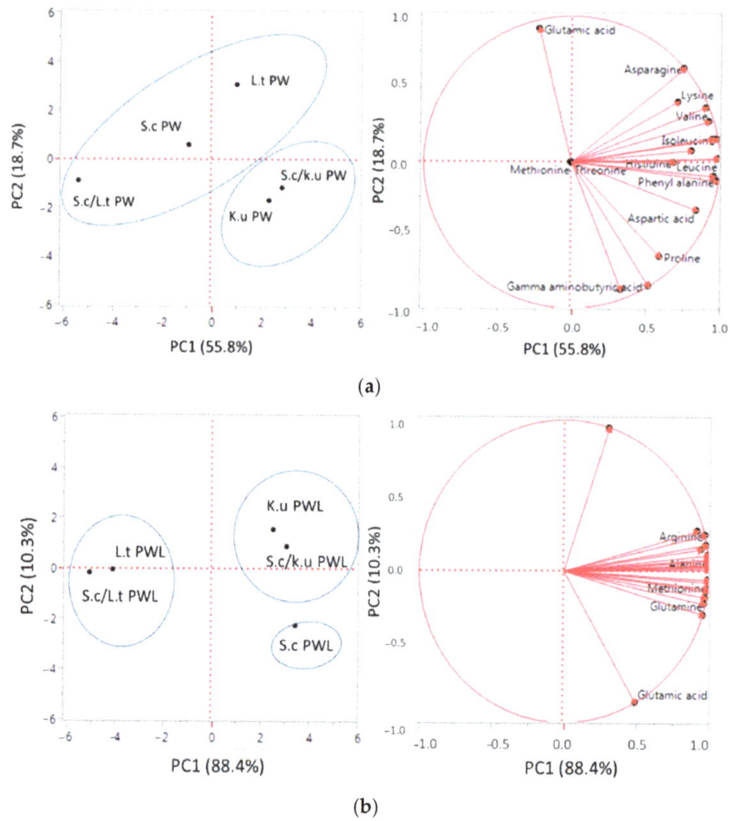

Figure 2. Principal component analysis of amino acid content of final beers. (**a**) Pils wort (PW): the variance explained by principal component analysis (PCA) analysis is PC 1 55.8% X-axis and PC 2 18.7% Y-axis. (**b**) Pils wort + lentil (PWL): the variance explained by principal component analysis (PCA) analysis is PC 1 88.4% X-axis and PC 2 10.3% Y-axis.

The trials carried out on PW (Figure 2a), *K. unispora* pure culture and mixed fermentations grouped in the lower right quadrant showed that *K. unispora* characterized the final amino acid content in mixed culture. Different behavior was exhibited by *L. thermotolerans* that in pure culture was in the right upper quadrant, in mixed (*L. thermotolerans/S. cerevisiae*) in the lower left quadrant, while *S. cerevisiae* pure culture was in an intermediate position highlighting a different effect of this yeast on the final amino acid content of the beer. The distribution of the trials carried out on PWL (Figure 2b) showed a more homogeneous distribution among pure and mixed fermentations. Indeed, *K. unispora* pure and mixed fermentations were in the right upper quadrant, *L. thermotolerans* pure and mixed trials on the line that separates the two left quadrants and *S. cerevisiae* in the lower right quadrant. These data indicated that *K. unispora* and *L. thermotolerans* affected the beer's final amino acid composition compared with *S. cerevisiae*.

3.3. By-Products and Volatile Profiles

The data for the main by-products and the volatile compounds in PW and PWL are reported in Table 3. The results of PW trials indicated that *L. thermotolerans* and *K. unispora* in mixed fermentation increased some aroma compounds compared to pure culture. *L. thermotolerans/S. cerevisiae* significantly increased the n-propanol, isobutanol, and amylic alcohol, while *K. unispora/S. cerevisiae* significantly increased ethyl butyrate in comparison with the other trials.

Table 3. The main by-products and volatile compounds (mg/L) of pure and mixed fermentation trials carried out in PW (pils wort) and PWL (pils wort with HRL). Data are means ± standard deviations from three independent experiments. Data with different superscript letters ($^{a, b, c, d}$) within each row between the fermentation trials tested on same wort (Duncan tests; $p < 0.05$). * OTVs (odor threshold value; mg/L).

	PW					PWL				
The Main By-Products (OTVs *)	*S. cerevisiae*	*L. thermotolerans*	*K. unispora*	*S. cerevisie/L. thermotolerans*	*S. cerevisiae/K. unispora*	*S. cerevisiae*	*L. thermotolerans*	*K. unispora*	*S. cerevisie/L. thermotolerans*	*S. cerevisiae/K. unispora*
Ethyl butyrate (0.14–0.37)	0.117 ± 0.025 ^{b}	0.067 ± 0.021 ^{c}	0.053 ± 0.007 ^{c}	0.083 ± 0.068 ^{b,c}	0.268 ± 0.016 ^{a}	0.335 ± 0.029 ^{a}	0.016 ± 0.009 ^{c}	0.055 ± 0.028 ^{b}	0.087 ± 0.037 ^{b}	0.060 ± 0.006 ^{b}
Ethyl acetate (7.5–31)	4.04 ± 0.52 ^{c}	7.08 ± 0.55 ^{a}	6.01 ± 0.89 ^{b,c}	7.74 ± 0.75 ^{a}	6.26 ± 0.47 ^{a,b}	15.81 ± 1.04 ^{a,b}	15.64 ± 1.08 ^{a,b}	20.54 ± 0.88 ^{a}	15.29 ± 3.43 ^{a,b}	11.73 ± 2.38 ^{b}
Linalool (0.0006–0.001)	0.047 ± 0.029 ^{a}	0.046 ± 0.022 ^{a}	0.037 ± 0.004 ^{a}	0.028 ± 0.020 ^{a}	0.031 ± 0.012 ^{a}	0.055 ± 0.033 ^{a}	0.031 ± 0.008 ^{a,b}	0.024 ± 0.002 ^{b}	0.032 ± 0.004 ^{a,b}	0.028 ± 0.00 ^{a,b}
Ethyl hexanoate (0.17–0.20)	0.027 ± 0.009 ^{a}	0.025 ± 0.005 ^{a}	0.028 ± 0.004 ^{a}	0.031 ± 0.00 ^{a}	0.020 ± 0.007 ^{a}	0.025 ± 0.004 ^{a}	0.023 ± 0.001 ^{a}	0.014 ± 0.007 ^{a}	0.019 ± 0.004 ^{a}	0.032 ± 0.010 ^{a}
Isoamyl acetate (0.30–0.72)	0.40 ± 0.05 ^{a}	0.11 ± 0.18 ^{b}	0.26 ± 0.22 ^{a,b}	0.32 ± 0.03 ^{a,b}	0.40 ± 0.01 ^{a}	0.820 ± 0.143 ^{b}	0.527 ± 0.021 ^{c}	1.514 ± 0.349 ^{a}	1.070 ± 0.063 ^{a}	0.851 ± 0.065 ^{b}
n-propanol (0.8–5.0)	18.09 ± 1.12 ^{b}	16.17 ± 1.01 ^{b}	17.01 ± 1.13 ^{b}	22.85 ± 1.69 ^{a}	18.35 ± 0.96 ^{b}	25.85 ± 2.18 ^{a}	18.28 ± 2.43 ^{c}	24.64 ± 1.70 ^{a,b}	20.92 ± 2.59 ^{b,c}	23.37 ± 3.20 ^{a,b,c}
Isobutanol (3.2–14.5)	8.838 ± 0.542 ^{a,b}	7.425 ± 0.510 ^{c}	9.937 ± 0.628 ^{b,c}	12.70 ± 2.36 ^{a}	11.19 ± 0.40 ^{a,b}	24.51 ± 3.69 ^{a}	13.56 ± 1.64 ^{b}	20.88 ± 2.48 ^{a}	12.52 ± 2.33 ^{b}	23.37 ± 3.20 ^{a}
Amylic alcohol (0.32–15.0)	6.722 ± 0.572 ^{a,b}	5.336 ± 0.104 ^{c}	6.404 ± 1.048 ^{b,c}	7.756 ± 0.491 ^{a}	7.689 ± 0.239 ^{a}	15.67 ± 2.01 ^{a}	6.727 ± 0.098 ^{c}	10.65 ± 0.92 ^{b}	7.822 ± 1.657 ^{c}	8.140 ± 1.755 ^{b,c}
Isoamylic alcohol (0.77–16.8)	46.74 ± 1.90 ^{a}	38.04 ± 1.81 ^{b,c}	36.47 ± 3.40 ^{c}	45.84 ± 3.15 ^{a}	39.45 ± 0.73 ^{b}	75.45 ± 0.19 ^{a}	53.50 ± 2.86 ^{c}	66.22 ± 0.62 ^{b}	53.51 ± 2.08 ^{c}	58.20 ± 3.03 ^{b,c}
β-phenyl ethanol (1.0–1.88)	2.384 ± 0.044 ^{a}	1.335 ± 0.472 ^{a,b}	0.794 ± 0.098 ^{b}	1.359 ± 0.382 ^{a,b}	1.093 ± 0.456 ^{a,b}	0.533 ± 0.004 ^{a,b}	0.450 ± 0.043 ^{a,b,c}	0.323 ± 0.030 ^{b,c}	0.551 ± 0.179 ^{a}	0.320 ± 0.034 ^{c}
Acetaldehyde (0.02–0.12)	49.46 ± 1.42 ^{a,b}	56.14 ± 4.99 ^{a}	52.14 ± 3.28 ^{a,b}	54.07 ± 8.72 ^{a}	35.08 ± 2.43 ^{c}	8.959 ± 1.216 ^{d}	144.46 ± 13.96 ^{a}	28.11 ± 0.53 ^{c}	5.039 ± 0.536 ^{d}	77.15 ± 21.80 ^{b}

A different trend was shown in PWL, where the *K. unispora* pure culture increased ethyl acetate, while isobutanol and isoamyl acetate were comparable to that exhibited by *K. unispora/S. cerevisiae,* and *L. thermotolerans/S. cerevisiae,* respectively. Moreover, *L. thermotolerans/S. cerevisiae* led to a significant increase in β-phenyl ethanol compared to the other trials. *L. thermotolerans* showed a significant enhancement of acetaldehyde content, while *S. cerevisiae* was characterized by ethyl butyrate, ethyl acetate, and amylic and isoamylic alcohol.

To assess the overall effects of yeast strains, modalities of inoculum, and different worts used, the data of the by-products, volatile compounds were analyzed by PCA (Figure 3).

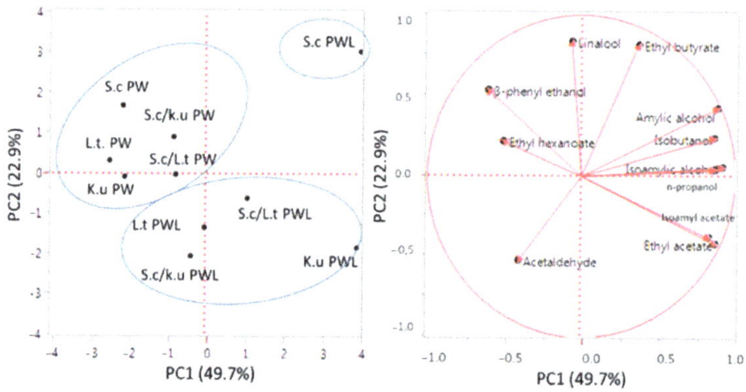

Figure 3. Principal component analysis for the main by-products and volatile compounds of craft beer obtained by different yeast strains in PW and PWL. The variance explained by principal component analysis (PCA) analysis is PC 1 49.7% X-axis and PC 2 22.9% Y-axis.

The graphic representation of PCA of the fermentation products responsible for beer's aroma showed a clear separation of beers obtained by two different worts. The PW trials were in the upper left quadrant while PWL trials were in the lower right quadrant, showing a less homogeneous distribution. Indeed, the fermentation with *S. cerevisiae* pure culture was in the upper right quadrant. The two non-*Saccharomyces* in pure and mixed fermentations were in the opposite quadrant (lower right quadrant). Furthermore, PCA analysis showed a separation between pure and mixed culture, highlighting an effect of non-*Saccharomyces* strains and their inoculation modality on the volatile composition of beer. In PWL NCY (pure and mixed fermentations) strongly characterized the aromatic profile resulting in more distance in the graphical distribution by *S. cerevisiae.*

3.4. Vitality Assay after 3 Months

The vitality assay of all strain tested (data not shown) showed a good vitality after 3 months of bottling, exhibiting a viable cell count c.a $\geq$ 6.5 log CFU/mL in both substrates and compared with the initial inoculum (c.a. 6 log CFU/mL). This is a promising result since the vitality after storage is very closed to that fixed for probiotic bacteria claim.

3.5. Sensorial Analysis

The beers obtained by pure and mixed fermentations on PW and PWL underwent sensory analysis, and the results are reported in Figure 4.

All beers analyzed showed differences for their main aromatic notes. The fermentation carried out with *K. unispora* pure fermentation on PW (Figure 4a) showed a significant difference in the perception of alcoholic solvent, malt note, sweetness and persistence. This last feature was also significantly emphasized in S. cerevisiae pure culture, which also exhibited a perception of the descriptor "other sulfide". *L. thermotolerans* showed malty and fruity/citric notes. Overall, all the beers are characterized by distinctive and characterizing notes.

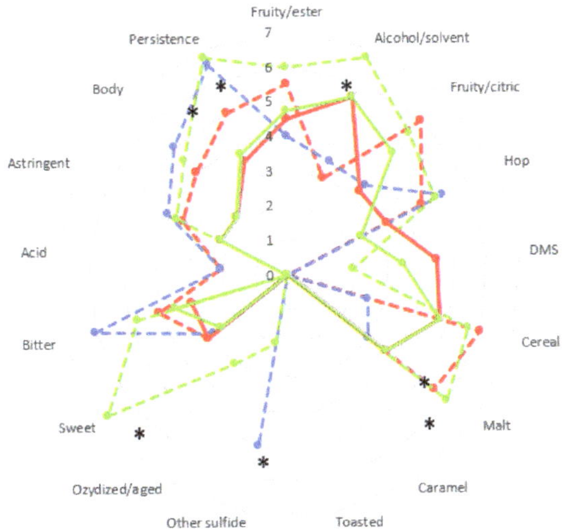

(a)

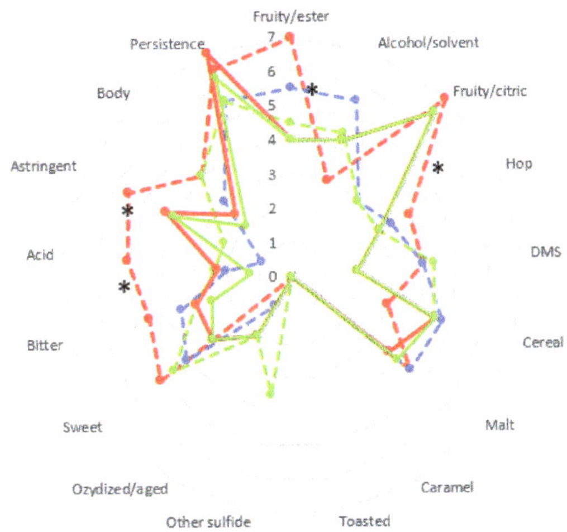

(b)

Figure 4. Sensory analysis of the beer produced by pure and mixed fermentation on pils wort (**a**) and pils wort added lentil (**b**). Pure culture of *S. cerevisiae* (- - -), *L. thermotolerans* (- - -) and *K. unispora* (- - -), and mixed fermentation of *S. cerevisiae*/*L. thermotolerans* (——) and *S. cerevisiae*/*K. unispora* (——). * significantly different (Fisher ANOVA; *p*-value 0.05). DMS, Dimethyl sulphide. Score 0: absence of the descriptor analyzed.

Regarding beer brewed with PWL (Figure 4b), in *L. thermotolerans* pure cultures the presence of lentils increased the perception of fruity esters, fruity/citric, astringency, and acidity. Additionally, *K. unispora*/*S. cerevisiae* emphasized the fruity/citric notes. Moreover, all fermentations exhibited persistence, and in all single fermentations, an increase in DMS was observed

4. Discussion

In recent years, the attention of researchers has been focused not only on the exploitation of new raw materials to obtain a distinctive beer but also on the potential health benefits for consumers [28]. Indeed, beer is rich in bioactive compounds coming from traditional ingredients (barley, hop, and yeast) but also from special ingredients such as spices, cereal, herbs, fruits, and legumes, which can affect the nutritional composition of the final product [23,29]. Although the health benefits of fermented beverages are well established in the scientific field, this is not equally perceived in public opinion. Nevertheless, the positive impact of yeast on consumer health can be related to several aspects: providing a source of probiotic microbes; providing prebiotic metabolites through the secondary metabolism of compounds derived from the grain, hops, or other ingredients; and the production of antimicrobial compounds [30,31].

In this work, selected yeasts with functional aptitude [23,24] were evaluated in pure and mixed culture on PW and PWL added with HRL. The goal of craft beer with functional properties and strengthen the sensory profile was obtained through a double strategy: on the one hand to provide for the increase in vegetable protein and on the other for the use of probiotic fermenting yeasts.

As expected, the addition 20% of HRL determined an enhancement of total protein concentration of craft beer supporting the recent trend of consumers demanding protein-enriched foods. Moreover, the hydrolysis process enriches the wort with essential amino acids that contribute to improving the nutritional profile of the final product. On the other hand, this addition increased the fermentation performance of yeasts both in pure and mix cultures, highlighting the possible use of this substrate in the brewing process to improve the overall fermentation process confirming previous preliminary findings [24]. Furthermore, changes in the concentrations of amino acids in the fortified wort, influencing the nitrogen metabolism, led to a greater availability of amino acids determining an enhancement of the fermentation activity, specifically of *L. thermotolerans* and *K. unispora* strains. In this regard, Krasnikova et al. [32] found an enhancement of enzymatic activity in *S. cerevisiae* due to the supplementation of wort with nitrogen sources from lentils.

The addition of HRL showed a limited increase in maltose concentration with a consequent increase in ethanol content. Over the last decade, plant-based beverages have gained popularity amongst aware consumers seeking alternative and environmentally sustainable options to traditional drinks. Recently, Nawaz et al. [33] reviewed the involvement of fermented yeasts in alternative substrates in the emerging segment of the functional legume-based beverages. They highlighted the effective opportunities to broaden and diversify new products, characterized by legume addition, which may offer better nutrition content and distinctive taste.

In brewing, the composition of wort is an essential part of beer flavor. In this regard, although wort is complex and not thoroughly characterized, the content of the amino acid indubitably affected the production of some minor metabolic products of fermentation which contribute to the flavor of the beer. The addition of a protein source such as HRL determined an effective improvement of aromatic compounds such as higher alcohols, esters, carbonyl and sulfur compounds. Except for the preliminary screening work [23], no other published work has investigated the use of hydrolyzed lentils to produce functional craft beer. The distinctive footprint found in pure and mixed fermentation with PWL could be partially related to the yeast catabolism of increased availability of amino acids. During wort fermentation, amino acid utilization by yeast is closely linked to flavour profile [34] such as higher alcohols and related esters. Thus, an improved understanding of amino acid uptake and assimilation is essential to generate defined amounts of metabolites to regulate specific sensory perception in fermented beverages.

Despite the high levels of nitrogen and/or free amino acids coming from HRL, in PWL trials the final flavors were characterized by specific amino acids, such as glutamate, aspartate, and asparagine. In line these results, Black and co-workers [35] highlighted the positive effect of fava bean that changed the proportion of unfavorable amino acids in brewing. On the other hand, the use of NCY in pure and mixed cultures positively influenced the aromatic and sensorial profile, particularly in fortified beers. The interest in NCY isolated from various food and the environmental source is another strategy to bring innovation in the brewing sector. Brewers widely seek the ability to improve the analytical and aromatic profiles during alcoholic fermentation [3,36–40]. In this work we evaluated the dual role of yeasts: metabolizing the amino acids characterizing the vegetal notes (unwanted) and using them as precursors of aromatic compounds.

Differently from *L. thermotolerans*, yeast species extensively investigated in the production of beer [36,38–41], *K. unispora* was only recently proposed [23]. This yeast is usually encountered at low frequency in natural environments; the genus *Kazachstania* was first discovered in 1971 with *Kazachstania viticola* isolated from fermenting grapes. Subsequently, other species such as *K. unispora* (formerly *Saccharomyces unisporus*) were found on grape must in different countries [42], although its metabolic footprint remains widely unknown. Mixed fermentations with NCY lead final products with increased persistence and astringency and a general exalted fruity note [23]. Moreover, all strains tested showed a good vitality in both substrates after bottle re-fermentation and storage. The viability found here was close to 7 Log/mL(claim for probiotic bacteria) indicating the potential functional trait of NCY *K. unispora* and *L. thermotolerans* here tested. Further investigations are necessary to confirm the positive role of these non-conventional yeasts at the human-gut-level, combining consumer acceptability.

5. Conclusions

This study showed that the combined use of selected NCY and the addition of HRL in pils wort could be a suitable strategy to manage a specialty craft beer with enhanced functional properties and a promising sensory profile.

The different amino acid metabolism of NCY tested led the aromatic profile of the fortified beers with distinctive and positive sensory notes.

The promising results indicate a possible market exploitation of such innovative fermented beverages. Indeed, the conjunction of a legume fortified beer and the presence of functional yeasts could represent a great opportunity to put an innovative product on the market.

Author Contributions: L.C., A.A., E.Z., M.C. and F.C. contributed equally to the manuscript. All the authors participated in the design and discussion of the research. L.C. carried out the experimental part of the work and carried out the analysis of the data. All authors have read and agreed to the published version of the manuscript.

Funding: The project SYSTEMIC "an integrated approach to the challenge of sustainable food systems: adaptive and mitigatory strategies to address climate change and malnutrition", Knowledge hub on Nutrition and Food Security, has received funding from national research funding parties in Belgium (FWO), France (INRA), Germany (BLE), Italy (MIPAAF), Latvia (IZM), Norway (RCN), Portugal (FCT), and Spain (AEI) in a joint action of JPI HDHL, JPI-OCEANS and FACCE-JPI launched in 2019 under the ERA-NET ERA-HDHL (n° 696295).

Institutional Review Board Statement: Not applicable.

Informed Consent Statement: Informed consent was obtained from all subjects involved in the study.

Data Availability Statement: The data presented in this study are available on request from the corresponding author.

Acknowledgments: The authors wish to thank the Birra dell'Eremo Microbrewery (Assisi, Italy) for providing the facilities and the wort for the trials. Moreover, thanks go to the UNIONBIRRAI association (Milano, Italy) and the trained testers belonging to UNINOBIRRAI BEER TASTERS (UBT Marche region) for helping the authors to complete the study for an industrial application.

Conflicts of Interest: The authors declare no conflict of interest.

References

1. Donadini, G.; Fumi, M.D.; Kordialik-Bogacka, E.; Maggi, L.; Lambri, M.; Sckokai, P. Consumer interest in specialty beers in three European markets. *Food Res. Int.* **2016**, *85*, 301–314. [CrossRef] [PubMed]
2. Euromonitor International. Beer in Brazil. 2018. Available online: https://www.euromonitor.com/beer-inbrazil/report (accessed on 17 November 2018).
3. Callejo, M.J.; Navas, J.G.; Alba, R.; Escott, C.; Loira, I.; González, M.C.; Morata, A. Wort fermentation and beer conditioning with selected non-*Saccharomyces* yeasts in craft beers. *Eur. Food Res. Technol.* **2019**, *245*, 1229–1238. [CrossRef]
4. Pires, E.J.; Teixeira, J.A.; Brányik, T.; Vicente, A.A. Yeast: The soul of beer's aroma—a review of flavour-active esters and higher alcohols produced by the brewing yeast. *Appl. Microbiol. Biotechnol.* **2014**, *98*, 1937–1949. [CrossRef]
5. Verstrepen, K.J.; Derdelinckx, G.; Dufour, J.P.; Winderickx, J.; Thevelein, J.M.; Pretorius, I.S.; Delvaux, F.R. Flavor-active esters: Adding fruitiness to beer. *J. Biosci. Bioeng.* **2003**, *96*, 110–118. [CrossRef]
6. Canonico, L.; Comitini, F.; Ciani, M. Dominance and influence of selected *Saccharomyces cerevisiae* strains on the analytical profile of craft beer refermentation. *J. Inst. Brew.* **2014**, *120*, 262–267. [CrossRef]
7. Nachel, M.; Ettlinger, S. Getting to know the mother beer categories: Ales, lagers, and more. In *Beer for Dummies*; Nachel, M., Ettlinger, S., Eds.; Wiley: Hoboken, NJ, USA, 2012.
8. Yeo, H.Q.; Liu, S.Q. An overview of selected specialty beers: Developments, challenges and prospects. *Int. J. Food Sci.* **2014**, *49*, 1607–1618. [CrossRef]
9. Basso, R.F.; Alcarde, A.R.; Portugal, C.B. Could non-*Saccharomyces* yeasts contribute on innovative brewing fermentations? *Food Res. Int.* **2016**, *86*, 112–120. [CrossRef]
10. Suzuki, K. 125th anniversary review: Microbiological instability of beer caused by spoilage bacteria. *J. Inst. Brew.* **2011**, *117*, 131–155. [CrossRef]
11. Vriesekoop, F.; Krahl, M.; Hucker, B.; Menz, G. 125th Anniversary Review: Bacteria in brewing: The good, the bad and the ugly. *J. Inst. Brew.* **2012**, *118*, 335–345. [CrossRef]
12. Etienne-Mesmin, L.; Livrelli, V.; Privat, M.; Denis, S.; Cardot, J.M.; Alric, M.; Blanquet-Diot, S. Effect of a new probiotic *Saccharomyces cerevisiae* strain on survival of *Escherichia coli* O157: H7 in a dynamic gastrointestinal model. *Appl. Environ. Microbiol.* **2011**, *77*, 1127–1131. [CrossRef]
13. Kelesidis, T.; Pothoulakis, C. Efficacy and safety of the probiotic *Saccharomyces boulardii* for the prevention and therapy of gastrointestinal disorders. *Therap. Adv. Gastroenterol.* **2012**, *5*, 111–125. [CrossRef] [PubMed]
14. Capece, A.; Romaniello, R.; Pietrafesa, A.; Siesto, G.; Pietrafesa, R.; Zambuto, M.; Romano, P. Use of *Saccharomyces cerevisiae* var. *boulardii* in co-fermentations with *S. cerevisiae* for the production of craft beers with potential healthy value-added. *Int. J. Food Microbiol.* **2018**, *284*, 22–30. [CrossRef] [PubMed]
15. de Paula, B.P.; de Souza Lago, H.; Firmino, L.; Júnior, W.J.F.L.; Corrêa, M.F.D.; Guerra, A.F.; Pereira, K.S.; Coelho, M.A.Z. Technological features of *Saccharomyces cerevisiae* var. *boulardii* for potential probiotic wheat beer development. *LWT* **2021**, *135*, 110233. [CrossRef]
16. Mulero-Cerezo, J.; Briz-Redón, Á.; Serrano-Aroca, Á. *Saccharomyces cerevisiae* var. *boulardii*: Valuable probiotic starter for craft beer production. *Appl. Sci.* **2019**, *9*, 3250. [CrossRef]
17. Boye, J.I.; Aksay, S.; Roufik, S.; Ribéreau, S.; Mondor, M.; Farnworth, E.; Rajamohamed, S.H. Comparison of the functional properties of pea, chickpea and lentil protein concentrates processed using ultrafiltration and isoelectric precipitation techniques. *Food Res. Int.* **2010**, *43*, 537–546. [CrossRef]
18. Tharanathan, R.N.; Mahadevamma, S. Grain legumes—A boon to human nutrition. *Trends Food Sci. Technol.* **2003**, *14*, 507–518. [CrossRef]
19. Yadav, D.N.; Sharma, M.; Chikara, N.; Anand, T.; Bansal, S. Quality characteristics of vegetable-blended wheat–pearl millet composite pasta. *Agric. Res.* **2014**, *3*, 263–270. [CrossRef]
20. de Almeida Costa, G.E.; da Silva Queiroz-Monici, K.; Reis, S.M.P.M.; de Oliveira, A.C. Chemical composition, dietary fibre and resistant starch contents of raw and cooked pea, common bean, chickpea and lentil legumes. *Food Chem.* **2006**, *94*, 327–330. [CrossRef]
21. Tosh, S.M.; Yada, S. Dietary fibres in pulse seeds and fractions: Characterization, functional attributes, and applications. *Food Res. Int.* **2010**, *43*, 450–460. [CrossRef]
22. Micioni Di Bonaventura, M.V.; Cecchini, C.; Vila-Donat, P.; Caprioli, G.; Cifani, C.; Coman, M.M.; Cresci, A.; Fiorini, D.; Ricciutelli, M.; Silvi, S.; et al. Evaluation of the hypocholesterolemic effect and prebiotic activity of a lentil (*Lens culinaris* Medik) extract. *Mol. Nutr. Food Res.* **2017**, *61*, 1700403. [CrossRef]
23. Canonico, L.; Zannini, E.; Ciani, M.; Comitini, F. Assessment of non-conventional yeasts with potential probiotic for protein-fortified craft beer production. *LWT* **2021**, *145*, 111361. [CrossRef]
24. Agarbati, A.; Canonico, L.; Marini, E.; Zannini, E.; Ciani, M.; Comitini, F. Potential probiotic yeasts sourced from natural environmental and spontaneous processed foods. *Foods* **2020**, *9*, 287. [CrossRef] [PubMed]
25. Bensadoun, A.; Weinstein, D. Assay of proteins in the presence of interfering materials. *Anal. Biochem.* **1976**, *70*, 241–250. [CrossRef]
26. Canonico, L.; Comitini, F.; Ciani, M. Influence of vintage and selected starter on *Torulaspora delbrueckii*/*Saccharomyces cerevisiae* sequential fermentation. *Eur. Food Res. Technol.* **2015**, *241*, 827–833. [CrossRef]

27. Analytica EBC. European Brewery Convention. 13 October 1997. Available online: http://www.europeanbreweryconvention.org/ (accessed on 25 March 2022).
28. Mellor, D.D.; Hanna-Khalil, B.; Carson, R. A review of the potential health benefits of low alcohol and alcohol-free beer: Effects of ingredients and craft brewing processes on potentially bioactive metabolites. *Beverages* **2020**, *6*, 25. [CrossRef]
29. Ambra, R.; Pastore, G.; Lucchetti, S. The Role of Bioactive Phenolic Compounds on the Impact of Beer on Health. *Molecules* **2021**, *26*, 486. [CrossRef]
30. Korcz, E.; Kerényi, Z.; Varga, L. Dietary fibers, prebiotics, and exopolysaccharides produced by lactic acid bacteria: Potential health benefits with special regard to cholesterol-lowering effects. *Food Funct.* **2018**, *9*, 3057–3068. [CrossRef]
31. De Vuyst, L.; Weckx, S. The cocoa bean fermentation process: From ecosystem analysis to starter culture development. *J. Appl. Microbiol.* **2016**, *121*, 5–17. [CrossRef]
32. Krasnikova, E.S.; Krasnikov, A.V.; Babushkin, V.A. The influence of composite flour mixtures on saccharomyces cerevisiae biotechnological properties and bread quality. *IOP Conf. Ser. Earth Environ. Sci.* **2020**, *521*, 022008. [CrossRef]
33. Nawaz, M.A.; Tan, M.; Øiseth, S.; Buckow, R. An emerging segment of functional legume-based beverages: A review. *Food Rev. Int.* **2022**, *38*, 1064–1102. [CrossRef]
34. Procopio, S.; Brunner, M.; Becker, T. Differential transcribed yeast genes involved in flavour formation and its associated amino acid metabolism during brewery fermentation. *Eur. Food Res. Technol.* **2014**, *239*, 421–439. [CrossRef]
35. Black, K.; Tziboula-Clarke, A.; White, P.J.; Iannetta, P.P.; Walker, G. Optimised processing of faba bean (*Vicia faba* L.) kernels as a brewing adjunct. *J. Inst. Brew.* **2021**, *127*, 13–20. [CrossRef]
36. Canonico, L.; Galli, E.; Ciani, E.; Comitini, F.; Ciani, M. Exploitation of three non-conventional yeast species in the brewing process. *Microorganisms* **2019**, *7*, 11. [CrossRef] [PubMed]
37. Canonico, L.; Ciani, E.; Galli, E.; Comitini, F.; Ciani, M. Evolution of aromatic profile of *Torulaspora delbrueckii* mixed fermentation at microbrewery plant. *Fermentation* **2020**, *6*, 7. [CrossRef]
38. Domizio, P.; House, J.F.; Joseph, C.M.L.; Bisson, L.F.; Bamforth, C.W. *Lachancea thermotolerans* as an alternative yeast for the production of beer. *J. Inst. Brew.* **2016**, *122*, 599–604. [CrossRef]
39. Holt, S.; Mukherjee, V.; Lievens, B.; Verstrepen, K.J.; Thevelein, J.M. Bioflavoring by non-conventional yeasts in sequential beer fermentations. *Food Microbiol.* **2018**, *72*, 55–66. [CrossRef] [PubMed]
40. Osburn, K.; Amaral, J.; Metcalf, S.R.; Nickens, D.M.; Rogers, C.M.; Sausen, C.; Caputo, R.; Miller, J.; Hongde, L.; Tennesen, J.M.; et al. Primary souring: A novel bacteria-free method for sour beer production. *Food Microbiol.* **2018**, *70*, 76–84. [CrossRef]
41. Zdaniewicz, M.; Satora, P.; Pater, A.; Bogacz, S. Low lactic acid-producing strain of *Lachancea thermotolerans* as a new starter for beer production. *Biomolecules* **2020**, *10*, 256. [CrossRef]
42. Bhattacharya, I.; Yan, S.; Yadav, J.S.S.; Tyagi, R.; Surampalli, R.Y. *Saccharomyces unisporus*: Biotechnological potential and present status. *Compr. Rev. Food Sci. Food Saf.* **2013**, *12*, 353–363. [CrossRef]

foods

MDPI

Article

Antihypertensive Peptides from Ultrafiltration and Fermentation of the Ricotta Cheese Exhausted Whey: Design and Characterization of a Functional Ricotta Cheese

Erica Pontonio [1], Marco Montemurro [1], Gina Valeria De Gennaro [1], Valerio Miceli [2] and Carlo Giuseppe Rizzello [3,*]

1 Department of Soil, Plant and Food Science, University of Bari Aldo Moro, via Giovanni Amendola 165/A, 70126 Bari, Italy; erica.pontonio@uniba.it (E.P.); marco.montemurro@uniba.it (M.M.); g.vale94@libero.it (G.V.D.G.)
2 ENEA Research Centre, Department for Sustainability, 72100 Brindisi, Italy; valerio.miceli@enea.it
3 Department of Environmental Biology, Sapienza University of Rome, Piazzale Aldo Moro 5, 00185 Rome, Italy
* Correspondence: carlogiuseppe.rizzello@uniroma1.it; Tel.: +39-0805442945

Citation: Pontonio, E.; Montemurro, M.; De Gennaro, G.V.; Miceli, V.; Rizzello, C.G. Antihypertensive Peptides from Ultrafiltration and Fermentation of the Ricotta Cheese Exhausted Whey: Design and Characterization of a Functional Ricotta Cheese. *Foods* **2021**, *10*, 2573. https://doi.org/10.3390/foods10112573

Academic Editor: Ioan Cristian Trelea

Received: 4 October 2021
Accepted: 22 October 2021
Published: 25 October 2021

Publisher's Note: MDPI stays neutral with regard to jurisdictional claims in published maps and institutional affiliations.

Abstract: Aiming at valorizing the ricotta cheese exhausted whey (RCEW), one of the most abundant by-products from the dairy industry, a biotechnological protocol to obtain bioactive peptides with angiotensin-I-converting enzyme (ACE)—inhibitory activity was set up. The approach was based on the combination of membrane filtration and fermentation. A *Lactobacillus helveticus* strain selected to be used as starter for the fermentation of the ultrafiltration protein-rich retentate (R-UF) obtained from RCEW. The fermented R-UF was characterized by a high anti-ACE activity. Peptides responsible for the bioactivity were purified and identified through nano-LC–ESI–MS/MS. The sequences identified in the purified active fractions of the fermented R-UF showed partial or complete overlapping with previously reported κ-casein antihypertensive fragments. The fermented R-UF was spray-dried and used to enrich ricotta cheese at different fortification level (1 and 5% *w*/*w*). An integrated approach including the assessment of the microbiological, chemical, functional, textural, and sensory properties was used to characterize the fortified products. A significantly higher anti-ACE activity was found in the ricotta cheese fortified with fermented R-UF as compared to the control and to the samples obtained with the unfermented R-UF fraction at the same levels of fortification. In particular, a 100 g portion of the ricotta cheese produced at 5% fortification level contained circa 30 mg of bioactive peptides. The fortification led to a moderate acidification, increased hardness and chewiness, and decreased the milk odor and taste of the ricotta cheese as compared to the control, while flavor persistence and sapidity improved.

Keywords: ricotta cheese exhausted whey; anti-ACE activity; fermentation; bioactive peptides; food by-products

1. Introduction

The management of food waste and by-products, due to the global increase of population and food consumption, is a challenge for the agri-food industry that faces growing economic costs for the treatment and/or disposal of waste and stringent environmental regulations [1].

Among different production activities [1], the dairy industry annually produces millions of tons of by-products, mainly represented by cheese whey (CW), which corresponds to the liquid fraction remaining after milk coagulation. Around 9–10 L of whey results from the production of 1 kg of cheese. Due to its consistent organic load [2,3], if discarded without treatment, CW creates significant problem for the environment. Besides disposal, CW is used for feed, and to a small extent as substrate for the recovery or the synthesis (through physical, chemical, and biotechnological processes) of molecules having nutritional and pharmaceutical potential. However, in many countries such as Portugal, Spain, Italy, and Turkey, CW is employed to produce whey-derived cheeses (e.g., requeijão, requesón, ricotta, and lor, respectively) [4–6]. Acidification and heating at 85–90 °C for 20–30 min of the CW, to allow coagulation and subsequent precipitation of whey proteins and separation of whey cheese mass, are used in the production of ricotta [7–9].

Although depending on the origin of the whey and the process employed, the whey cheese yield is lower than 4%, unless whey is previously concentrated. The liquid remaining after whey cheese separation ("cheese exhausted whey", CEW) represents more than 90% of the original whey and it is the main by-product of the whey cheese production chain. CEW obtained from the ricotta production is called "Scotta" in Italy, and commonly defined as ricotta cheese exhausted whey (RCEW).

As is the case of CW, since it is characterized by high values of BOD (biochemical oxygen demand) and COD (chemical oxygen demand) (circa 50 and 80 g/L, respectively), CEW is considered highly polluting [9,10]. Indeed, if discarded into water sources, it reduces the dissolved oxygen, and represents a risk to aquatic life, as well as to the environment and human health [11]. Due to the high content of lactose (35–50 g/L), which cannot be digested by most animals without suffering from digestive disorders [9,12], the valorization of CEW represents a considerable problem. Moreover, small- and medium-size dairy industries often lack dimension to make the necessary investments for CEW valorization [2,8]. Nevertheless, CEW is considered as a source of functional and bioactive compounds, especially proteins and peptides [13–15].

The concentration processes required by the relatively low content of proteins in CEW and needed to ensure high hydrolysis yields, are some of the main problems impairing its industrial reuse [16]. Several methods, such as ultrafiltration, diafiltration, nanofiltration, ion exchange chromatography, electrophoresis, crystallization, and precipitation have been proposed to concentrate and separate proteins from other components contained in CW and CEW. These techniques could be applied for the recovery of the major compounds (β-lactoglobulin and α-lactalbumin), but also for minor compounds, such as lactoperoxidase, lactoferrin, immunoglobulins, as well as protein derivatives such as the biologically active peptides obtained by fermentation with either selected proteolytic microorganisms or the enzymatic or chemical hydrolysis of proteins [17].

In this work, a biotechnological protocol for obtaining bioactive peptides with ACE-inhibitory activity was set up by using a protein-rich fraction obtained from RCEW through a membrane ultrafiltration (UF) process [18] as substrate. A proper lactic acid bacteria (LAB) strain was selected to be used as a starter for the fermentation of RCEW. Bioactive peptides were purified and identified. The fermented RCEW active fraction was spray-dried and used to enrich ricotta cheese at different fortification level (1 and 5% *w/w*). Functional ricotta cheese, produced at a pilot plant level, was characterized by an integrated approach including the assessment of the microbiological, chemical, functional, textural, and sensory properties.

2. Materials and Methods

2.1. Microrganisms

The 5 strains of lactic acid bacteria (LAB) used in this study are listed in Table S1. Strains were previously isolated from cheeses: "Flor di capra" (sheep milk cheese) [19], Pecorino Umbro (sheep milk cheese) [19], Caciocavallo Pugliese [20], Parmigiano Reggiano [21], and fermented milk [22] and identified by partial sequencing of the 16S rRNA, recA, pheS, and rpoA genes. *Lacticaseibacillus casei*, *Lactobacillus delbrueckii* subsp. *bulgaricus*, *Lactobacillus helveticus*, and *Lactococcus lactis* strains were cultured on De Man, Rogosa, and Sharpe (MRS, Oxoid, Basingstoke, Hampshire, UK) at 37 °C for 24 h. *Streptococcus thermophilus* strain was grown on M17 containing 0.5% (*w/v*) lactose (Oxoid) at 37 °C for 24 h. All strains belong to the culture collection of the Department of Soil, Plant and Food Sciences (DiSSPA) of the University of Bari (Italy) and were maintained as stocks in 15% (*v/v*) glycerol at −80 °C.

2.2. RCEW Fermentation

Caseificio dei Colli Pugliesi (Santeramo, Italy) kindly provided the RCEW that was used as substrate for the starter selection. The RCEW (aliquot of 200 L) was pasteurized (72 °C for 15 s) immediately after ricotta cheese production using an industrial plate pasteurizer (Alfa Laval, Lund, Sweden). After pasteurization, the RCEW was cooled down at 4 °C in less than 16 min. No protein flocculation occurred during thermal treatment. The RCEW was then stored at 4 °C and fermented within 24 h from pasteurization.

Strains were singly used to ferment the RCEW. Cells were harvested by centrifugation (10,000× *g*, 10 min, 4 °C), washed twice in 50 mM phosphate buffer, pH 7.0, and re-suspended in RCEW at a final cell density of circa 7.0 log10 cfu/mL of sample. All the samples were incubated for 24 h at 30 or 37 °C, according to the strain used.

Values of pH prior and after the fermentation were measured using a pH meter M.507 (Crimson, Milan, Italy). The cell density of presumptive LAB was enumerated using either the MRS or M17 medium according to the strain inoculated (Table S1).

Plates were incubated under anaerobiosis (AnaeroGen and AnaeroJar, Oxoid) at 30 or 37 °C, according to the strain used, for 48 h. The concentration of peptides in RCEW and fermented RCEW was determined through the o-phtaldialdehyde (OPA) method as described by Church et al. [23]. Aiming at removing proteins and free amino acids, treatments with trifluoroacetic acid (TFA, 0.05% *w*/*v*) and dialysis (cutoff 500 Da) were performed, respectively. The total free amino acids (TFAA) concentration was determined using a Biochrom 30 series amino acid analyzer (Biochrom Ltd., Cambridge Science Park, UK) with a Li-cation-exchange column (0.46 cm internal diameter), by post column derivatization with ninhydrin, as described by Rizzello et al. [24]. Fermentations and analysis were carried out in triplicate.

2.3. RCEW ACE-Inhibitory Activity

ACE-inhibitory activity in the RCEW was analyzed prior and after the fermentation by the method reported first by Cushman and Cheung [25] and Wu and Ding [26] and modified by Rizzello et al. [27]. Compared to the original method, the new analytical procedure does not include the ethyl acetate extraction. Previously, it was successfully used for the analysis of protein hydrolysates [28]. Two hundred microliters of 5 mM hippuryl-L-histidyl-L-leucine (HHL) solution (obtained in a Na-borate buffer, pH 8.3, supplemented with 300 mM NaCl) were mixed with 60 μL of the peptides mixture and 40 μL of 100 mU/mL ACE (in a phosphate buffer at 10 mM, pH 7.0, supplemented with 500 mM NaCl). Water was used in the negative control instead of samples. Incubation of the reaction mixture was carried out at 37 °C for 60 min; to stop the reaction, 250 μL of 1 M HCl were added at the end of the incubation. The mixture was then analyzed by HPLC, using a Resource RPC C18 column (6.4 × 100 mm, particle size 15 μm; GE Healthcare Bio-Sciences AB), aiming at quantifying separately the product and hippuric acid (HA) from HHL. In detail, elution was carried out by using a mobile phase composed of water and acetonitrile (CH_3CN) containing 0.05% TFA. The flow rate was 1 mL/min, while the detection was carried out at 228 nm. The analysis was carried out by setting a linear CH_3CN gradient from 5 to 46% (between 16 and 62 min). The inhibition activity was calculated using the following equation:

$$\text{inhibition activity (\%)} = ((\text{Pc} - \text{Ps})/(\text{Pc} - \text{Pb}))100, \quad (1)$$

where Pc is the HA-peak area of the control, Ps is the HA-peak area of the reaction mixture (sample), and Pb is the HA-peak area of the reaction mixture without ACE.

2.4. RCEW Fractionation and Characterization

RCEW was fractionated at the pilot plant of the ENEA (Brindisi, Italy), delivered under refrigerated conditions (4 ± 2 °C), and processed within 24 h from production [29].

A multistep fractionation process based on a separative membrane process previously proposed by Raho et al. [18] was applied to RCEW, aiming at collecting separately a protein-rich fraction, a lactose-rich fraction, and ultrapure water. From each fractionation step, two fractions were obtained: (i) aqueous permeate (P) containing the molecules able to cross the membrane and (ii) retentate (R) containing a residual part of these and all the molecules unable to cross the membrane. The retentate of the first fractionation step (R-UF), rich in proteins, was collected and used in this work. The ultra-filtration (UF) was carried out by a prototypal system equipped with a spiral wound PESH (polyethersulfone) membrane (30 kDa cutoff) ISUH030 4040 C1 (Microdyn-Nadir, Wiesbaden, Germany), with a spacer of 44 mil (1.117 mm) and an area of 6 m^2. The process was carried out at the flow rate of 3000 L/h, at 14 °C. The transmembrane pressure (TMP) corresponded to 3.47 bar while permeate flow was 15.4 $L/h{\cdot}m^2$.

Fraction recover was expressed as % *v*/*v*, while the volume concentration ratio (VCR) was calculated as the ratio between initial feed volume and retentate volume. Before analysis, all the RCEW fractions obtained were stored at −20 °C. All the analyses were performed in triplicate.

The pH of the RCEW and R-UF was determined as reported above. A standard ISO 8968-1 Kjeldahl-based method [30] was used to quantify the total nitrogen (TN). The calculation of total proteins was carried out by using 6.38 as the conversion factor. Sugars (glucose, galactose, and lactose) were determined by HPLC analysis, using an ÄKTA purifier HPLC (GE Healthcare) as reported by Verni et al. [31]. Commercial standards of lactose, glucose, and galactose (Sigma Aldrich, Milano, Italy) were used for the identification of the sugars and for obtaining the calibration curves. The TFAA concentration was determined using a Biochrom 30 series amino acid analyzer as described above. The international standards methods Gerber [32] and AOAC 945.46 [33] were used to quantify fat and ash, respectively. The sodium chloride content in the R-UF was measured by a Sherwood 926 chloride-analyzer (Sherwood Scientific, Cambridge, UK).

2.5. R-UF Fermentation

The R-UF was fermented with *L. helveticus* PR4, selected for its ability to increase ACE-inhibitory activity of the RCEW in the preliminary screening. In details, the R-UF was inoculated as described above (final cell density of circa 7.0 log10 cfu/mL) and fermentation was carried out at 37 °C for 24 h. Lactose, protein and peptide concentrations, and ACE-inhibitory activity were determined as described above.

2.6. Spray-Drying

Unfermented and fermented R-UF were spray-dried using a pilot plant spray dryer (Mini Spray Dryer B-290, Büchi, Switzerland) at a drying rate of 1.0 L/h. The spray-drying system included a fluid nozzle (0.7 mm diameter) to atomize liquid feed into fine droplets, and a drying chamber (16.5 cm diameter, 45 cm height) in which atomized liquid was dried by a flow of hot air. Two cyclone separators were used for collecting powder. The first separator collects coarser particles, and the fine and ultrafine particles were recovered by the first and the second separators, respectively. A peristaltic pump was used to feed the system with a controlled flow rate. The inlet and outlet temperatures of the spray dryer system were 200 and 180 °C, respectively. For each spray-drying experiment, 50 to 100 mL of sample (preconditioned at 25 °C) was pumped with a feed flow rate fixed at 10 mL/min. Pressure ranged from 5 to 8 bar. Dried powders were collected in the glass bottle connected to the separators and then stored until further analysis in airtight containers.

2.7. Purification and Identification of the Bioactive Peptides

An aliquot of fermented R-UF corresponding to 15 mg of peptides was automatically fractionated (2 mL per fraction, 33 fractions for each run) by reversed-phase fast performance liquid chromatography (RP-FPLC), using a Resource RPC column and ÄKTA FPLC equipment, with the UV detector operating at 214 nm (GE Healthcare Bio-Sciences AB,

Uppsala, Sweden). Elution was carried out using a mobile phase composed of water and acetonitrile (CH_3CN), containing 0.05% TFA (1 mL/min flow rate). The analysis was carried out using a gradient elution (CH_3CN concentration in mobile phase was increased from 5 to 46% between 16 and 62 min, and from 46 to 100% between 62 and 72 min).

Fractions were freeze-dried to remove solvents and re-dissolved in sterile water to determine the peptide concentration through the OPA method. Each fraction was also subjected to the ACE-inhibitory assay as reported above.

The peptides contained in the fractions with the highest ACE-inhibitory activity were further purified and identified. The analysis was performed by nano-LC–ESI–MS/MS (nano-liquid chromatography–electrospray ionization–mass spectra/mass spectra), by using a ion trap mass spectrometer (Finnigan LCQ Deca XP Max, Life Technologies GmbH, Darmstadt, Germany, ThermoElectron). A nano-ESI interface was used. MS spectra were automatically recorded through Xcalibur software (Life Technologies GmbH, Darmstadt, Germany, ThermoElectron), in positive ion mode, following the manufacturer's instrument settings. The software program BioWorks 3.2 (Life Technologies GmbH, Darmstadt, Germany, ThermoElectron) was used for MS/MS spectra processing. Peptides were identified through the Mascot search engine (Matrix Science, London, UK) using the NCBIProt database (National Centre for Biotechnology Information, Bethesda, MD, USA). Settings used for peptides identification were: instrument type, "ESI-trap"; enzyme, "none"; peptide mass tolerance, ±0.1%; fragment mass tolerance, ±0.5 Da. Results were screened as described by Chen et al. [34]. Validated peptide sequences explained all the major peaks in the MS/MS spectrum.

2.8. Ricotta Production

Experimental ricotta cheese was produced at industrial level at "Caseificio dei Colli Pugliesi" (Santeramo, Bari, Italy) from December 2019 to March 2020. The whey, collected from mozzarella making, was strained and then heated at 85–90 °C under stirring conditions. When proteins started to flocculate, stirring was stopped to promote the formation of aggregates. The curd was then scooped, moved into perforate molds, and kept for 30 min in a cool room (4 °C) for draining. Then, 5 kg portions of the clot were gently mixed with the spray-dried R-UF and fermented R-UF (fR-UF). Five different types of ricotta cheese were obtained: ricotta cheeses supplemented with 1% and 5% (*w*/*w*) of the spray-dried fermented fR-UF (RCf1 and RCf5); ricotta cheeses supplemented with 1% and 5% (*w*/*w*) of the spray dried unfermented R-UF (RC1 and RC5), and a control sample produced without a R-UF addition. After R-UF addition, ricotta cheeses were divided into 100 g portions, placed in perforate molds, and stored at 4 °C until analysis. Three ricotta productions were done on three different days.

2.9. Ricotta Characterization

The pH values of ricotta samples were determined as reported above. Ricotta samples were homogenized in a home pulverizing machine. The ash contents of whey samples were determined according to AOAC [33]. The determination of fat (% *w*/*w*) was carried out using the Gerber–Van Gulik method [32]. The solid non-fat content in the whey was determined by using the following formula: total solid% − fat%. The total nitrogen was determined using the Kjeldahl method and converted into protein percentage using the conversion factor 6.38 (the nitrogen-to-protein conversion factor for milk and dairy products) [30]. The moisture content and total dry matter of ricotta were determined by drying samples on a BRASIMET Model ESE 35 stove at 105 °C for 12 h to achieve a constant weight [35].

For the microbial analysis, 10 g of each ricotta cheese were suspended in 90 mL of sterile sodium chloride (0.9% *w*/*v*) solution and homogenized in a BagMixer 400P (Interscience, St Nom, France) at room temperature to enumerate the microbial cell number. Serial 10-fold dilutions were then plated into Plate Count Agar (PCA, Oxoid, Basingstoke, Hampshire, UK) supplemented with cycloheximide (0.1 g/L), MRS (Oxoid) supplemented with cycloheximide (0.1 g/L), Violet Red Bile Glucose Agar (VRBGA, Oxoid), and Slanetz and Bartley Agar (Oxoid) used to enumerate total thermophilic bacteria, presumptive thermophilic LAB, *Enterobacteriaceae* and enterococci, respectively. Except for *Enterobacteriaceae*, which were enumerated after 24 and 48 h, all microbial groups were counted after 48 h incubation at 37 °C. Yeasts and molds were enumerated on malt extract (Oxoid) and wort agar (Oxoid), respectively, after incubation for 48 h at 25 °C.

Water-soluble extracts of the ricotta samples were prepared according to the method of Kuchroo and Fox [36] with some modifications. Ricotta samples were mixed with a phosphate buffer (0.05 M and pH 7) in a ratio 1:2 (*w*/*v*) and homogenized in a BagMixer 400P (Interscience, St Nom, France) at room temperature using a stomacher (PBI International, Milano, Italia). Then, samples were incubated at 40 °C for 1 h under stirring condition (75 rpm) and centrifuged at 4500 rpm for 30 min at 4 °C. The supernatant was filtered (0.22 μm), the pH adjusted with 0.1 M HCl to pH 4.6, and centrifuged at 11,200× *g* for 10 min at 4 °C.

The 70% ethanol-soluble extract was prepared from the water-soluble extract by adding absolute ethanol to a final ethanol concentration of 700 mL/L. The mixture was held for 30 min at 20 °C and then centrifuged (3000× *g* for 30 min). The supernatant was filtered through Whatman no. 1 filter paper. Rotary evaporation under vacuum (model no. RE100, Bibby Sterlin Ltd., Stone, UK) at 30 °C was used for removing ethanol from the extract. The ethanol-soluble and -insoluble extracts were resuspended in water, lyophilized, and stored at 4 °C until further analyses.

The peptide profile of the ethanol-soluble fraction was determined by RP-HPLC using a Waters 626 system equipped with a Waters 600 controller and a Waters 717 plus autosampler (Waters Corp., Milford, MA, USA). Guard and analytical columns used were Nucleosil C8 (5 μm particle, 300 Å pore size), with 4.6 × 10 mm × mm and 4.6 × 250 mm diameter, respectively (Macherey-Nagel GmbH, Duren, Germany). A Varian 9050 UV–Vis Detector (Varian Inc., Walnut Creek, CA, USA) was used for detection at 214 nm. Chromatograms were recorded by the Millennium software program (Waters). The chromatographic conditions were as follows: solvent A, 1 mL/L TFA (Sigma, St. Louis, MO, USA) in deionized, HPLC-grade water (Milli-Q system, Waters Corp.); solvent B, 1 mL/L TFA in CH_3CN (HPLC-grade, Labscan Ltd., Dublin, Ireland). Samples (4 mg/mL) were dissolved in solvent A and filtered through 0.45 μm cellulose acetate filters (Sartorius GmbH, Gottingen, Germany). Aliquots of 40 μL were loaded onto the column and eluted (0.75 mL/min flow rate) using the gradient previously proposed by Shakeel-Ur-Rheman et al. [37]

An ACE-inhibitory assay was performed on the ethanol-soluble fraction as described above.

2.10. Sensory Analysis

Sensory analysis of ricotta samples was carried out by 10 semi-trained panelists, with an equal distribution of men and women ranging in age between 25 and 40 years. Before the analysis, 30 panelists underwent a minimum of 5 h of training and were chosen based on their capacity to distinguish and describe the sensory attributes listed in Table S2.

The 5 h training was performed as follows: (i) the assessors attended a short class on the type of ricotta cheese to be evaluated; (ii) a clear definition of sensory attributes was provided along with the presentation and evaluation of the different physical references (Table S2). Physical and written definitions were available to panelists during all sensory evaluations.

The lexicon consisted of nineteen attributes and followed the Lawless and Heymann [38] guidelines. Ricotta samples were randomly coded and served at 18 to 20 °C in portions of 20 g, together with non-salted table biscuits and still water, to panelists placed separately in rooms for unbiased evaluation of sensory attributes. Ricotta samples were scored from 1 (very unpleasant) to 7 (excellent and fresh) where the value 4 was set as the minimum threshold of sensory acceptability, in agreement with what is reported in the literature for the sensory analysis of dairy products [39,40]. The study protocol followed the ethical guidelines of the sensory laboratory. A written informed consent was obtained from each participant.

2.11. Texture Profile Analysis of Ricotta

The textural properties of ricotta were evaluated at room temperature (18 ± 2 °C) by using a FRTS-100N texture analyzer (Imada, Toyohashi, Japan) equipped with a cylinder probe FR-HA-30J on ricotta cheese specimens of 60 ± 1 g. The instrument settings were the following: test speed of 1 mm/s, 30% deformation of the sample, and two compression cycles while the parameters evaluated were hardness (expressed as the maximum force at first compression), cohesiveness (ratio of the areas of the second and the first compression peak), springiness (height of the product on the second compression divided by the height of the first peak), and chewiness (hardness × cohesiveness × springiness).

2.12. Statistical Analysis

All the chemical, microbiological, and physical analyses were carried out in triplicate for each batch of sample. Data were subjected to one-way ANOVA; pair-comparison of treatment means was achieved by Tukey's procedure at $p < 0.05$, using the statistical software package Statistica 12.5 (StatSoft Inc., Tulsa, OK, USA).

3. Results

3.1. LAB Selection

The five strains of LAB were singly inoculated into the RCEW. The acidification and proteolytic activity (quantified as concentration of peptides released during fermentation) were used as selection criteria together with the capability to enhance the ACE-inhibitory activity of the fermented RCEW. The capability of growth of the strain was also assessed through the enumeration of the presumptive LAB. Overall, the pH of fermented RCEW was from 0.5 to 1 unit lower than the CT (Table 1).

Table 1. Characteristics of the RCEW inoculated (7.0 log10 cfu/mL) with selected LAB.

	RCEW					
	CT	**FC13**	**B15Z**	**CR12**	**PR4**	**DIBCA2**
pH	4.87 ± 0.05 ^{a}	4.10 ± 0.04 ^{c}	3.92 ± 0.04 ^{d}	4.02 ± 0.05 ^{d}	3.76 ± 0.05 ^{e}	4.26 ± 0.05 ^{b}
LAB density (log10 cfu/mL)	3.9 ± 0.1 ^{d}	8.5 ± 0.2 ^{a}	7.9 ± 0.2 ^{b}	8.1 ± 0.1 ^{b}	9.0 ± 0.2 ^{a}	7.5 ± 0.1 ^{c}
Peptides (mg/L)	364 ± 4 ^{e}	703 ± 10 ^{c}	775 ± 15 ^{b}	819 ± 9 ^{a}	795 ± 12 ^{b}	525 ± 6 ^{d}
ACE-inhibitory activity (%)	7.2 ± 0.2 ^{d}	16.9 ± 0.2 ^{c}	26.5 ± 0.2 ^{b}	7.9 ± 0.4 ^{d}	68.4 ± 0.6 ^{a}	7.5 ± 0.3 ^{d}

The data are the means of three independent experiments ± standard deviations ($n = 3$). $^{a–e}$ Values in the same row with different superscript letters differ significantly ($p < 0.05$). Fermentation was carried out at 30 °C (*L. delbrueckii* subsp. *bulgaricus* B15Z and *L. lactis* DIBCA2) or 37 °C (*L. casei* FC13, *L. helveticus* PR4, and *S. thermophilus* CR12). An uninoculated but incubated (30 °C) RCEW sample was used as control (CT).

Growth from 0.5 (*L. lactis* DIBCA2) to 2 (*L. helveticus* PR4) log10 cfu/mL were observed during the incubation of the RCEW. A significant ($p < 0.05$) and markedly lower cell density was found in CT (3.9 ± 0.1 log10 cfu/mL). Increases from circa 40 (*L. lactis* DIBCA2) to 125% (*L. helveticus* PR4 and *S. thermophilus* CR12) were found in fermented RCEW (Table 1). Nevertheless, the highest peptide concentration did not correspond to the highest ACE-inhibitory activity. Indeed, the RCEW fermented with *L. lactis* DIBCA2 and *S. thermophilus* CR12 did not show ($p > 0.05$) significant increases of the ACE-inhibitory activity compared to the CT (Table 1). Significantly ($p < 0.05$) higher values (2 and 4 times) were found for *L. casei* FC13 and *L. delbrueckii* subsp. *bulgaricus* B15Z, while an activity circa 10 times higher than CT was found for the RCEW fermented with *L. helveticus* PR4. According to the data reported above, *L. helveticus* PR4 was selected and used for further experiments.

3.2. RCEW Fractionation

The RCEW was subjected to ultrafiltration, as previous described by Raho et al. [18]. The RCEW's initial pH was 5.2 (Table 2).

Table 2. Proximal composition of ricotta cheese exhausted whey (RCEW), ultrafiltration protein-rich retentate (R-UF), and ultrafiltration permeate (P-UF).

	RCEW	R-UF	P-UF
pH	5.2 ± 0.1 [a]	5.2 ± 0.2 [a]	5.2 ± 0.1 [a]
Total Nitrogen (mg/L)	60.02 ± 0.54 [b]	290.02 ± 4.05 [a]	1.34 ± 0.02 [c]
Proteins * (% *w*/*v*)	0.38 ± 0.06 [b]	1.85 ± 0.05 [a]	0.08 ± 0.01 [c]
Total Free Amino Acids (mg/L)	417 ± 5 [b]	52 ± 2 [c]	516 ± 9 [a]
Lactose (% *w*/*v*)	3.80 ± 0.10 [b]	5.19 ± 0.22 [a]	3.42 ± 0.11 [c]
Glucose (% *w*/*v*)	<0.01	<0.01	<0.01
Galactose (% *w*/*v*)	<0.01	<0.01	<0.01
Fat (% *w*/*v*)	0.16 ± 0.10 [b]	0.75 ± 0.04 [a]	<0.01
Ash (% *w*/*v*)	1.08 ± 0.12 [b]	0.23 ± 0.08 [c]	1.35 ± 0.11 [a]

The data are the means of three independent experiments ± standard deviations (n = 3). [a–c] Values in the same row with different superscript letters differ significantly ($p < 0.05$). * Protein content is calculated as total nitrogen × 6.38.

Most of the nitrogen and fat of the RCEW was retained in the protein-rich retentate (R-UF), as a result of the separation process. R-UF corresponded to the 20.3% *v*/*v* of the processed RCEW, with a volume concentration ratio (VCR) of 4.92.

The protein concentration of the fraction, estimated based on the TN content (290 ± 4 mg/L) corresponded to 1.85 ± 0.05% *w*/*v*, while fat was 0.75 ± 0.04% *w*/*v*. The retentate was moreover characterized by a relevant lactose content (5.19 ± 0.22% *w*/*v*) (Table 2).

As expected, a low TN concentration characterized the UF permeate (P-UF). The remaining TN corresponded to organic compounds having molecular mass lower than 30 kDa (mainly peptides and free amino acids, FAA). In particular, the TFAA concentration in P-UF corresponded to 516 mg/L. The ash concentration of the permeate was significantly ($p < 0.05$) higher than that of the RCEW since fractionation and separation of the R-UF allowed a partial water removal.

3.3. Fermentation and Characterization of the Protein-Rich RCEW Fraction

Aiming at releasing the highest concentration of bioactive peptides, *L. helveticus* PR4 was used to ferment the R-UF fraction obtained from the RCEW. The main characteristics of the R-UF (prior the fermentation) and fR-UF (after the fermentation) are summarized in Table 3.

Table 3. Characteristics of the RCEW fractions R-UF (prior fermentation) and fR-UF (after fermentation with *Lactobacillus helveticus* PR4 at 37 °C for 24 h).

	R-UF	fR-UF
pH	5.2 ± 0.1 [a]	3.82 ± 0.2 [b]
LAB density (log10 cfu/mL)	7.21 ± 0.2 [b]	8.82 ± 0.4 [a]
Peptides (mg/L)	65 ± 5 [b]	3230 ± 25 [a]
TFAA (mg/L)	50 ± 2 [b]	896 ± 12 [a]
Lactose (%)	5.2 ± 0.2 [a]	3.5 ± 0.1 [b]
ACE-inhibitory activity (%)	nd	88.2 ± 1.1

LAB, thermophilic presumptive lactic acid bacteria; TFAA, total free amino acids. The data are the means of three independent experiments ± standard deviations (n = 3). [a,b] Values in the same row, with different superscript letters, differ significantly ($p < 0.05$).

A total bacteria cell density of 2.65 log10 cfu/mL characterized the R-UF before the inoculum. During fermentation, *L. helveticus* PR4 reached a cell density of 8.8 ± 0.4 log cfu/mL, while the pH decreased by circa 1.5 units (Table 3). The fR-UF was characterized by concentrations of peptides and TFAA 50 and 18 times higher than the R-UF. The concentration of the lactose was also significantly ($p < 0.05$) lower (circa 33%). The fR-UF was characterized by an ACE-inhibitory activity of 88.2 ± 1.1% (Table 3).

The spray-drying process allowed to recover circa 6.95 ± 0.10 and 7.80 ± 0.08% (*w*/*v*) powder from the R-UF and fR-UF, respectively, both having circa 7% moisture. Spray-dried

R-UF and fR-UF were characterized by similar TN (3.54 ± 0.12 and 3.56 ± 0.09 g/kg, respectively). According to the conventional calculation (applying the conversion factor 6.38 to TN), the protein concentration theoretically corresponded to 22.58% and 22.71% (*w*/*w*) of the spray-dried R-UF and fR-UF, respectively. Nevertheless, this estimation is approximate, and the analysis revealed peptides and TFAA concentrations markedly higher in the fR-UF (46.44 vs. 0.83 g/kg of peptides and 12.80 vs. 0.62 g/kg of TFAA), due to the LAB proteolysis occurring during fermentation. As expected, the spray dried R-UF and fR-UF were characterized by relevant concentration of lactose (65.5 and 50.04%, respectively) and fat (circa 9.0% for both). The LAB density in the R-UF and fR-UF was <10^4 cfu/g.

3.4. ACE-Inhibitory Peptides

The fR-UF was fractionated by RP-FPLC; then, fractions were freeze-dried, resuspended in 1 mL of sterilized water, and assayed for assessing the ACE-inhibitory activity. Four of the thirty fractions (Figure 1) showed ACE-inhibitory activity, corresponding to 68.11 (fraction A), 54.49 (fraction B), 35.05 (fraction C), and 36.55% (fraction D) (Figure 1).

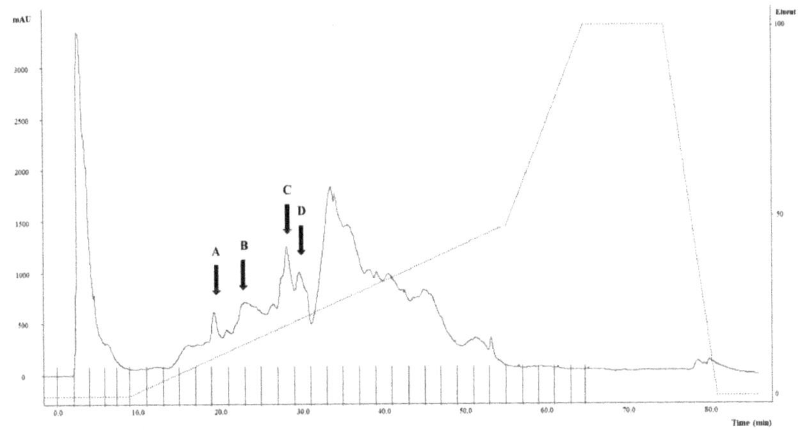

Figure 1. RP-FPLC chromatogram (UV detector 214 nm) showing the WSE fractions obtained from the purification of the fermented R-UF. The gradient of eluent B is represented by the dashed line. A–D refer to the fractions with anti-ACE activity.

The peptides concentration of the fractions was 0.32, 0.43, 0.52, and 0.92 mg/mL for A, B, C, and D, respectively. Two out of four extract fractions contained a mixture of peptides. Indeed, fractions A and D contained four and two peptides, respectively. The identified peptides, having 12–25 amino acid residues, were characterized by molecular mass ranging from 1304.4 (AQPTDASAQFIR) to 2657.9 Da (NQDKTEIPTINTIASGEPTSTPTIE) and hydrophobic ratio between 32 (DETHLEAQPTDASAQ) and 49% (VIESPPEINTVQVTSTAV and AQPTDASAQFIR). The net charges at pH 7 (antimicrobial peptide calculator and Predictor-APD3) [41] was negative for all peptides except for RHPYFYAPELLYYANK and AQPTDASAQFIR having positive and neutral values, respectively (Table 4).

3.5. Ricotta Cheeses Characterization

Microbiological properties of the ricotta cheeses were also investigated. Overall, no significant ($p > 0.05$) differences were found among samples. The microbial cell density ranged from circa 2.5 ± 0.3 (yeast, molds, and *Enterococcus*) to 3.5 ± 0.2 (LAB and *Enterobacteriaceae*) log10 ufc/g. Total bacteria cell density was 3.8 ± 0.3 log10 ufc/g.

Biochemical and nutritional properties of the ricotta cheeses are summarized in Table 5.

Table 4. List of the peptides identified in the partially purified peptide fractions obtained from the fR-UF through RP-FPLC. Fermentation was carried out with *Lactobacillus helveticus* PR4 at 37 °C for 24 h.

Fraction	Sequence	Mass (Da)	Length (aa)	Net Charge *	Hydrophobic Ratio (%)	NCBI Accession n° (Protein)
A	TIASGEPTSTPTTEA	1462.532	15	−2	38	CAF03625.1 (kappa-casein)
	DAFLGSFLYEYSR	1567.721	13	−1	44	P02769.4 (serum albumin)
	RHPYFYAPELLYYANK	2045.327	16	1.24	42	
	DETHLEAQPTDASAQ	1612.63	15	−3.75	32	ABY26541.1 (glycosylation-dependant cell adhesion molecule 1)
B	ILNKPEDETHLEAQPT	1835.001	16	−2.75	36	AAB27385.1 (PP3 homolog)
C	VIESPPEINTVQVTSTA	1884.118	17	−2	49	CAF03625.1 (kappa casein)
D	NQDKTEIPTINTIASGEPTSTPTIE	2657.869	25	−3	36	ACF15188.1 (kappa casein)
	AQPTDASAQFIR	1304.428	12	0	49	AAB27385.1 (PP3 homolog)

* The net charge was calculated at pH 7.

Table 5. Biochemical, nutritional, and textural characteristics of the ricotta cheeses.

	Ricotta cheeses				
	RC	RC1	RC5	RCf1	RCf5
	Biochemical and nutritional characteristics				
pH	5.02 ± 0.03 ^{a}	5.02 ± 0.02 ^{a}	5.02 ± 0.03 ^{a}	4.95 ± 0.03 ^{b}	4.88 ± 0.04 ^{b}
TTA (mL NaOH 0.1 M)	8.1 ± 0.2 ^{b}	8.1 ± 0.2 ^{b}	8.2 ± 0.3 ^{b}	8.5 ± 0.1 ^{b}	8.9 ± 0.3 ^{a}
Moisture (%)	76.6 ± 0.5 ^{a}	76.5 ± 0.7 ^{a}	73.1 ± 0.5 ^{b}	76.2 ± 0.8 ^{a}	73.0 ± 0.5 ^{b}
Proteins [1] (% *w*/*v*)	9.8 ± 0.5 ^{b}	10.3 ± 0.6 ^{ab}	10.9 ± 0.7 ^{a}	10.3 ± 0.4 ^{ab}	11.0 ± 0.6 ^{a}
Fat (% *w*/*v*)	10.2 ± 0.3 ^{a}	10.3 ± 0.4 ^{a}	10.7 ± 0.4 ^{a}	10.3 ± 0.3 ^{a}	10.7 ± 0.5 ^{a}
Carbohydrates (% *w*/*v*)	3.5 ± 0.3 ^{b}	3.9 ± 0.4 ^{b}	6.3 ± 0.2 ^{a}	3.7 ± 0.3 ^{b}	5.8 ± 0.3 ^{a}
Ash (% *w*/*v*)	2.4 ± 0.2 ^{a}	2.5 ± 0.3 ^{a}	2.6 ± 0.2 ^{a}	2.5 ± 0.1 ^{a}	2.6 ± 0.3 ^{a}
ACE-inhibitory activity	7.2 ±0.2 ^{c}	7.5 ± 0.3 ^{c}	7.5 ± 0.2 ^{c}	25.1 ± 0.2 ^{b}	64.3 ± 0.2 ^{a}
	Textural Profile Analysis				
Hardness (g)	244.9 ± 11.0 ^{b}	244.1 ± 23.5 ^{b}	357.3 ± 32.0 ^{a}	242.5 ± 14.5 ^{b}	253.2 ± 19.3 ^{b}
Cohesiveness	0.519 ± 0.015 ^{b}	0.468 ± 0.016 ^{c}	0.566 ± 0.034 ^{a}	0.584 ± 0.027 ^{a}	0.601 ± 0.028 ^{a}
Springiness	0.716 ± 0.016 ^{a}	0.694 ± 0.011 ^{b}	0.608 ± 0.036 ^{c}	0.709 ± 0.010 ^{ab}	0.627 ± 0.021 ^{c}
Chewiness (g)	91.0 ± 3.5 ^{a}	79.9 ± 5.2 ^{bc}	100.4 ± 6.9 ^{a}	63.6 ± 10.4 ^{c}	85.0 ± 11.8 ^{ab}

[1] Protein concentration is calculated as total nitrogen × 6.38. The data are the means of three independent experiments ± standard deviations ($n = 3$). $^{a-c}$ Values in the same row, with different superscript letters, differ significantly ($p < 0.05$). RC, control ricotta cheese (without fortification); RC1 and RC5, ricotta cheeses fortified with 1 and 5% (*w*/*w*) spray-dried unfermented R-UF, respectively; RCf1 and RCf5, ricotta cheeses fortified with 1 and 5% (*w*/*w*) fR-UF, respectively.

Overall, the fortification affected some of the characteristics of the samples with the RCf showing the highest differences. As expected, the pH and TTA values were significantly ($p < 0.05$) lower in RCf1 and RCf5 compared to the RC samples, with the lowest value found in RCf5 (Table 5). The highest level of fortification (5% *w*/*w*), regardless the fermentation, led to lower moisture (circa 4%) and carbohydrates concentration significantly ($p < 0.05$) higher than the corresponding samples fortified with 1% (*w*/*w*) of either the R-UF or fR-UF (Table 5). In addition, the protein content was slightly affected by the fortification, with RCf5 showing the highest concentration (circa 12% higher than RC).

A weak ACE-inhibitory activity was found in RC, RC1, and RC5 (from 7.2 to 7.5%), while significant ($p > 0.05$) higher values were found in RCf1 and RCf5 (Table 5). The highest anti-ACE activity (circa nine times higher than RC), was found in RCf5.

Textural profile analysis showed slight differences among samples. The fortification with 1% R-UF or fR-UF did not cause significant ($p > 0.05$) differences in the hardness, compared to control RC. Nevertheless, when 5% (*w*/*w*) R-UF was added to the formulation (RC5), a marked increase in the hardness, cohesiveness, and chewiness was found (Table 5).

RCf5 had a similar ($p < 0.05$) hardness compared to that of RC and a lower hardness and chewiness compared to those of the corresponding RC5 (Table 5).

3.6. Sensory and Textural Characteristics of Ricotta Cheese

The sensory data of the ricotta cheeses are summarized in Table 6.

Table 6. Sensory analysis of ricotta cheeses.

Attributes	RC	RC1	RC5	RCf1	RCf5
			Visual aspect		
Color intensity	5.1 ± 0.2 [b]	5.3 ± 0.4 [b]	5.3 ± 0.3 [b]	5.3 ± 0.4 [b]	6.0 ± 0.4 [a]
Color homogeneity	4.5 ± 0.2 [a]	4.8 ± 0.3 [a]	4.0 ± 0.3 [b]	4.8 ± 0.3 [a]	4.8 ± 0.3 [a]
			Odor		
Milk	5.0 ± 0.3 [a]	4.8 ± 0.4 [ab]	4.8 ± 0.5 [ab]	4.5 ± 0.2 [ab]	3.8 ± 0.4 [c]
Acidic	1.1 ± 0.4 [ab]	1.1 ± 0.4 [ab]	1.1 ± 0.2 [b]	1.5 ± 0.3 [ab]	1.8 ± 0.4 [a]
			Flavor		
Sapidity	3.5 ± 0.3 [b]	3.5 ± 0.4 [b]	3.5 ± 0.4 [b]	4.0 ± 0.3 [ab]	4.3 ± 0.4 [a]
Bitterness	0.5 ± 0.3 [b]	0.5 ± 0.2 [b]	0.5 ± 0.2 [b]	0.5 ± 0.2 [b]	1.1 ± 0.4 [a]
Acidity	4.0 ± 0.3 [c]	4.3 ± 0.3 [c]	4.3 ± 0.3 [c]	5.3 ± 0.4 [b]	6.5 ± 0.4 [a]
Sweetness	2.8 ± 0.5 [a]	2.1 ± 0.2 [b]	2.0 ± 0.4 [b]	2.0 ± 0.3 [b]	2.0 ± 0.3 [b]
Milk	3.0 ± 0.4 [a]	2.8 ± 0.4 [a]	2.8 ± 0.5 [a]	3.0 ± 0.2 [a]	2.7 ± 0.3 [a]
Butter	3.3 ± 0.6 [a]	2.7 ± 0.4 [a]	2.7 ± 0.4 [a]	2.7 ± 0.4 [a]	2.7 ± 0.5 [a]
Off-flavors	0.7 ± 0.3 [a]	0.7 ± 0.3 [a]	0.8 ± 0.4 [a]	0.8 ± 0.4 [a]	0.8 ± 0.2 [a]
Persitency	3.2 ± 0.5 [b]	3.0 ± 0.3 [b]	3.2 ± 0.4 [b]	3.8 ± 0.5 [b]	6.0 ± 0.7 [a]
			Texture		
Adhesiveness	2.7 ± 0.5 [ab]	3.0 ± 0.2 [a]	3.0 ± 0.4 [a]	3.3 ± 0.3 [a]	2.3 ± 0.0 [b]
Graininess	2.5 ± 0.3 [b]	2.8 ± 0.5 [ab]	3.2 ± 0.4 [a]	2.5 ± 0.4 [b]	2.7 ± 0.4 [ab]
Friability	3.2 ± 0.4 [a]	3.3 ± 0.4 [a]	3.3 ± 0.2 [a]	3.3 ± 0.5 [a]	3.3 ± 0.4 [a]
Wetness	5.0 ± 0.2 [a]	5.0 ± 0.4 [ab]	4.7 ± 0.2 [ab]	4.7 ± 0.3 [ab]	4.5 ± 0.2 [b]

The data are the means of three independent experiments ± standard deviations ($n = 3$). [a–c] Values in the same row, with different superscript letters, differ significantly ($p < 0.05$). RC, control ricotta cheese (without fortification); RC1 and RC5, ricotta cheeses fortified with 1 and 5% (w/w) spray-dried unfermented R-UF, respectively; RCf1 and RCf5, ricotta cheeses fortified with 1 and 5% (w/w) fR-UF, respectively.

According to the visual aspect, the fortification led to higher values of both color intensity and homogeneity, especially when the fR-UF was used. Moreover, the fortification led to a decrease of the milk odor, while increasing the acidic flavor. Similar data were found when the flavor was analyzed. Bitterness and off-flavors were almost not found in any of the sample analyzed. However, the persistence of flavor and sapidity was higher in fortified ricotta cheeses than in the RC, with the highest value found in RCf5. When tested, the main difference found between fortified and control ricotta cheeses was in terms of wetness, which was significantly lower than that of the control. The difference was higher when the fR-UF was used (Table 6).

4. Discussion

The dairy industry produces RCEW as one of its most abundant by-products, which is an inexpensive substrate rich in nutrients such as proteins (mainly in a denatured state), soluble peptides, oligosaccharides, lactose, non-protein nitrogen, hydro-soluble vitamins, several minerals, and free amino acids [18,42,43]. Although RCEW has a similar composition to CW, the concentration of its constituents is markedly lower, except for the ash content, often affected by acid and salt added to improve the whey proteins flocculation and aggregation and the ricotta yield [43]. Due to the same reasons (e.g., the addition of sodium bicarbonate to promote flocculation), the RCEW pH is higher than that of whey. Fat is almost completely removed with ricotta production [43]. Here, the multi-step fractionation allowed to retain the fat in the UF retentate and the separation of two fractions: one rich in protein (1.85%, w/v), which can be easily subjected to the recovery of whey proteins to be used as food and feed supplements [18], and one rich in lactose.

RCEW proteins have a globular structure, with a uniform distribution of non-polar, polar, and charged groups; they include β-lactoglobulin, α-lactalbumin, immunoglobulins, serum albumin, lactoferrin, lactoperoxidase, and several enzymes. Other protein components such as glycomacropeptide or caseinomacropeptide, which are released from κ-casein in the first step of enzymatic curdling are also present [1].

Food proteins usually contain, encrypted into their sequences, several biologically active peptides, inactive until bound into the primary structure [44]. Once released through enzymatic proteolysis and/or microbial fermentation, the free forms of the peptides demonstrate health effects in the gut or after systemic absorption into blood circulation.

Bioactive peptides have been reported to lower blood sugar, serum cholesterol, blood pressure, inhibit microbial growth, and hinder cancer development [45]. Among the high number of active sequences identified and characterized from plant and animal sources, antihypertensive peptides deriving from caseins and whey proteins were deeply investigated [46,47]. It was indeed reported that existing synthetic antihypertensive drugs have several side effects, thus pushing the scientific community towards the research of antihypertensive peptides as alternative therapeutics to control systemic blood pressure and to prevent cardiovascular diseases [48]. The key enzyme involved in blood pressure regulation is angiotensin-I-converting enzyme (ACE), a transmembrane metallopeptidase presents in many tissues (lung, thoracic aorta, heart, kidney, and liver) and biological fluids [49]. ACE hydrolyses the decapeptide angiotensin I into the octapeptide angiotensin II, which binds to AT1 receptors on vascular smooth muscles and endothelial cells leading to vasoconstriction and an increase of the blood pressure. Angiotensin II also inactivates the endothelium-dependent vasodilator bradykinin [50]. The inhibition of the ACE activity is therefore the target for antihypertensive agents. Whey proteins such as β-lactoglobulin, α-lactalbumin, bovine serum albumin, and immunoglobulin exhibit diverse physiological functions and their hydrolysates have been shown to have ACE inhibitory activity [46,51]. It was moreover demonstrated that RCEW is a good source of anti-ACE peptides, that can be successfully released by different proteolytic enzymes [52].

Many LAB have been shown to degrade milk proteins to release bioactive peptides during the fermentation occurring in dairy processes. Indeed, LAB are characterized by an efficient proteolytic system able to hydrolyze proteins of the growth matrix into small peptides (2 to 40 amino acid chains) and free amino acids [53,54]. Thanks to their proteolytic activity, LAB strains of the species *Lactobacillus acidophilus, Levilactobacillus brevis, Ligilactobacillus animalis, Lactococcus lactis, Lactobacillus helveticus,* and *Lactiplantibacillus plantarum* were successfully used as microbial catalysts for producing ACE-inhibitory peptides [22,55–57].

In this work, LAB strains previously isolated from dairy products and already selected for their capability to release ACE-inhibitory molecules [19–22,57] were used as starters for the fermentation of the RCEW. Among the five strains used, *L. helveticus* PR4 was selected since it was able to improve the ACE-inhibitory activity of the fermented RCEW, although *S. thermophilus* CR12 led to the release of a higher total amount of peptides.

Due to its high proteolytic activity, lactic acid production, rate of milk acidification, and ability to improve the flavor and texture, strains of *L. helveticus* are often used to produce fermented milk beverages and hard cheeses [58]. The *L. helveticus* strains exhibit strong extracellular proteinase activity and capacity to release relevant amount and types of different peptides during milk fermentation [57,59]. The number of cell envelope proteinases (CEP) varies among *L. helveticus* strains ranging from one to four enzymes [60], such as the pattern of the cytoplasmic peptidases [47]. Several *L. helveticus* strains were previously selected for their capability to release anti-ACE peptides from caseins [58].

In this research work, a *L. helveticus* PR4 strain was employed as selected starter for the fermentation of the R-UF fraction derived from RCEW through membrane separation, confirming the potential of the substrate to be enriched in antihypertensive peptides. Aiming at identifying the peptides responsible for the bioactivity, the fermented R-UF was fractionated by FPLC and the active subfractions characterized by the highest anti-ACE activity, subjected to a purification and nano-LC–ESI–MS/MS analysis.

Eight peptides were identified in the four active fractions, all characterized by a molecular mass ranging from 1.4 to 2.6 kDa. Two of the purified fractions contained mixtures of different peptides. Three of the identified peptides derived from the hydrolysis of the κ-casein. TIASGEPTSTPTTEA and NQDKTEIPTINTIASGEPTSTPTIE, which were identified in partially purified fractions A (eluted in 12% eluent B) and D (eluted in 22% eluent B), respectively, showed partial or complete overlapping with previously reported κ-casein antihypertensive fragments (DQTEIPT, DKTEIPTINTIA, KTEIPTINTIA, TEIPTIN, EIPTINTIA, EIPTINTIAS, PTIN, and PTINTIASSGEP) [58]. Such peptides were purified from a κ-casein hydrolysate obtained using *L. helveticus* CEP cell envelope proteinases. Moreover, the other κ-casein-derived peptide, VIESPPEINTVQVTSTA, was previously reported as antihypertensive, together with other sequences sharing the epitopes VIESPPEIN and TVQVTSTA [46,58,61].

To the best of our knowledge, the two peptides deriving from serum albumin (DAFLGSFLYEYSR and RHPYFYAPELLYYANK), were not previously reported as anti-ACE peptides; nevertheless, serum album hydrolysates produced with different enzymes, such as papain and proteinase-K, were already reported as sources of different antihypertensive sequences [62,63].

In addition, the cell adhesion molecule 1 (GlyCAM-1), known as proteose-peptone 3 (PP3) [64], corresponding to the C-terminal-truncated variant of lactophorin [65] was already reported as source of angiotensin-converting enzyme (ACE)-inhibitory peptides. Three different GlyCAM-1 fragments were identified in the active subfractions of the fermented R-UF, and in particular, a large similarity was found between DETHLEAQPTDASAQ and a previously identified bioactive sequence isolated in RCEW hydrolysates obtained with commercial proteases [52].

Aiming at RCEW valorization through the reuse as food ingredient, the fermented R-UF fraction, enriched in bioactive peptides, was spray-dried and used as functional supplement for making fortified ricotta cheese, at a pilot plant level. The fortified ricotta cheese was characterized by a marked increase of the anti-ACE activity compared to the control and to the ricotta cheese samples obtained with the unfermented R-UF fraction at the same levels of fortification.

Based on the concentration found in the purified active fractions, it can be estimated that bioactive peptides corresponded to circa 14% (*w*/*w*) of the total peptide concentration of the spray-dried fermented R-UF. Consequently, circa 30 mg of bioactive peptides can be found in a 100 g portion of the ricotta cheese produced at a 5% fortification level.

Although the most recent literature shows a positive antihypertensive in vivo effect when the ACE-inhibitory peptides are used at doses of 3.8–52.5 mg/kg [66,67], an analytical quantification of the bioactive peptides in fortified ricotta cheese is needed to make further considerations. Moreover, in vivo and human studies are necessary to elucidate the dose-effect and the mechanisms involved in the activity of these sequences.

Compared to the use of unfermented spray-dried R-UF fraction, the fermentation reflected in a moderate acidification of the final product, especially at the highest level of inclusion in the ricotta cheese formulation (5% *w*/*w*). Obviously, the moisture of the ricotta cheeses was also affected by the addition of the dried RCEW fraction at the highest level of fortification. The supplementation also caused a relevant increase of the protein (calculated on the basis of the total nitrogen) and lactose contents, the latter was still abundant in both unfermented and fermented R-UF, although at a lower level. Probably due to the extensive proteolysis, the use of the fermented spray-dried R-UF attenuated the increase in hardness and chewiness that characterized the ricotta cheese fortified with the 5% unfermented R-UF.

Since peptides generally have a taste, covering the entire range of established taste modalities: sweet, bitter, umami, sour and salty [68], the sensory properties of the fortified ricotta cheese needed to be investigated. The fortification modified the overall characteristics of the ricotta cheese by decreasing the milk odor and taste, while increasing the acidic one. Moreover, the persistence of flavor and sapidity was also improved with the fortification.

5. Conclusions

A biotechnological protocol to obtain bioactive peptides with angiotensin-I-converting enzyme (ACE)-inhibitory activity was set up. A *Lactobacillus helveticus* selected strain was used as starter for the fermentation of the ultrafiltration protein-rich retentate (R-UF) obtained from the RCEW. The fermented R-UF was characterized by a high anti-ACE activity and the peptides responsible for the bioactivity showed partial or complete overlapping with previously reported κ-casein antihypertensive fragments.

When the freeze-dried fermented R-UF was used to enrich ricotta cheese at different fortification levels (1 and 5% *w*/*w*), a high anti-ACE activity was found. Indeed, a 100 g portion of the ricotta cheese produced at 5% fortification level containing circa 30 mg of bioactive peptides.

The fortification led to a moderate acidification, increased hardness and chewiness, and decreased the milk odor and taste of the ricotta cheese as compared to the control, while flavor persistence and sapidity improved.

The biotechnological protocol proposed here for the enrichment in antihypertensive bioactive peptides represent an easily scalable process that can be employed for the upcycle of the RCEW and its components, providing novel supplements for the design of innovative functional foods, with ricotta cheese as one of the many potential applications.

Supplementary Materials: The following are available online at https://www.mdpi.com/article/10.3390//foods10112573/s1, Table S1: List of microorganisms used in this study. The source of isolation and corresponding references are also reported, Table S2: List and definition of the attributes used for the sensory analysis carried out on ricotta cheese samples.

Author Contributions: Conceptualization, C.G.R., V.M. and E.P.; methodology, C.G.R., V.M. and E.P.; formal analysis, G.V.D.G. and M.M.; investigation, G.V.D.G. and M.M.; resources, C.G.R.; data curation, E.P.; writing—original draft preparation, C.G.R. and E.P.; writing—review and editing, C.G.R. and E.P.; supervision, E.P.; project administration, C.G.R.; funding acquisition, C.G.R. All authors have read and agreed to the published version of the manuscript.

Funding: This research was developed under the project "BIOCOSÌ-Sviluppo di tecnologie e processi innovativi per la produzione di imballaggi 100% BIOdegradabili e COmpostabili per un'industria Sostenibile, circolare ed Intelligente", project ID O2N6AE7), funded by Regione Puglia (Italy)-POR Puglia FESR-FSE 2014–2020, INNONETWORK call.

Acknowledgments: Vincenzo Pontrandolfo and Caseificio dei Colli Pugliesi (Santeramo in Colle, Bari, Italy) are gratefully thanked for having provided the RCEW and for their support on the production of the ricotta cheeses. Vito Emanuele Carofiglio and Domenico Centrone (Eggplant Srl, Italy), and Mariantonietta Porcelli and Alessandra Taddeo (Agriplan, Italy) are gratefully thanked for their technical, economic, and administrative support during the project "BIOCOSì".

Conflicts of Interest: The authors declare no conflict of interest. The funders had no role in the design of the study; in the collection, analyses, or interpretation of data; in the writing of the manuscript, or in the decision to publish the results.

References

1. Pires, A.F.; Marnotes, N.G.; Rubio, O.D.; Garcia, A.C.; Pereira, C.D. Dairy by-products: A review on the valorization of whey and second cheese whey. *Foods* **2021**, *10*, 1067. [CrossRef] [PubMed]
2. Castelli, H.; Du Vale, L. *Handbook on Cheese: Production, Chemistry and Sensory Properties*; Nova Science Publishers Inc.: Hauppauge, NY, USA, 2013.
3. Jelen, P. Whey processing. Utilization and products. In *Encyclopedia of Dairy Sciences*; Fuquay, J., Fox, P., McSweeney, P., Eds.; Academic Press: London, UK, 2003.
4. Ryan, M.P.; Walsh, G. The biotechnological potential of whey. *Rev. Environ. Sci. Bio/Technol.* **2016**, *15*, 479–498. [CrossRef]
5. Pires, A.F.; Marnotes, N.G.; Bella, A.; Viegas, J.; Gomes, D.M.; Henriques, M.H.; Pereira, C.J. Use of ultrafiltrated cow's whey for the production of whey cheese with Kefir or probiotics. *J. Sci. Food Agric.* **2021**, *101*, 555–563. [CrossRef] [PubMed]

6. Prazeres, A.R.; Carvalho, F.; Rivas, J. Cheese whey management: A review. *J. Environ. Manag.* **2012**, *110*, 48–68. [CrossRef]
7. Madureira, A.; Pereira, C.; Truszkowska, K.; Gomes, A.; Pintado, M.; Malcata, F. Survival of probiotic bacteria in a whey cheese vector submitted to environmental conditions prevailing in the gastrointestinal tract. *Int. Dairy J.* **2005**, *15*, 921–927. [CrossRef]
8. Sommella, E.; Pepe, G.; Ventre, G.; Pagano, F.; Conte, G.M.; Ostacolo, C.; Manfra, M.; Tenore, G.C.; Russo, M.; Novellino, E. Detailed peptide profiling of "Scotta": From a dairy waste to a source of potential health-promoting compounds. *Dairy Sci. Technol.* **2016**, *96*, 763–771. [CrossRef]
9. Maragkoudakis, P.; Vendramin, V.; Bovo, B.; Treu, L.; Corich, V.; Giacomini, A. Potential use of scotta, the by-product of the ricotta cheese manufacturing process, for the production of fermented drinks. *J. Dairy Res.* **2016**, *83*, 104. [CrossRef]
10. Macedo, A.; Duarte, E.; Pinho, M. The role of concentration polarization in ultrafiltration of ovine cheese whey. *J. Membr. Sci.* **2011**, *381*, 34–40. [CrossRef]
11. Yadav, J.S.S.; Yan, S.; Pilli, S.; Kumar, L.; Tyagi, R.D.; Surampalli, R.Y. Cheese whey: A potential resource to transform into bioprotein, functional/nutritional proteins and bioactive peptides. *Biotechnol. Adv.* **2015**, *33*, 756–774. [CrossRef]
12. Argenta, A.B.; Scheer, A.D.P. Membrane separation processes applied to whey: A review. *Food Rev. Int.* **2019**, *36*, 499–528. [CrossRef]
13. Secchi, N.; Giunta, D.; Pretti, L.; García, M.R.; Roggio, T.; Mannazzu, I.; Catzeddu, P. Bioconversion of ovine scotta into lactic acid with pure and mixed cultures of lactic acid bacteria. *J. Ind. Microbiol. Biotechnol.* **2012**, *39*, 175–181. [CrossRef]
14. Sansonetti, S.; Curcio, S.; Calabro, V.; Iorio, G. Bio-ethanol production by fermentation of ricotta cheese whey as an effective alternative non-vegetable source. *Biomass Bioenergy* **2009**, *33*, 1687–1692. [CrossRef]
15. Macedo, A.; Duarte, E.; Fragoso, R. Assessment of the performance of three ultrafiltration membranes for fractionation of ovine second cheese whey. *Int. Dairy J.* **2015**, *48*, 31–37. [CrossRef]
16. Tavares, T.G.; Malcata, F.X. Whey proteins as source of bioactive peptides against hypertension. In *Bioactive Food Peptides in Health and Disease*; Hernández-Ledesma, B., Hsieh, C.C., Eds.; IntechOpen: London, UK, 2013; p. 75.
17. Pihlanto, A. Whey proteins and peptides: Emerging properties to promote health. *Nutrafoods* **2011**, *10*, 29–42. [CrossRef]
18. Raho, S.; Carofiglio, V.E.; Montemurro, M.; Miceli, V.; Centrone, D.; Stufano, P.; Schioppa, E.; Pontonio, E.; Rizzello, C.G. Production of the polyhydroxyalkanoate PHBV from ricotta cheese exhausted whey by *Haloferax mediterranei* fermentation. *Foods* **2020**, *9*, 1459. [CrossRef]
19. Siragusa, S.; De Angelis, M.; Di Cagno, R.; Rizzello, C.G.; Coda, R.; Gobbetti, M. Synthesis of γ-aminobutyric acid by lactic acid bacteria isolated from a variety of Italian cheeses. *Appl. Environ. Microbiol.* **2007**, *73*, 7283–7290. [CrossRef]
20. Gobbetti, M.; Morea, M.; Baruzzi, F.; Corbo, M.R.; Matarante, A.; Considine, T.; Di Cagno, R.; Giunee, T.; Fox, P.F. Microbiological, compositional, biochemical and textural characterisation of Caciocavallo Pugliese cheese during ripening. *Int. Dairy J.* **2002**, *12*, 511–523. [CrossRef]
21. Gatti, M.; Lazzi, C.; Rossetti, L.; Mucchetti, G.; Neviani, E. Biodiversity in *Lactobacillus helveticus* strains present in natural whey starter used for Parmigiano Reggiano cheese. *J. Appl. Microbiol.* **2003**, *95*, 463–470. [CrossRef]
22. Nejati, F.; Rizzello, C.G.; Di Cagno, R.; Sheikh-Zeinoddin, M.; Diviccaro, A.; Minervini, F.; Gobbetti, M. Manufacture of a functional fermented milk enriched of Angiotensin-I Converting Enzyme (ACE)-inhibitory peptides and γ-amino butyric acid (GABA). *LWT-Food Sci. Technol.* **2013**, *51*, 183–189. [CrossRef]
23. Church, F.C.; Swaisgood, H.E.; Porter, D.H.; Catignazni, G.L. Spectrophotometric assay using *o*-phthaldialdehyde for determination of proteolysis in milk and isolated milk proteins. *J. Dairy Sci.* **1983**, *66*, 1219–1227. [CrossRef]
24. Rizzello, C.G.; Nionelli, L.; Coda, R.; De Angelis, M.; Gobbetti, M. Effect of sourdough fermentation on stabilisation, and chemical and nutritional characteristics of wheat germ. *Food Chem.* **2010**, *119*, 1079–1089. [CrossRef]
25. Cushman, D.W.; Cheung, H.S. Spectrophotometric assay and properties of the angiotensin-converting enzyme of rabbit lung. *Biochem. Pharmacol.* **1971**, *20*, 1637–1648. [CrossRef]
26. Wu, J.; Ding, X. Characterization of inhibition and stability of soy-protein-derived angiotensin I-converting enzyme inhibitory peptides. *Food Res. Int.* **2002**, *35*, 367–375. [CrossRef]
27. Rizzello, C.G.; Cassone, A.; Di Cagno, R.; Gobbetti, M. Synthesis of angiotensin I-converting enzyme (ACE)-inhibitory peptides and γ-aminobutyric acid (GABA) during sourdough fermentation by selected lactic acid bacteria. *J. Agric. Food Chem.* **2008**, *56*, 6936–6943. [CrossRef] [PubMed]
28. Xu, X.; Singh, T.K.; Hillier, A.J. An HPLC assay for the determination of ACE inhibitory activity of food protein derived peptides. *Aust. J. Dairy Technol.* **2006**, *61*, 228.
29. Miceli, V.; Camassa, A.; Pizzichini, D.; Russo, C. Il Trattamento Sostenibile Dei Reflui Provenienti Dai Comparti Olivicolo, Viti-Vinicolo e Lattiero Caseario al Centro Del Progetto Del Progetto Bio-Energy Smart Source System, ENEA. 2016. Available online: https://iris.enea.it/retrieve/handle/20.500.12079/6766/515/RT-2016-39-ENEA.pdf (accessed on 30 September 2021).
30. ISO/FIL. *Milk and Milk Products: Determination of Nitrogen Content. Part 1: Kjeldahl Principle and Crude Protein Calculation*; Joint Standard ISO 8968-1/IDF 20-1:2001; International Standardisation Organisation: Geneva, Switzerland, 2001.
31. Verni, M.; De Mastro, G.; De Cillis, F.; Gobbetti, M.; Rizzello, C.G. Lactic acid bacteria fermentation to exploit the nutritional potential of Mediterranean faba bean local biotypes. *Food Res. Int.* **2019**, *125*, 108571. [CrossRef]
32. ISO/IDF. *Milk. Determination of Fat Content*; Joint Standard ISO 2446/IDF 226:2008; International Standardisation Organisation: Geneva, Switzerland, 2008.

33. AOAC. Ash of milk-Gravimetric method, method 945.46. In *Official Methods of Analysis of AOAC International*, 20th ed.; Latimer, G.W., Ed.; AOAC International: Gaithersburg, MD, USA, 2016; p. 10.
34. Chen, Y.; Kwon, S.W.; Kim, S.C.; Zhao, Y. Integrated approach for manual evaluation of peptides identified by searching protein sequence databases with tandem mass spectra. *J. Proteome Res.* **2005**, *4*, 998–1005. [CrossRef]
35. AOAC Official methods of analysis of AOAC international. In *Official method 926.08. Moisture in Cheese*, 16th ed.; AOAC International: Rockville, MD, USA, 1995; Volume 2, p. 58.
36. Kuchroo, C.N.; Fox, P.F. Soluble nitrogen in Cheddar cheese: Comparison of extraction procedures. *Milchwissenschaft* **1982**, *37*, 331–335.
37. Rehman, S.-U.; McSweeney, P.L.H.; Fox, P.F. Protocol for the manufacture of miniature cheeses. *Lait* **1998**, *78*, 703–720.
38. Lawless, H.T.; Heymann, H. *Sensory Evaluation of Food: Principles and Practices*; Springer: New York, NY, USA, 2010; Volume 2.
39. Lacivita, V.; Conte, A.; Manzocco, L.; Plazzotta, S.; Zambrini, V.A.; Del Nobile, M.A.; Nicoli, M.C. Surface UV-C light treatments to prolong the shelf-life of Fiordilatte cheese. *Innov. Food Sci. Emerg. Technol.* **2016**, *36*, 150–155. [CrossRef]
40. Chen, C.; Wolle, D.; Sommer, D. Mozzarella. In *The Sensory Evaluation of Dairy Products*, 2nd ed.; Clark, M., Costello, M., Drake, M., Bodyfelt, F., Eds.; Springer Science + Business Media: Dordrecht, the Netherlands, 2009; pp. 468–469.
41. Wang, G.; Li, X.; Wang, Z. APD3: The antimicrobial peptide database as a tool for research and education. *Nucleic Acids Res.* **2016**, *44*, 1087–1093. [CrossRef]
42. Minhalma, M.; Magueijo, V.; Queiroz, D.P.; de Pinho, M.N. Optimization of "Serpa" cheese whey nanofiltration for effluent minimization and by-products recovery. *J. Environ. Manag.* **2007**, *82*, 200–206. [CrossRef]
43. Monti, L.; Donati, E.; Zambrini, A.V.; Contarini, G. Application of membrane technologies to bovine Ricotta cheese exhausted whey (scotta). *Int. Dairy J.* **2018**, *85*, 121–128. [CrossRef]
44. Daliri, E.B.M.; Oh, D.H.; Lee, B.H. Bioactive peptides. *Foods* **2017**, *6*, 32. [CrossRef]
45. Sánchez, A.; Vázquez, A. Bioactive peptides: A review. *Food Qual. Saf.* **2017**, *1*, 29–46. [CrossRef]
46. Daliri, E.B.M.; Lee, B.H.; Park, B.J.; Kim, S.H.; Oh, D.H. Antihypertensive peptides from whey proteins fermented by lactic acid bacteria. *Food Sci. Biotechnol.* **2018**, *27*, 1781–1789. [CrossRef]
47. Brown, L.; Pingitore, E.V.; Mozzi, F.; Saavedra, L.; Villegas, J.M.; Hebert, E.M. Lactic acid bacteria as cell factories for the generation of bioactive peptides. *Protein Pept. Lett.* **2017**, *24*, 146–155. [CrossRef]
48. Daliri, E.B.M.; Lee, B.H.; Oh, D.H. Current trends and perspectives of bioactive peptides. *Crit. Rev. Food* **2018**, *58*, 2273–2284. [CrossRef]
49. Nakamura, K.; Naramoto, K.; Koyama, M. Blood-pressure-lowering effect of fermented buckwheat sprouts in spontaneously hypertensive rats. *J. Funct. Foods* **2013**, *5*, 406–415. [CrossRef]
50. Manzanares, P.; Salom, J.B.; Garcı´a-Tejedor, A.; Fernández-Musoles, R.; Ruiz-Giménez, P.; Gimeno-Alcañíz, J.V. Unraveling the mechanisms of action of lactoferrin-derived antihypertensive peptides: ACE inhibition and beyond. *Food Funct.* **2015**, *6*, 2440–2452. [CrossRef]
51. Ibrahim, H.R.; Ahmed, A.S.; Miyata, T. Novel angiotensin-converting enzyme inhibitory peptides from caseins and whey proteins of goat milk. *J. Adv. Res.* **2017**, *8*, 63–71. [CrossRef] [PubMed]
52. Monari, S.; Ferri, M.; Russo, C.; Prandi, B.; Tedeschi, T.; Bellucci, P.; Zambrini, A.V.; Donati, E.; Tassoni, A. Enzymatic production of bioactive peptides from scotta, an exhausted by-product of ricotta cheese processing. *PLoS ONE* **2019**, *14*, e0226834. [CrossRef] [PubMed]
53. Venegas-Ortega, M.G.; Flores-Gallegos, A.C.; Martínez-Hernández, J.L.; Aguilar, C.N.; Nevárez-Moorillón, G.V. Production of bioactive peptides from lactic acid bacteria: A sustainable approach for healthier foods. *Compr. Rev. Food Sci. Food Saf.* **2019**, *18*, 1039–1051. [CrossRef] [PubMed]
54. Gobbetti, M.; Minervini, F.; Rizzello, C.G. Angiotensin I-converting-enzyme-inhibitory and antimicrobial bioactive peptides. *Int. J. Dairy Technol.* **2004**, *57*, 173–188. [CrossRef]
55. Brzozowski, B.; Lewandowska, M. Prolyl endopeptidase—Optimization of medium and culture conditions for enhanced production by *Lactobacillus acidophilus*. *Electron. J. Biotechnol.* **2014**, *17*, 204–210. [CrossRef]
56. Hayes, M.; Stanton, C.; Slattery, H.; O'Sullivan, O.; Hill, C.; Fitzgerald, G.; Ross, R. Casein fermentate of *Lactobacillus animalis* DPC6134 contains a range of novel propeptide angiotensin-converting enzyme inhibitors. *Appl. Environ. Microbiol.* **2007**, *73*, 4658–4667. [CrossRef]
57. Minervini, F.; Algaron, F.; Rizzello, C.G.; Fox, P.F.; Monnet, V.; Gobbetti, M. Angiotensin I-converting-enzyme-inhibitory and antibacterial peptides from *Lactobacillus helveticus* PR4 proteinase-hydrolyzed caseins of milk from six species. *Appl. Environ. Microbiol.* **2003**, *69*, 5297–5305. [CrossRef]
58. Skrzypczak, K.; Gustaw, W.; Szwajgier, D.; Fornal, E.; Waśko, A. κ-casein as a source of short-chain bioactive peptides generated by *Lactobacillus helveticus*. *J. Food Sci. Technol.* **2017**, *54*, 3679–3688. [CrossRef]
59. Pihlanto, A. Lactic fermentation and bioactive peptides. In *Lactic Acid Bacteria-R & D for Food, Health and Livestock Purposes*; Kongo, M., Ed.; InTechOpen Limited: London, UK, 2013; pp. 309–333.
60. Sadat-Mekmene, L.; Genay, M.; Atlan, D.; Lortal, S.; Gagnaire, V. Original features of cell-envelope proteinases of *Lactobacillus helveticus*: A review. *Int. J. Food Microbiol.* **2011**, *146*, 1–13. [CrossRef]

61. Tavares, T.; del Mar Contreras, M.; Amorim, M.; Pintado, M.; Recio, I.; Malcata, F.X. Novel whey-derived peptides with inhibitory effect against angiotensin-converting enzyme: In vitro effect and stability to gastrointestinal enzymes. *Peptides* **2011**, *32*, 1013–1019. [CrossRef]
62. Abubakar, A.; Saito, T.; Kitazawa, H.; Kawai, Y.; Itoh, T. Structural analysis of new antihypertensive peptides derived from cheese whey protein by proteinase K digestion. *J. Dairy Sci.* **1998**, *81*, 3131–3138. [CrossRef]
63. Lafarga, T.; Aluko, R.E.; Rai, D.K.; O'Connor, P.; Hayes, M. Identification of bioactive peptides from a papain hydrolysate of bovine serum albumin and assessment of an antihypertensive effect in spontaneously hypertensive rats. *Food Res. Int.* **2016**, *81*, 91–99. [CrossRef]
64. Valk-Weeber, R.L.; Nichols, K.; Dijkhuizen, L.; Bijl, E.; van Leeuwen, S.S. Variations in N-linked glycosylation of glycosylation-dependent cell adhesion molecule 1 (GlyCAM-1) whey protein: Intercow differences and dietary effects. *J. Dairy Sci.* **2021**, *104*, 5056–5068. [CrossRef]
65. Ganzorig, K.; Urashima, T.; Fukuda, K. Exploring potential bioactive peptides in fermented bactrian camel's milk and mare's milk made by mongolian nomads. *Foods* **2020**, *9*, 1817. [CrossRef]
66. Marques, C.; Manuela Amorim, M.; Odila Pereira, J.; Estevez Pintado, M.; Moura, D.; Calhau, C.; Pinheiro, H. Bioactive peptides —Are there more antihypertensive mechanisms beyond ACE inhibition? *Curr. Pharm. Des.* **2012**, *18*, 4706–4713. [CrossRef]
67. Pei, J.; Hua, Y.; Zhou, T.; Gao, X.; Dang, Y.; Wang, Y. Transport, in vivo antihypertensive effect, and pharmacokinetics of an Angiotensin-Converting Enzyme (ACE) inhibitory peptide LVLPGE. *J. Agric. Food Chem.* **2021**, *69*, 2149–2156. [CrossRef]
68. Temussi, P.A. The good taste of peptides. *J. Pept. Sci.* **2012**, *18*, 73–82. [CrossRef]

foods

MDPI

Article

Changes in Functionality of *Tenebrio molitor* Larvae Fermented by *Cordyceps militaris* Mycelia

Neul-I Ha [1,2], Seul-Ki Mun [1], Seung-Bin Im [2], Ho-Yeol Jang [3], Hee-Gyeong Jeong [2], Kyung-Yun Kang [3], Kyung-Wuk Park [3], Kyoung-Sun Seo [2], Seung-Eon Ban [2], Kyung-Je Kim [2] and Sung-Tae Yee [1,*]

[1] Department of Pharmacy, Sunchon National University, Jungang-Ro, Suncheon 549-742, Korea
[2] Jangheung Research Institute for Mushroom Industry, Jangheung 59338, Korea
[3] Suncheon Research Center for Bio Health Care, Jungang-Ro, Suncheon 57922, Korea
* Correspondence: sungtae@sunchon.ac.kr; Tel.: +82-61-750-3752; Fax: +82-61-750-3708

Abstract: The Food and Agriculture Organization (FAO) has been estimating the potential of insects as human food since 2010, and for this reason, *Tenebrio molitor* larvae, also called mealworms, have been explored as an alternative protein source for various foods. In this study, in order to increase nutrient contents and improve preference as an alternative protein source, we fermented the *T. molitor* larvae by *Cordyceps militaris* mycelia. *T. molitor* larvae were prepared at optimal conditions for fermentation and fermented with *C. militaris* mycelia, and we analyzed *T. molitor* larvae change in functionality with proximate composition, β-glucan, cordycepin, adenosine, and free amino acids content. *T. molitor* larvae fermented by *C. militaris* mycelia showed higher total protein, total fiber, and β-glucan content than the unfermented larvae. In addition, the highest cordycepin content (13.75 mg/g) was observed in shaded dried *T. molitor* larvae fermented by *C. militaris* mycelia. Additionally, the isolated cordycepin from fermented *T. molitor* larvae showed similar cytotoxicity as standard cordycepin when treated with PC-9 cells. Therefore, we report that the optimized methods of *T. molitor* larvae fermented by *C. militaris* mycelia increase total protein, total fiber, β-glucan and produce the amount of cordycepin content, which can be contributed to healthy food and increase *T. molitor* larvae utilization.

Keywords: edible insects; *Tenebrio molitor* larvae; functionality change; fermentation; *Cordyceps militaris* mycelia

Citation: Ha, N.-I.; Mun, S.-K.; Im, S.-B.; Jang, H.-Y.; Jeong, H.-G.; Kang, K.-Y.; Park, K.-W.; Seo, K.-S.; Ban, S.-E.; Kim, K.-J.; et al. Changes in Functionality of *Tenebrio molitor* Larvae Fermented by *Cordyceps militaris* Mycelia. *Foods* **2022**, *11*, 2477. https://doi.org/10.3390/foods11162477

Academic Editors: Carlo Giuseppe Rizzello and Palmira De Bellis

Received: 1 August 2022
Accepted: 15 August 2022
Published: 17 August 2022

1. Introduction

As the world's population steadily increases, the International Feed Industry Federation (IFIF) predicted that it will reach 10 billion by 2050 [1]. This means that it is necessary to increase food production to meet consumer demands. Mass breeding of livestock has a negative impact on the environment by producing a large amount of greenhouse gases and ammonia and spending considerable water, energy, and land [2]. A previous study reported that insects are significantly decreased the production of am-monia and greenhouse gases (CO_2, N_2O, and CH_4) and also reported that less land-dependent than livestock [3,4]. In addition, insects are estimated to be able to supply food to approximately two billion people. For these reasons, edible insects have received worldwide attention as potential sources of alternative proteins [5,6].

Among the edible insects, *Tenebrio molitor* larvae, also called mealworms, are commercially reared in many countries as food for animals or even for human consumption. For example, *T. molitor* larvae are the most common insect produced in China, and the amount is more than 1000 tons per year [7,8]. In a previous study, *T. molitor* larvae were reported to have approximately 50% crude protein content [9] and contain essential amino acids, which are recommended for adult consumption by the World Health Organization (WHO) [7,10]. For these reasons, producers who cultivate *T. molitor* larvae and their production amounts have steadily increased. However, *T. molitor* larvae are usually produced in terms of living status and dried status, whereas consumers do not prefer them because of their disgusting appearance [11].

Therefore, it is necessary to develop a method that leads to an increase in *T. molitor* larvae consumption and utilization.

Cordyceps militaris, called Dong-Chong-Xia-Cao, which includes means of winter worm transforming into summer grass, is one of the entomogenous fungi that form a fruiting body mainly on pupae or larvae [12,13]. It has been traditionally used as an herbal medicine in Korea and China to enhance longevity and vitality [14] and contains many types of biologically active compounds such as cordycepin, adenosine, cordycepic acid, sterols, nucleosides, and polysaccharides [15]. Among them, cordycepin (3′-deoxyadenosine), a derivative of adenosine, was the first compound to be isolated from *C. militaris* [16]. Cordycepin has been reported to have antitumor, antiangiogenic, antimetastatic, and antiproliferative effects, and it induces apoptosis in cancer cells [17].

Biofortification is a very important method to improve the shortage of food and provide nutrients for those who rarely have access to foods [18]. Among those, fermentation is a very economical and useful method of improving nutrient ingredients by producing active ingredients [19,20]. Fermentation is a process that is subjected to the action of microorganisms or enzymes, and it can cause biochemical changes in food [21]. These changes may result in sensory qualities and improved nutritional value [22]. A previous study that fermented the *T. molitor* larvae performed lactic acid bacteria and soy sauce fermentation, but there is not enough study fermented with *C. millitaris* myceila using their entomogenous character. On account of this reason, in this study, *T. molitor* larvae fermented by *C. militaris* mycelia were performed to increase their utilization and progressing appearance. In addition, we analyzed changes in functionality to confirm the potential of nutritional components as food.

2. Materials and Methods

2.1. Samples

T. molitor larvae used in this study were purchased from Myeong Pum, Inc. (Jangseong, Korea). *T. molitor* larvae were reared for 9 weeks and did not feed for two days before harvest to remove feces. Harvested *T. molitor* larvae were separated from wheat and excretions using steel mesh. *T. molitor* larvae were separated into two test groups to compare the nutrient composition and functional ingredient and to find optimal fermentation conditions. One test group of *T. molitor* larvae was spread in a steel tray and dried under shade for 20 h at room temperature (shade dried, SD). The other test group of *T. molitor* larvae was also spread in a steel tray and dried in an oven for 20 h at 80 °C and boiled for 30 min (30 min boil after hot air dried, 30BHAD).

2.2. Chemicals and Standards

Analytical-standard cordycepin and adenosine were purchased from Sigma-Aldrich (St. Louis, MO, USA). Standards of amino acids, OPA was purchased from Agilent (Agilent Technologies, Santa Clara, CA, USA). Certified HPLC-grade solvents methanol and acetonitrile were purchased from J. T. Baker Chemicals (Center Valley, Coopersburg, PA, USA). All other reagents used for analysis were analytical grade. Potato dextrose agar (PDA) and potato dextrose broth (PDB) were purchased from BD DIFCO (Detroit, MI, USA).

2.3. Fermentation

The *C. militaris* strain (No. 40226) was obtained from the Korean Agricultural Culture Collection (KACC, Suwon, Korea) and used in this study. To prepare the inoculum, *C. militaris* mycelia in the PDA plates were punched using a cork borer (diameter 5 mm), and five discs were transplanted into 500 mL PDB and incubated for 10 days at 25 °C with 150 rpm in a shaking incubator. The 500 mL liquid media were transferred to 1 L PDB, incubated for 15 days at 25 °C, and homogenized for transplantation. *T. molitor* larvae of two test groups (SD, 30BHAD) were filled with each polypropylene bottle and sterilized in an autoclave for 30 min at 121 °C. The polypropylene bottle was cooled to room temperature on a clean bench and inoculated with an equal ratio of inoculum (*v*/*w*,

1:2). The polypropylene bottles containing *T. molitor* larvae and *C. militaris* mycelia were harvested after incubation for 60 days at 25 °C and freeze-dried for use in the analysis.

2.4. Proximate Composition Determination

The contents of proximate compositions were measured according to the official methods of the Association of Official Analytical Chemists (AOAC, Rockville, Maryland, 2005). The moisture content was determined by oven-drying methods at 105 °C for 12 h. Crude protein (N × 6.25) in the samples was determined using the Kjeldahl method. Crude fat was extracted using a Soxhlet apparatus. To measure the ash content, samples were burned at 600 °C for 3 h. The crude fiber was extracted as digesting with dilute H_2SO_4 and KOH solutions.

2.5. β-Glucan Determination

The β-glucan content was measured using a Megazyme yeast and mushroom kit (Megazyme, Bray, Ireland), and β-glucan content was calculated as the total glucan content minus the α-glucan content.

2.6. Analysis of Free Amino Acids

To determine free amino acids, samples were extracted for 3 h in 60 °C water bath with distilled water. The extraction was filtered using a syringe filter of 0.45 μm (Advantec Dismicr, Tokyo, Japan) and incubated at 4 °C after adding sulfosalicylic acid. After 4 h, water extract was separated by centrifugation at 10,000× *g* for 10 min and used for analysis. The free amino acids were quantified by high-performance liquid chromatography (HPLC; 1200 Series HPLC system, Agilent Technologies, Santa Clara, CA, USA) after derivatization with o-phthaldialdehyde. The column size was 2.1 × 150 mm, with a 4 μm particle size (Agilent Technologies, Santa Clara, CA, USA) and a 340 nm detection wavelength at 40 °C. Mobile phase A was 10 mM sodium phosphate and 10 mM sodium tetraborate (*v*/*v*, 1:1), pH 8.2, and mobile phase B was acetonitrile, methanol, and distilled water (*v*/*v*/*v*, 45:45:10) with a flow rate 0.35 mL/min.

2.7. Analysis of Cordycepin and Adenosine

To determine cordycepin and adenosine, samples were extracted using a sonicator for 90 min in distilled water. The water extract was separated by centrifugation at 2264× *g* for 15 min. The supernatant was filtered using a syringe filter of 0.45 μm (Advantec Dismicr, Tokyo, Japan). HPLC analysis was performed on a 1200 Series HPLC system (Agilent Technologies, Santa Clara, CA, USA) with an Agilent ZORBAX Eclipse XDB-C_{18} column (Agilent Technologies, Santa Clara, CA, USA). The parameters used were as follows: flow rate of 0.8 mL/min, injection volume of 10 μL, wavelength of 260 nm, and column temperature of 30 °C.

2.8. Separation and Purification Cordycepin

Separation and purification were performed to determine the cell viability effect of cordycepin isolated from SD *T. molitor* larvae fermented by *C. militaris* mycelia. Semi-HPLC was used for separation. The composition ratio of the mobile phase was LC-water and methanol (*v*/*v*, 70:30) to obtain a sub-fraction by loading the extract of SD *T. molitor* larvae fermented by *C. militaris* mycelia into the cartridge column. Among them, the main fraction that included the cordycepin peak was collected and concentrated under freeze drying and gained 46.5 mg of dried cordycepin peak fraction. HPLC-MS/ELSD was used for purification. The dried peak fraction and cordycepin standard were dissolved in methanol and analyzed using an Agilent Poroshell 120 SB—C_{18}. Detection was performed under the condition that the composition ratio of the mobile phase was water: Acetonitrile (*v*/*v*, 70:30), and the flow rate was 0.4 mL/min. The injection volume was 10 μL, and the detection wavelength was 260 nm.

2.9. Cell Culture and Viability Assay

PC-9 cells (human lung adenocarcinoma cell line) were obtained from the European Collection of Authenticated Cell Cultures (ECACC). The cells were cultured in RPMI-1640 medium containing 10% fetal bovine serum, 1% penicillin/streptomycin, and 0.1%

2-mercaptoethanol. Cultures were maintained at 37 °C and 5% CO_2, and the media were changed every three days. Cells were seeded in a 96-well plate (6 × 10^4 cells/well) and subcultured for 24 h, and 10, 30, 100, and 300 μM isolated cordycepin and standard cordycepin were treated to each well. The cells were incubated in 5% CO_2 at 37 °C for 48 h. Subsequently, 100 μL of the solution was removed from each well, and 10 μL of CCK-8 reagent was added. After 1 h, the absorbance was measured at 450 nm using a microplate reader (Versamax, Molecular Devices, San Jose, CA, USA).

2.10. Statistical analysis

Data are presented as mean ± SD of three replicates. Statistical differences between groups were compared using the Student's *t*-test. Probability values less than 0.05, were considered significant (*p* values * < 0.05, ** < 0.01, and *** < 0.001). One-way ANOVA and Tukey's post hoc test were performed using Graphpad 7.0 (Graphpad Software, San Dieogo, CA, USA).

3. Results

3.1. Proximate Composition and β-Glucan Contents

Proximate compositions were analyzed to determine the usefulness of *T. molitor* larvae fermented by *C. militaris* mycelia as a functional food source. Table 1 shows the proximate composition of *T. molitor* larvae fermented with or without *C. militaris* mycelia. The proximate composition of SD *T. molitor* larvae and fermented by *C. militaris* mycelia contained moisture (3.5% and 9.15%), crude protein (44.01% and 59.34%), crude fat (32.02% and 18.54%), crude ash (3.21% and 3.84%), and crude fiber (5.54% and 9.48%, respectively). In addition, 30BHAD *T. molitor* larvae fermented by *C. militaris* mycelia contained moisture (2.25% and 8.87%), crude protein (47.49% and 64.24%), crude fat (36.9% and 32.02%), crude ash (2.63% and 2.21%), and crude fiber (12.97% and 15.62%, respectively). By fermentation of *T. molitor* larvae with *C. militaris* mycelia, they substantially increased the crude protein and crude fiber content and reduced the crude fat content. In a previous study, it was reported that bioactive compounds such as ergosterol, the precursor of vitamin D_2, are integrally linked with mycelial growth [23]. The protein and fat contents of several mushrooms grown in various media have been described previously [24]. Microbial proteins can be derived from a variety of microorganisms, both unicellular and multicellular, including bacteria, yeast, and fungi [25]. In this study, *T. molitor* larvae fermented by *C. militaris* mycelia produced more crude protein than unfermented *T. molitor* larvae. We suggest that these results are affected by microbial proteins and can be more attractive as alternative protein sources than unfermentation.

Table 1. Proximate composition and β-glucan content of shade dried (SD), and 30 min boil after hot air dried (30BHAD) *Tenebrio molitor* larvae before and after fermentation.

Variables (%)	SD		30BHAD	
	Unfermented	Fermented	Unfermented	Fermented
Moisture	3.5 ± 0.24 ^{a1}	9.15 ± 0.79 ^{b}	2.25 ± 0.09 ^{c}	8.87 ± 0.15 ^{b}
Crude protein	44.01 ± 0.27 ^{a}	59.34 ± 0.12 ^{b}	47.49 ± 0.19 ^{c}	64.24 ± 0.52 ^{d}
Crude fat	32.02 ± 0.21 ^{a}	18.54 ± 1.24 ^{b}	36.9 ± 0.58 ^{c}	32.02 ± 0.21 ^{d}
Crude ash	3.21 ± 0.04 ^{a}	3.84 ± 0.19 ^{a}	2.63 ± 0.42 ^{ab}	2.21 ± 0.15 ^{b}
Crude fiber	5.54 ± 0.24 ^{a}	9.48 ± 0.2 ^{b}	12.97 ± 1.44 ^{c}	15.62 ± 0.19 ^{d}
β-Glucan	0.34 ± 0.17 ^{a}	4.77 ± 0.26 ^{b}	0.61 ± 0.29 ^{a}	8.78 ± 0.33 ^{c}

[1] All values are means ± SD (N = 3). $^{a–d}$ Groups that do not share a common letter indicate a significant difference ($p < 0.05$).

β-Glucan contents of *T. molitor* larvae fermented by *C. militaris* mycelia are shown in Table 1. First, different content of β-glucan was shown between SD *T. molitor* larvea (0.34%) and fermented SD *T. molitor* larvea (4.77%). The β-glucan content of 30BHAD *T. molitor* larvae was 0.61%, and when 30BHAD *T. molitor* larvae fermented by *C. militaris* mycelia, it contained 8.78%, which the highest β-glucan content in the samples. The β-glucan content of *T. molitor* larvae fermented by *C. militaris* mycelia was substantially higher than those of the non-fermented *T. molitor* larvae. It is well known that mushrooms contain many bioactive compounds that have diverse beneficial impacts on human health [26]. In particular, β-glucan is the main polysaccharide normally present in the cell walls of fungi and yeast. It has been reported that β-glucan affects immune system activation, as well as antimicrobial, antioxidant, and antitumor effects [27,28]. According to these results, we expect the use of *T. molitor* larvae as functional food.

3.2. Free Amino Acids Content

The free amino acids content of *T. molitor* larvae before and after fermentation are shown in Table 2. In all of the samples, a total of 15 amino acids, which include eight essential and seven non-essential amino acids, were detected. Total free amino acids content of SD *T. molitor* larvae was 2587.73 mg/100 g and after fermentation contained 1626.05 mg/100 g. Moreover, total free amino acids content of 30BHAD *T. molitor* larvae was 1080.14 mg/100 g and after fermentation contained 3018.25 mg/100 g. In case of SD *T. molitor* larvae, total free amino acids content was 0.62-fold decreased after fermentation, but 30BHAD *T. molitor* larvae was 2.79-fold significantly increased. Total essential amino acids content of SD *T. molitor* larvae was 1138.64 mg/100 g but decreased after fermentation to 607.33 mg/100 g (0.53-fold). On the other hand, the total essential amino acids content of 30BHAD *T. molitor* larvae was 490.88 mg/100 g and after fermentation increased to 1262.5 mg/100 g (2.57-fold). *C. militaris* is entaentomogenous fungi that form a fruiting body mainly on pupae or larvae and use insects as a nutrient source for grown [12,13]. For these reasons, it seems that the free amino acids content of SD *T. molitor* larvae fermented by *C. militaris* mycelia was decreased [29]. On the other hand, the total free amino acids in 30BHAD *T. molitor* larvae was increased after fermentation. These free amino acids content differences require additional experiments. Other previous studies showed that when *Pleurotus eryngii* and *Bacillus subtilis* were fermented in *Gryllus bimaculatus*, total free amino acids increased due to protease [29,30]. Additionally, aspartic acid content of 30BHAD *T. molitor* larvae was 28.72 mg/100 g and after fermentation increased to 149.61 mg/100 g. The glutamic acid content of 30BHAD *T. molitor* larvae was 143.68 mg/100 g and after fermentation increased to 588.89 mg/100 g. These free amino acids are known for being highly responsible for the taste properties and development of flavor-forming reactions [31]. Aspartic acid and glutamic acid are related to umami [32], umami content of 30BHAD *T. molitor* larvae was significantly increased when fermented by *C. militaris* mycelia.

Table 2. Free amino acid contents of shade dried (SD), and 30 min boil after hot air dried (30BHAD) *Tenebrio molitor* larvae before and after fermentation.

Variables (mg/100 g)	SD		30BHAD	
	Unfermented	Fermented	Unfermented	Fermented
Aspartic acid	91.22 ± 11.49 [a1]	84.8 ± 14.16 [a]	28.72 ± 4.33 [b]	149.61 ± 7.44 [c]
Glutamic acid	224.22 ± 37.16 [a]	292.31 ± 44.24 [a]	143.68 ± 30.04 [a]	588.89 ± 67.85 [b]
Serine	81.47 ± 8.26 [a]	59.53 ± 7.67 [ab]	40.66 ± 7.24 [b]	88.61 ± 11.54 [a]
Histidine	165.14 ± 16.02 [a]	81.7 ± 10.16 [b]	52.18 ± 11.6 [b]	128.87 ± 19.16 [a]
Glycine	48.02 ± 3.56 [a]	25.04 ± 2.94 [b]	21.45 ± 2.95 [b]	36.38 ± 5.56 [c]
Threonine	77.35 ± 8.61 [a]	52.28 ± 5.87 [b]	38.44 ± 6.85 [b]	112.77 ± 9.37 [c]
Arginine	456.76 ± 42.2 [a]	385.09 ± 35.44 [a]	172.39 ± 32.27 [b]	602.94 ± 85.19 [c]
Alanine	239.22 ± 31.98 [a]	56.68 ± 13.94 [b]	122.08 ± 25.07 [b]	98.04 ± 38 [b]
Tyrosine	308.19 ± 42.63 [a]	115.27 ± 12.17 [b]	60.27 ± 11.24 [b]	191.28 ± 24.32 [c]
Valine	189.16 ± 30.24 [a]	113.28 ± 11.29 [b]	91.54 ± 16.9 [b]	221.11 ± 18.24 [a]
Methionine	63.82 ± 5.39 [a]	33.36 ± 5.16 [b]	18.48 ± 2.09 [c]	26.16 ± 1.9 [bc]
Phenylalanine	120.66 ± 12.59 [a]	49.37 ± 5.59 [b]	51.32 ± 9.62 [b]	102.14 ± 11.98 [a]
Isoleucine	124.7 ± 16.84 [a]	48.77 ± 5.07 [b]	59.51 ± 10.6 [b]	70.32 ± 4.85 [b]
Leucine	143.62 ± 16.6 [a]	57.47 ± 5.58 [b]	91.01 ± 17.92 [bc]	117.56 ± 16.43 [ac]
Lysine	254.18 ± 70.69 [a]	171.1 ± 4.31 [a]	88.41 ± 33.91 [a]	483.57 ± 114.86 [b]
Total free AA	2587.73 ± 330.84 [a]	1626.05 ± 160.85 [b]	1080.14 ± 219.62 [b]	3018.25 ± 419.4 [a]
Total EAA	1138.64 ± 162.42 [a]	607.33 ± 41.08 [b]	490.88 ± 107.59 [b]	1262.5 ± 188.49 [a]

[1] All values are means ± SD (N = 3). [a–c] Groups that do not share a common letter indicate a significant difference ($p < 0.05$).

3.3. Cordycepin and Adenosine Contents

Table 3 shows the cordycepin and adenosine contents of *T. molitor* larvae fermented with or without *C. militaris* mycelia. Cordycepin and adenosine contents of unfermented *T. molitor* larvae were not detected, and when SD *T. molitor* larvae fermented by *C. militaris* mycelia, it contained 13.75 mg/g of cordycepin and 0.85 mg/g of adenosine. Additionally, 30BHAD *T. molitor* larvae fermented by *C. militaris* mycelia contained 9.19 mg/g of cordycepin and 1.66 mg/g of adenosine. The cordycepin and adenosine contents of *T. molitor* larvae were produced according to fermentation.

Table 3. Cordycepin and adenosine contents of various *Cordyceps militaris*, including *Tenebrio molitor* larvae before and after fermentation.

Samples		Variables (mg/g)	
		Cordycepin	Adenosine
SD	Unfermented	ND [1]	ND
	Fermented	13.75 ± 0.03 [a3]	0.85 ± 0.03 [a]
30BHAD	Unfermented	ND	ND
	Fermented	9.19 ± 0.02 [b]	1.66 ± 0.05 [b]
CCMBR1 [2]	Fruiting body	0.78 ± 0.02 [c]	2.77 ± 0.03 [c]
	Substrate	0.26 ± 0.01 [d]	0.4 ± 0.01 [d]
CCMBR2	Fruiting body	1.16 ± 0.02 [e]	3.24 ± 0.03 [e]
	Substrate	0.63 ± 0.01 [f]	0.23 ± 0.01 [f]
CCMP	Fruiting body	1.06 ± 0.04 [g]	4.1 ± 0.09 [g]
	Substrate	6.45 ± 0.02 [h]	1.7 ± 0.12 [b]

[1] ND—not detected. [2] CCMBR1—commercial *C. militaris* grown on brown rice 1; CCMBR2—commercial *C. militaris* grown on brown rice 2; CCMP—commercial *C. militaris* grown on the pupae. CCMBR1,2 were purchased from distinct farmer market. [3] All values are means ± SD (N = 3). [a–h] Groups that do not share a common letter indicate a significant difference ($p < 0.05$).

Additionally, we purchased commercial *C. militaris* to compare cordycepin and adenosine contents with SD and 30BHAD *T. molitor* larvae fermented by *C. militaris* mycelia. Commercial *C. militaris* were separated from the fruiting body and substrate for analysis. Table 3 also shows the cordycepin and adenosine contents of commercial *C. militaris*. The contents of cordycepin in commercial *C. militaris* were found to be highest in order to CCMP substrate (6.45 mg/g), CCMP fruiting body (1.06 mg/g), CCMBR2 fruiting body (1.16 mg/g). Furthermore, the adenosine contents in commercial *C. militaris* were found to be highest in order to CCMP fruiting body (4.1 mg/g), CCMBR2 fruiting body (3.24 mg/g), CCMBR1 fruiting body (2.77 mg/g). In the case of CCMBR1, 2 cordycepin and adenosine contents of fruiting bodies were higher than those of the substrate. Additionally, the adenosine content of the CCMP fruiting body was higher than the substrate, but cordycepin content was higher in the fruiting body. Among the different substrates of *C. militaris*, the SD *T. molitor* larvae had the highest cordycepin content (13.75 mg/g). Cordycepin, which was first isolated from *C. militaris*, has been reported to have antitumor, antiangiogenic, antimetastatic, and antiproliferative effects [16,17]. Previous studies have reported cordycepin content in *C. militaris* fruiting bodies of brown rice were 5.62 mg/g [33], 1.2–3.5 mg/g [34], and in a substrate of brown rice was 2.27–3.97 mg/g [35]. A previous study reported that the differences between cordycepin contents are due to differences in the *C. militaris* strain [35]. This study showed that *T. molitor* larvae produce amount of cordycepin when fermented by *C. militaris* mycelia. According to a previous study, cordycepin is well known for strong antitumor effects. Therefore, we expect an additional bio-active effect of *T. molitor* larvae fermented by *C. militaris* mycelia.

3.4. Cell Viability Assay

We used the human lung cancer cell line PC-9 to investigate the cell viability of the isolated cordycepin produced in *T. molitor* larvae. Standard cordycepin and isolated cordycepin were administered at 10, 30, 100, and 300 μM for 48 h. The CCK-8 assay showed a dose-dependent decrease in the growth of PC-9 cells. The toxicity of standard and isolated cordycepin at 100 μM was 42.4% and 61%, respectively; at 300 μM, it was 31% and 42.7%, respectively (Figure 1). Cordycepin is well known for cytotoxicity in various cancers, and PC-9 cells are more sensitive to lung cancer [36]. In this study, we confirmed the cordycepin isolated from SD *T. molitor* larvae fermented by *C. militaris* mycelia reduces PC-9 cell viability as well as standard cordycepin.

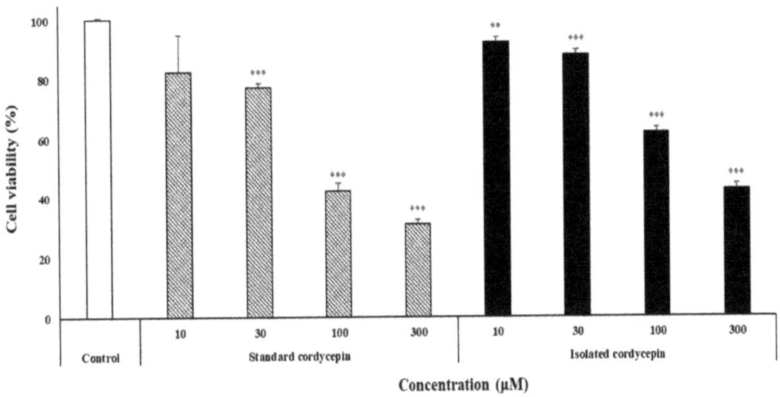

Figure 1. Effect of standard cordycepin and isolated cordycepin on PC-9 human lung cancer cell viability. Cells were seeded at a density of 6×10^3 cells/well in a 96-well plate and then treated with different concentrations of standard cordycepin and isolated cordycepin (10, 30, 100, and 300 μM) for 48 h. The culture supernatant was removed, and cell counting kit-8 was added. All data are expressed as mean ± SD of three independent experiments. ** $p < 0.01$, *** $p < 0.001$ compared to the control.

4. Conclusions

This study was performed to analyze the functional change of *T. molitor* larvae according to fermentation by *C. militaris* mycelia. The functional changes of *T. molitor* larvae were investigated using two drying methods (SD and 30BHAD) and fermentation with C. militaris mycelia. The results showed not only did *T. molitor* larvae increase crude protein, crude fiber, β-glucan, and free amino acids (only 30BHAD *T. molitor*), but it also reduced crude fat when fermented with *C. militaris* mycelia. Moreover, *T. molitor* produced cordycepin and adenosine after fermentation, and also SD *T. molitor* larvae fermented by *C. militaris* mycelia produced the highest contents of cordycepin compared with commercial *C. militaris*. Then, we isolated cordycepin from *T. molitor* larvae, which has the highest content, and analyzed its cell viability activity; it inhibited the growth of PC-9 cancer cells. In this study, we only performed cytotoxicity effects of isolated cordycepin from fermented *T. molitor* larvae, but if additional anti-cancer effects are confirmed, it can be possible to develop functional foods using fermented *T. molitor* larvae. These results suggest that fermentation with *C. militaris* mycelia can increase the functionality of *T. molitor* larvae. In addition, fermentation using *C. militaris* mycelia to *T. molitor* larvae is considered to will be useful for the development of high-protein and eco-friendly alternative foods.

Author Contributions: Conceptualization, S.-T.Y.; data curation, N.-I.H.; funding acquisition, K.-S.S. and K.-J.K.; investigation, N.-I.H., S.-K.M., S.-B.I., H.-Y.J. and H.-G.J.; project administration, S.-T.Y.; resources, K.-Y.K., K.-W.P. and S.-E.B.; supervision, S.-T.Y.; writing—original draft preparation, N.-I.H.; writing—review and editing, S.-T.Y. and K.-S.S. All authors have read and agreed to the published version of the manuscript.

Funding: This research received external funding from Jangheung Mushroom Industry Research (Grant Number: JMI 2022-01).

Institutional Review Board Statement: Not applicable.

Informed Consent Statement: Not applicable.

Data Availability Statement: Data are contained within the article.

Acknowledgments: This study was carried out with the support of the 'R&D Program for Forest Science Technology (Project No. 2020198A00-2022-BA01)' provided by the Korea Forest Service (Korea Forestry Promotion Institute).

Conflicts of Interest: The authors declare no conflict of interest.

References

1. Hong, J.; Han, T.; Kim, Y.Y. Mealworm (*Tenebrio molitor* Larvae) as an alternative protein source for monogastric animal: A review. *Animals* **2020**, *10*, 2068. [CrossRef]
2. Grau, T.; Vilcinskas, A.; Joop, G. Sustainable farming of the mealworm *Tenebrio molitor* for the production of food and feed. *Z. Nat. C* **2017**, *72*, 337–349. [CrossRef]
3. Oonincx, D.G.; Van Itterbeeck, J.; Heetkamp, M.J.; Van Den Brand, H.; Van Loon, J.J.; Van Huis, A. An exploration on greenhouse gas and ammonia production by insect species suitable for animal or human consumption. *PLoS ONE* **2010**, *5*, e14445. [CrossRef]
4. Nowak, V.; Persijn, D.; Rittenschober, D.; Charrondiere, U.R. Review of food composition data for edible insects. *Food Chem.* **2016**, *193*, 39–46. [CrossRef]
5. Makkar, H.P.; Tran, G.; Heuzé, V.; Ankers, P. State-of-the-art on use of insects as animal feed. *Anim. Feed Sci. Technol.* **2014**, *197*, 1–33. [CrossRef]
6. McLeod, A. *World Livestock 2011—Livestock in Food Security*; Food and Agriculture Organization of the United Nations (FAO): Rome, Italy, 2011.
7. Purschke, B.; Brüggen, H.; Scheibelberger, R.; Jäger, H. Effect of pre-treatment and drying method on physico-chemical properties and dry fractionation behaviour of mealworm larvae (*Tenebrio molitor* L.). *Eur. Food Res. Technol.* **2018**, *244*, 269–280. [CrossRef]
8. Dossey, A.T.; Morales-Ramos, J.A.; Rojas, M.G. *Insects as Sustainable Food Ingredients: Production, Processing and Food Applications*; Academic Press: Cambridge, MA, USA, 2016.
9. Barker, D.; Fitzpatrick, M.P.; Dierenfeld, E.S. Nutrient composition of selected whole invertebrates. *Zoo Biol. Publ. Affil. Am. Zoo Aquar. Assoc.* **1998**, *17*, 123–134. [CrossRef]
10. Rumpold, B.A.; Schlüter, O.K. Nutritional composition and safety aspects of edible insects. *Mol. Nutr. Food Res.* **2013**, *57*, 802–823. [CrossRef]
11. Hwang, J.; Choe, J.Y. How to enhance the image of edible insect restaurants: Focusing on perceived risk theory. *Int. J. Hosp. Manag.* **2020**, *87*, 102464. [CrossRef]
12. Won, S.-Y.; Park, E.-H. Anti-inflammatory and related pharmacological activities of cultured mycelia and fruiting bodies of *Cordyceps militaris*. *J. Ethnopharmacol.* **2005**, *96*, 555–561. [CrossRef]
13. Nam, K.S.; Jo, Y.S.; Kim, Y.H.; Hyun, J.W.; Kim, H.W. Cytotoxic activities of acetoxyscirpenediol and ergosterol peroxide from *Paecilomyces tenuipes*. *Life Sci.* **2001**, *69*, 229–237. [CrossRef]
14. Park, S.J.; Jang, H.-J.; Hwang, I.-H.; Kim, J.M.; Jo, E.; Lee, M.-G.; Jang, I.-S.; Joo, J.C. *Cordyceps militaris* extract inhibits the NF-κB pathway and induces apoptosis through MKK7-JNK signaling activation in TK-10 human renal cell carcinoma. *Nat. Prod. Commun.* **2018**, *13*, 465–470. [CrossRef]
15. Yue, K.; Ye, M.; Zhou, Z.; Sun, W.; Lin, X. The genus *Cordyceps*: A chemical and pharmacological review. *J. Pharm. Pharmacol.* **2013**, *65*, 474–493. [CrossRef]
16. Cunningham, K.; Manson, W.; Spring, F.; Hutchinson, S. *Cordycepin*, a metabolic product isolated from cultures of *Cordyceps militaris* (Linn.) Link. *Nature* **1950**, *166*, 949. [CrossRef]
17. Tuli, H.S.; Sandhu, S.S.; Sharma, A. Pharmacological and therapeutic potential of *Cordyceps* with special reference to *Cordycepin*. *3 Biotech* **2014**, *4*, 1–12. [CrossRef]
18. Jaiswal, D.K.; Krishna, R.; Chouhan, G.K.; de Araujo Pereira, A.P.; Ade, A.B.; Prakash, S.; Verma, S.K.; Prasad, R.; Yadav, J.; Verma, J.P. Bio-fortification of minerals in crops: Current scenario and future prospects for sustainable agriculture and human health. *Plant Growth Regul.* **2022**, *98*, 5–22. [CrossRef]
19. Marta, C.-K.; Beata, P. Changes in the Folate Content and Fatty Acid Profile in Fermented Milk Produced with Different Starter Cultures during Storage. *Molecules* **2021**, *26*, 6063.
20. Mrinal, S.; Rotimi, E.A.; Anil, K.P.; Tejpal, D. Enhancing Micronutrients Bioavailability through Fermentation of Plant-Based Foods: A Concise Review. *Fermentation* **2021**, *7*, 63.
21. Fox, P.F. Fermented Foods of the World. A Dictionary and Guide G. Campbell-Platt. *Ir. J. Food Sci. Technol.* **1987**, *11*, 199.
22. An, B.; Sara, B.; Sorel, T.S.; Harshadrai, R.; Oliver, K.S.; Van, C.L. Effect of Blanching Plus Fermentation on Selected Functional Properties of Mealworm (*Tenebrio molitor*) Powders. *Foods* **2020**, *9*, 917.
23. Gutarowska, B.; Żakowska, Z. Mathematical models of mycelium growth and ergosterol synthesis in stationary mould culture. *Lett. Appl. Microbiol.* **2009**, *48*, 605–610. [CrossRef] [PubMed]
24. Falanghe, H.; Smith, A.; Rackis, J. Production of fungal mycelial protein in submerged culture of soybean whey. *Appl. Microbiol.* **1964**, *12*, 330–334. [CrossRef] [PubMed]

25. Yabaya, A.; Ado, S. Mycelial protein production by *Aspergillus niger* using banana peels. *Sci. World J.* **2008**, *3*, 9–12. [CrossRef]
26. Lu, R.; Conrad, P.; Yacine, H. Antitumor activity of mushroom polysaccharides. *Food Funct.* **2012**, *3*, 1118–1130.
27. Zhu, F.; Du, B.; Bian, Z.; Xu, B. Beta-glucans from edible and medicinal mushrooms: Characteristics, physicochemical and biological activities. *J. Food Compos. Anal.* **2015**, *41*, 165–173. [CrossRef]
28. Wunjuntuk, K.; Ahmad, M.; Techakriengkrai, T.; Chunhom, R.; Jaraspermsuk, E.; Chaisri, A.; Kiwwongngam, R.; Wuttimongkolkul, S.; Charoenkiatkul, S. Proximate composition, dietary fibre, beta-glucan content, and inhibition of key enzymes linked to diabetes and obesity in cultivated and wild mushrooms. *J. Food Compos. Anal.* **2022**, *105*, 104226. [CrossRef]
29. Jang, H.W.; Choi, J.H.; Park, S.Y.; Park, B.R. Changes in Nutritional Composition of *Gryllus bimaculatus* Fermented by *Bacillus* sp. and Mycelium of Basidiomycetes. *J. Korean Soc. Food Cult.* **2019**, *34*, 785–792.
30. Baek, L.-M.; Park, L.-Y.; Park, K.-S.; Lee, S.-H. Effect of Starter Cultures on the Fermentative. *Korean J. Food Sci. Technol.* **2008**, *40*, 400–405.
31. Ionescu, R.E.; Fillit, C.; Jaffrezic-Renault, N.; Cosnier, S. Urease–gelatin interdigitated microelectrodes for the conductometric determination of protease activity. *Biosens. Bioelectron.* **2008**, *24*, 489–492. [CrossRef]
32. Thiansilakul, Y.; Benjakul, S.; Shahidi, F. Antioxidative activity of protein hydrolysate from round scad muscle using alcalase and flavourzyme. *J. Food Biochem.* **2007**, *31*, 266–287. [CrossRef]
33. Wen, T.; Li, G.; Kang, J.; Kang, C.; Hyde, K.D. Optimization of solid-state fermentation for fruiting body growth and cordycepin production by *Cordyceps militaris*. *Chiang Mai J. Sci.* **2014**, *41*, 858–872.
34. Li, J.F.; Hoang, V.A.; Ahn, J.C.; Yang, D.U.; Lee, D.W.; Yang, D.C. Isolation of New Strain of *Cordyceps militaris* HB8 and Optimal Condition for Production of Adenosine and Cordycepin in Fruit Body. *Korean J. Plant Resour.* **2020**, *33*, 696–706.
35. Chiang, S.-S.; Liang, Z.-C.; Wang, Y.-C.; Liang, C.-H. Effect of light-emitting diodes on the production of cordycepin, mannitol and adenosine in solid-state fermented rice by *Cordyceps militaris*. *J. Food Compos. Anal.* **2017**, *60*, 51–56. [CrossRef]
36. Wei, C.; Yao, X.; Jiang, Z.; Wang, Y.; Zhang, D.; Chen, X.; Fan, X.; Xie, C.; Cheng, J.; Fu, J. Cordycepin inhibits drug-resistance non-small cell lung cancer progression by activating AMPK signaling pathway. *Pharmacol. Res.* **2019**, *144*, 79–89. [CrossRef] [PubMed]

foods

MDPI

Review

Antifungal Preservation of Food by Lactic Acid Bacteria

Ahmad Nasrollahzadeh [1,2,*], Samira Mokhtari [3], Morteza Khomeiri [4] and Per E. J. Saris [3,*]

1 Department of Food Science and Technology, Urmia University, Urmia 49171, Iran
2 Nobonyad Nasr Food Industry Specialists Company, Tehran 11369, Iran
3 Department of Microbiology, Faculty of Agriculture and Forestry, University of Helsinki, 00014 Helsinki, Finland; samira.mokhtari@helsinki.fi
4 Department of Food Science and Technology, Gorgan University of Agricultural Sciences and Natural Resources, Gorgan 49189, Iran; khomeiri@gau.ac.ir
* Correspondence: a.nasrollahzadeh@urmia.ac.ir (A.N.); per.saris@helsinki.fi (P.E.J.S.); Tel.: +98-938-045-1013 (A.N.); +358-505-203-696 (P.E.J.S.)

Abstract: Fungal growth and consequent mycotoxin release in food and feed threatens human health, which might even, in acute cases, lead to death. Control and prevention of foodborne poisoning is a major task of public health that will be faced in the 21st century. Nowadays, consumers increasingly demand healthier and more natural food with minimal use of chemical preservatives, whose negative effects on human health are well known. Biopreservation is among the safest and most reliable methods for inhibiting fungi in food. Lactic acid bacteria (LAB) are of great interest as biological additives in food owing to their Generally Recognized as Safe (GRAS) classification and probiotic properties. LAB produce bioactive compounds such as reuterin, cyclic peptides, fatty acids, etc., with antifungal properties. This review highlights the great potential of LAB as biopreservatives by summarizing various reported antifungal activities/metabolites of LAB against fungal growth into foods. In the end, it provides profound insight into the possibilities and different factors to be considered in the application of LAB in different foods as well as enhancing their efficiency in biodetoxification and biopreservative activities.

Keywords: synthetic preservatives; preservation enhancement; metabolites; supplementation with LAB

Citation: Nasrollahzadeh, A.; Mokhtari, S.; Khomeiri, M.; Saris, P.E.J. Antifungal Preservation of Food by Lactic Acid Bacteria. *Foods* **2022**, *11*, 395. https://doi.org/10.3390/foods11030395

Academic Editors: Carlo Giuseppe Rizzello and Palmira De Bellis

Received: 30 December 2021
Accepted: 24 January 2022
Published: 29 January 2022

1. Introduction

Fungi are among the most serious food-spoiling micro-organisms threatening the quality and health of food, food products, and feed [1]. Fungal plant-pathogens destroy up to 30% of crop products, and spoiling fungi and their toxins contaminate about 25% of raw materials produced by agriculture worldwide [2]. It is estimated that the annual economic loss caused by the spoilage of bread by fungi will reach to more than EUR 200 million in Western Europe [3].

The disadvantages of using synthetic preservatives such as the formation of carcinogenic nitrosamines in food are well known, though mold species are also becoming resistant to them [4,5]. The biopreservation of food products by natural and biological compounds may be a satisfactory alternative to solving microbial spoilage of food and food products and its consequent economic loss, which will also contribute to reducing the incidence of foodborne illnesses [6].

According to extensive studies in recent decades, LAB being able to produce active compounds such as fatty acids, organic acids, hydrogen peroxide, peptides, and reuterin represent ideal biopreservatives for conventional chemical antifungal preservatives against spoilage and toxigenic compounds in food [7,8]. A total of 25% of Europe's diet and 60% of the diet of many developing countries is composed of fermented food, and LAB play a great role in the fermentation process [9,10]. In addition, LAB cultures isolated from native fermented food products with probiotic attributes and mycotoxin binding may be of immense value in decontaminating mycotoxins in food [11,12].

This review aimed to summarize the capability of LAB as green preservatives in different foods by highlighting their antifungal substances and mechanisms of their action. Moreover, foodborne diseases caused by pathogenic fungi as well as the hazards of synthetic preservatives for human health were outlined. Finally, a comprehensive insight into various aspects of the application of LAB as biopreservatives in foods was provided.

2. Foodborne Diseases

Foodborne diseases (also called foodborne infection or food poisoning) comprise a wide spectrum of diseases resulted from the ingestion of foodstuff spoilage or pathogen microorganisms and toxic chemicals. Foodborne diseases count as a considerable cause of morbidity and mortality, which subsequently pose a remarkable impediment to socioeconomic development all around the world [13]. Since many different pathogenic microorganisms can contaminate food, there is a wide variety of foodborne infections. The Centers for Disease Control and Prevention (CDC) estimated that each year in the United States of America, 48 million people become sick as a result of foodborne illness, 128,000 people are hospitalized, and 3000 people die [14]. According to the WHO, unsafe food causes 600 million cases of foodborne diseases and 420,000 deaths annually worldwide, of which 30% belong to children under 5 years of age. The WHO estimated that eating unsafe food leads to the loss of 33 million years of lives globally each year [15]. The production and release of mycotoxins in food is the most important and dangerous effect caused by fungi to human health [16].

3. Synthetic Preservatives and Hazards of Their Use

Synthetic preservatives are substances of chemical origin that inhibit the growth of spoilage microorganisms. Some examples are benzoates, sorbates, propionate, EDTA, nitrites, and sulfites [17]. The majority of preservatives used today are synthetic rather than natural, and several of them potentially pose life-threatening side effects over time for humans as well as negative impacts on the environment [18]. Researchers have reported that synthetic preservatives can cause serious health hazards such as cancer, allergy, asthma, hyperactivity, and damage to the nervous system [19,20]. A scientific report described the cumulative behavioral effects of bread preservative on children. Daily consumption of preservative in foods has the potential to cause irritability, restlessness, inattention, and sleep disturbance in children [21]. Table 1 shows the most common synthetic antifungal preservatives, their negative effect on human health, and fungi that have developed partial resistance to them.

Table 1. The most common synthetic antifungal preservatives, their negative effect on human health, and fungi that have developed partial resistance to them.

Preservatives	Food	Health Effects	Resistant Fungi	References
Benzoate	Fruit products	Neurotransmission and cognitive functioning		[19]
	Acidic foods	Hyperactivity and allergic reactions		[22]
	Margarine			[23]
	Cereals		*Zygosaccharomyces bailii*	[24]
	Meat	Genotoxic	*Aspergillus flavus*	[25]
	Carbonated drinks	Clastogenic intercalation in the DNA structure	*Aspergillus niger* and *Penicillium notatum*	[5]
			Elymus repens and *A. niger*	[26]
			Aspergillus conicus, Penicillium, Cladosporium and *Wallemia*	[27]
Propionate		Hypersensitivity		[19]
		Visual irritability	*E. repens* and *A. niger*	[26]
	Breads and other baked goods	Restlessness	*A. conicus, Penicillium, Cladosporium* and *Wallemia*	[27]
		Inattention		
		Sleep disturbance	*Penicillium expansum* and *Penicillium roqueforti*	[28]
			P. roqueforti	[29]

Table 1. *Cont.*

Preservatives	Food	Health Effects	Resistant Fungi	References
Sorbate	Syrups Dairy products Cakes Mayonnaise Margarine Processed meats	Cytotoxic and genotoxic effects DNA breakage Irritant to respiratory epithelium	*P. roquefortii*	[19] [30] [31] [32]
			A. flavus	[32]
			P. notatum and *A. niger*	[5]
			Rhizopus nigricans	[4]
			E. repens and *A. niger*	[26]
			A. conicus, Penicillium, Cladosporium and *Wallemia*	[27]

Benzoates mainly inhibit mold, yeasts, and bacteria in liquid environments such as acidic and soft drinks. Sodium benzoate is the most common salt of benzoate used in carbonated drinks, fruit juices, and some other foods with a pH of 3.6 or lower. It is established that benzoate can react with ascorbic acid in drinks and produce benzene, which is a carcinogen [17]. It is also reported to influence neurotransmission and cognitive functioning [33]. Although sodium benzoate is regarded as safe by major regulatory agencies, there are still questions over its adverse effects on human health. Sodium benzoate intake of above 5 mg/kg resulted in allergy and hyperactivity. Sodium benzoate has the potential to cause changes in the cell cycle and impairment in DNA as well as being considered as genotoxic and clastogenic [34].

Propionates inhibit mold growth in baked goods [17]. Although regarded as GRAS by FDA, there is still a lack of clarity on the metabolic effects of propionate in humans. Propionate may cause hyperinsulinemia, promoting adiposity and metabolic abnormalities over time [35]. Propionate preservatives are also reported to contribute to or cause visual irritability, restlessness, inattention, and sleep disturbance in some children [21].

Sorbates prevent mold/yeast growth in food products [17]. Even though sorbate is legally used in the food industry, it still has the potential to cause harmful side effects if consumed in quantities higher than the standard limits or if used long-term [36]. Various research results showed that the increased potassium sorbate intake above 25 mg/kg may lead to producing mutagenic compounds and inducing chromosome damage and DNA breakage and irritation to the respiratory epithelium [36,37].

Apart from negative impacts on health, synthetic preservatives may also adversely affect the organoleptic properties of the food. One of the serious problems in cheese preserved with sorbate is the decomposition of sorbic acid and potassium sorbate to trans-1, 3-pentadiene by resistant strains, and the consequent undesirable taste and odor in cheese, known as kerosene [38,39]. According to Ferrand et al. [40], sorbates might influence the taste of the food, though they are physiologically harmless and less toxic compared to benzoates [40].

Some fungi and yeasts have acquired the ability to resist chemical treatments and preservatives, which consequently creates the demand for a higher dose of the preservatives to be used. Frequent use of common antifungal agents is blamed for causing mutation in the target microorganisms and increasing their resistance [41,42]. It has been reported that some *Penicillium*, *Saccharomyces*, *Zygosaccharomyces*, *Rhizopus*, and *Yarrowia* strains can grow in the presence of potassium sorbate [4,5,43,44]. Additionally, *Z. bailii* and *P. roqueforti* isolates have been reported to be resistant to and even degrade benzoate, respectively [23,44]. These facts together with the demand for least processed foods and the potential hazards of synthetic preservative usage have directed the research sector for seeking alternatives for food preservation.

4. Lactic Acid Bacteria (LAB)

LAB homofermentatives are the species that produce lactic acid as the sole final product, while the heterofermentative ones produce lactic acid, CO_2, and ethanol or acetate. At least half of the final product carbon is a form of lactate [45].

For centuries, LAB have been employed as bacteria performing a central role in a diversity of fermented foods involving milk, vegetables, meats, and sourdough by inducing rapid acidification of the raw material [46,47]. When it is used regularly, LAB-fermented food confers health benefits by strengthening the body in the battle with pathogenic bacterial infections [48].

LAB have also received considerable attention as probiotics over the past few years. Improving health by the biotransformation of different compounds in the gastrointestinal tract into bioavailable ones such as vitamins and short-chain fatty acids by LAB have been reported [49,50]. Immune modulation, anticarcinogenic and antitumor activity, the reduction of cholesterol, alleviation of lactose intolerance, normalization of stool transit, hepatic encephalopathy, and treatment of peptic ulcers are a number of health benefits and indicate the safety of probiotics LAB. Additionally, some modes of action of probiotic LAB are acid tolerance, adhesion to mucus and epithelial cells, production of antimicrobial compounds, and immune stimulation [51–53].

4.1. LAB as Green Preservatives in Food Systems

Fermentation of some foods by LAB strains with antifungal properties has been demonstrated to reduce chemical preservative usage in the food. According to Axel et al. [54], the use of sourdough fermented with specific strains of antifungal LAB can reduce chemical preservatives in bakery products [54].

LAB can be used as natural compounds to replace the chemical preservatives and are associated with health-promoting and probiotic properties [55]. LAB strains with antifungal activity also have the potential to work in synergy with synthetic preservatives. A combination of propionate and sorbate with acetic acid was shown to represent synergistic effects against fungal species of *P. roqueforti* and *A. niger* [56]. In another study, sourdough fermented by antifungal *Lactiplantibacillus plantarum* strains was studied for inhibition activity against *Fusarium culmorum*, *A. niger*, or *P. expansum* spores. Strong synergistic activity was reported when a combination of calcium propionate and the sourdoughs fermented by *L. plantarum* into the bread formulation was applied. The reduced use of calcium propionate up to 1000 ppm maintained inhibition only when the antifungal sourdough was added. Additionally, the increase in shelf life was interestingly higher than that obtained using calcium propionate alone (3000 ppm) [28].

In some research, in situ addition of LAB into food and feed was proven to delay fungal growth. Some examples are in fruits and vegetables, sour cream and semi-hard cheese, quinoa, and rice bread [54,57,58].

In situ application of LAB strains with antifungal activity in some foods have proven potential to act better than synthetic preservatives and competency to replace them in the foods. Rice dough fermented by some LAB isolated from kimchi resisted against three fungal species of *Cladosporium* sp. YS1, *Penicillium crustosum* YS2, and *Neurospora* sp. YS3 much better than that of 0.3% calcium propionate [59]. One *Leuconostoc* and five *Lactobacillus* strains surface sprayed on bakery products were shown to delay the growth of some resistant and semi-resistant fungi to calcium propionate, potassium sorbate, and sodium benzoate [26]. In the study of Mandal, Sen, and Mandal [60], the antifungal compound of *Pediococcus acidilactici* LAB 5 at a high dilution (0.43 mg mL^{-1}) exerted a greater inhibition of *Curvularia lunata* conidia than sodium benzoate [60]. Valerio et al. [61] also reported that *Leuconostoc citreum*, *Weissella cibaria*, and *Lactobacillus rossiae* isolated from Italian durum wheat semolina inhibit fungal strains of *A. niger*, *P. roqueforti*, and *Endomyces fibuliger* to the same or a higher extent in comparison with calcium propionate. The results of the study indicated a potent inhibitory activity of the ten LAB strains used in their study compared to that obtained with calcium propionate (0.3% *w*/*v*) against the most widespread contaminant of bakery products, *P. roqueforti* [61].

4.2. Antifungal Activity Spectrum of LAB

LAB have a reported potential use as adjunct or starter cultures to inhibit fungi growth in the final products such as fruit and vegetables, dairy, and bakery. Twenty LAB isolates from fermented cassava were investigated against fungal pathogens associated with the spoilage of vegetables and fresh fruits. Strong inhibition of the radial growth and spores of the fungal pathogens was observed when the products were inoculated with the antifungal metabolites produced by the strains [57].

In cheese, *Lactobacillus amylovorus* DSM 19280 was used as an adjunct culture in a cheddar cheese model system contaminated with *P. expansum* spores. The presence of the strain resulted in a four-day delay in *Penicillium* growth on the cheddar cheese compared with the control [62]. *Lactobacillus rhamnosus* A238 was also shown alone or in combination with *Bifidobacterium animalis* to inhibit mold growth on cottage cheese for at least 21 days at 6 °C [63]. In another study, 12 selected *L. plantarum* isolates were inoculated into cottage cheese challenged with *Penicillium commune*. All the isolates were found to prevent the obvious *P. commune* growth on cottage cheese by between 14 and more than 25 days longer than the control [64]. *Lactobacillus brevis* and *Enterococcus faecium* isolated from "chal", a product from yogurt, reduced the growth of *Rhodotorula glutinis* in doogh, diluted yogurt, over 15 days of storage [65].

In sour cream and semi-hard cheeses, Lactobacillus paracasei CIRM-BIA1759 and L. rhamnosus CIRM-BIA1761 were tested as adjunct cultures. In situ assays showed that the strains postponed the growth of *P. commune*, *Rhodotorula mucilaginosa*, and *Mucor racemosus* on sour cream for 2–24 days and also delayed the growth of *P. commune* in semi-hard cheese for 1–6 days [58]. Ouiddir et al. [66] tested the antifungal activity of *L. plantarum* CH1, *L. paracasei* B20, and *Leuconostoc mesenteroides* L1 in sour cream and sourdough bread challenged with fungal spoilers. The strains delayed the growth of the *Aspergillus tubingensis*, *A. flavus*, *P. commune*, and *M. racemosus* for up to 5 days in sourdough bread. In sour cream, *L. plantarum* CH1 and *L. paracasei* B20 completely inhibited *P. commune* growth for 5 and 3 days, respectively [66].

In bakery products, in situ sprays of one *Leuconostoc* and five *Lactobacillus* strains delayed one or several fungal species growths. The incorporation of the same strains in milk-bread-roll preparation also delayed fungal growths [26]. In another study, two strains of *Lactobacillus* were used for sourdough fermentation of quinoa and rice flour. *L. reuteri* R29 and *L. brevis* R2Δ fermented sourdough bread reached a shelf life of quinoa and rice from 2 to 4 days, respectively [54]. A Chinese steamed bread manufactured with *L. plantarum* CCFM259 did not show any fungal contamination until 7 days of storage, a similar level of inhibition compared with that obtained by 0.25% (*w*/*w*) calcium propionate [67]. Fermenting rice dough with some LAB isolated from kimchi greatly retarded the growth of three fungal species from *Cladosporium*, *Neurospora*, and *Penicillium* genus in the rice cakes [59].

Different LAB isolates have the potential to synergically inhibit fungal growth in food. Seven strains of LAB were selected and tested for their anti-penicillium activity to prevent *Penicillium chrysogenum* growth in cottage cheese. They found that some of the strains act in synergy, and their combination has potential for use as bio-preservatives in fresh cheese [63].

Antifungal activity of LAB depends on the pH, temperature, growth media, incubation time, nutrients, antifungal compounds, production levels, and mode of action [68]. Mandal, Sen, and Mandal [60] observed that the production of antifungal compound(s) from *P. acidilactici* LAB 5 against pathogenic fungi showed a great dependency on media specifications. TGE, and TGE + Tween 80 media did not support the production of any antifungal compounds, while the fungal growth was completely restricted in MRS agar media [60]. Another study reported that supplementation of WFH media with 2.5% olive oil and 150 mM glycerol raised the antifungal activity of *L. brevis* Lu35 and *L. reuteri* 5529, respectively [26]. The addition of linoleic acid supported the antifungal activity of *Lactobacillus hammesii* [69]. Rouse et al. [70] reported that when grown in different carbon sources, the antifungal activity of the LAB strains tested was stable, although the quantity of metabolites produced varied depending on the carbon source. Among the sugars tested, for three out of four strains, glucose and lactose were the best and worst, respectively [70].

The incubation time has been observed to greatly influence the antifungal activity of LAB. Rouse et al. [70] observed that the four tested LAB cultures were unable to grow at 10 and 42 °C, and consequently, no growth was observed. Incubation between 21 and 37 °C, however, improved growth, and the bacteria presented different levels of antifungal activity with the optimal production of the antifungal compounds between 25 and 30 °C [70]. The antifungal activity of *Lactobacillus coryniformis* subsp. *coryniformis* strain Si3 was slightly better at 30 °C as compared to 25 °C for 40 h. The production of antifungal compounds by the strain was reported to begin in the log phase and reach a maximum level in the early stages of the stationary phase followed by a drop in activity [71].

The antifungal activity of LAB has also been found to be influenced by pH. The antifungal activity of *L. plantarum* K35 was reported to be pH-dependent and favorable to acidic conditions [72]. Antifungal properties of *L. coryniformis* subsp. *coryniformis* strain Si3 were also observed at maximum at pH values of between 3.0 and 4.5, with a decrease in pH between 4.5 and 6.0, and loss at higher pH values. Readjustment of the pH to 3.6 fully returned the activity [71]. In another study, antifungal attributes of the four LAB strains were found to be good at pH 3, moderate at pH 5, and low at pH 7, although poor fungal inhibition was maintained at pH 8 [73]. The effect of temperature and pH on the antifungal properties of the *L. plantarum* strain against *Aspergillus fumigatus* and *Rhizopus stolonifer* in temperatures ranging from 20 °C to 40 °C and pH ranging from 4.0 to 7.0 for 48 h of incubation was investigated. A combination of 30 °C and pH 6.5 °C presented optimum antifungal activity [74].

The role of the concentration of supernatant in the antifungal activity of LAB were also highlighted by Shehata et al. (2019), where they observed that increasing the supernatant concentration of *Lactobacillus* sp. RM1 decreased the growth of *Aspergillus parasiticus*, *A. flavus*, and *Aspergillus carbonarius* [18].

Fermentation time was also found to be effective in the antifungal activity of LAB. Longer fermentation times of barley malt substrate fermented by LAB resulted in higher carboxylic acids released by them against *F. culmorum* macroconidia. The maximal concentrations of the acids were obtained after 48 h of fermentation [75]. Among the four LAB strains studied by Muhialdin, Hassan, and Saari [73], the highest antifungal activity of *Lc. mesenteroides* and three *L. plantarum* occurred in different incubation times of 24 h and 48 h, respectively. They highlighted the significance of incubation time, growth stages, and temperature for the production of antifungal compounds. According to them, maximizing the production of inhibitory compounds could be obtained by determining the optimum growth conditions [73].

The inhibitory activities of LAB are strain specific. Selecting the best strain/combination of strains of LAB for biopreservation that would cause the minimum unfavorable changes in the product requires prior experiments. In a study, more than 200 yeast and 200 LAB strains were tested as biopreservatives against fungal growth during the cocoa fermentation process. The most promising candidates among all belonged to only four species of *Lactobacillus fermentum*, *L. plantarum*, *Saccharomyces cerevisiae*, and *Candida ethanolica* [76]. *Lactobacillus coryniformis* subsp. *coryniformis* strain Si3 was observed to have strong inhibitory activity against *A. fumigatus*, *P. Roqueforti*, *Aspergillus nidulans*, *Mucor hiemalis*, *Fusarium. graminearum*, *Talaromyces flavus*, *Fusarium poae*, *F. culmorum*, and *Fusarium sporotrichoides*. A weaker activity from the same strain was observed against *Kluyveromyces marxianus*, *Debaryomyces hansenii*, and *S. cerevisiae* while displaying no activity against *Sporobolomyces roseus*, *R. glutinis*, and *Pichia anomala* [71]. Further support for this is another study where the inhibitory percentage of *L. brevis* was stronger than *E. faecium* against *P. chrysogenum* [77].

The selected strains must be adapted or adaptable to the environmental conditions of target food, as well as the production process through which the food is prepared, so that their activity and release of antifungal compounds can be reasonably expected during storage. *L. reuteri* was added to a fermented milk product to inhibit pathogens and spoilage microorganisms. No change in the pH, acidity, soluble solids, color, or rheological aspects of the fermented milk product in the presence of reuterin was observed [78]. Fermentation of an oat-based beverage by *L. plantarum* UFG 121 also best preserved it against *F. culmorum*, causing no differences in terms of some qualitative features as compared to the control [79].

Bacterial metabolite profiles of LAB could sometimes be beneficially modulated, altering their spectrum of antifungal activity. Both the culture medium and the target fungal species determine the antifungal activity of LAB. The quality and quantity of the antifungal metabolites of *Lactobacillus pentosus* ŁOCK 0979 were reported to be dependent on the culture medium compounds. The presence of galactosyl-polyols and gal-erythritol improved the anticandidal properties of *L. pentosus* ŁOCK 0979. The addition of the culture medium of the strain conferred an inhibitory attribute against *Aspergillus brassicicola* and *A. niger.*

4.3. Antifungal Metabolites of LAB

LAB inhibitory compounds are secondary metabolites produced after 48 h of fermentation [70]. Shehata et al. [18] observed that the production of the antifungal metabolites of the LAB strain *Lactobacillus* sp. RM1 was initiated at the growth log phase (12–14 h), reached the highest at the strain stationary phase (24 h), and remained stable. The proposed mechanisms explaining the fungal inhibitory effect of LAB are competition over the available nutrients and space, clogging the pathogen's path through the matrix, and manipulation of the spore membrane, causing viscosity and permeability [80]. Sangmanee and Hongpattarakere [72] revealed that the mechanism of antifungal action of *L. plantarum* K35 supernatant causes damage to the cytoplasmic membrane and cell wall and consequent leakage of cytoplasmic content, the formation of membrane-bound vesicles followed by the destruction of mitochondria and nuclei [72].

Lactic acid, formic acid, acetic acid, caproic acid, and phenyllactic acid (PLA), as organic acids, as well as other metabolites from LAB such as carbon dioxide, hydroxyl fatty acids, hydrogen peroxide, diacetyl, ethanol, reuterin, cyclic dipeptides, protein compounds, reutericyclin, proteinaceous, acetoin, and volatile compounds such as diacetyl are natural antimicrobial and antifungal metabolites produced by LAB [7,58]. Table 2 summarizes the number of LAB studied for their antifungal metabolites, fungal spectrum of activity, and their in situ application in the last 10 years.

Table 2. Selected LAB studied for their antifungal metabolites, fungal spectrum of activity, and their in situ application in the last 10 years.

LAB Isolate	Antifungal Compound	Activity Spectrum	Food Product	Reference
L. pentosus G004 *L. fermentum* Te007 *L. paracasi* D5 *Pediococcus pentosaceus* Te010	Protein-like compounds	*A. niger* and *Aspergillus oryzae*	Bread Tomato Cheese	[81]
L. amylovorus DSM 19280	Acetic acid Lactic acid Hydrocinnamic acid Azelaic acid 4-Hydroxybenzoic acid	*P. expansum*	Cheddar cheese	[62]
L. plantarum LR/14	Antimicrobial peptides AMPs LR14	*A. niger, Rhizopus stolonifera, M. racemosus* and *P. chrysogenum*	Wheat grain	[82]

Table 2. *Cont.*

LAB Isolate	Antifungal Compound	Activity Spectrum	Food Product	Reference
L. plantarum	Phenolic acids Organic acids	*F. culmorum* *P. expansum* MUCL2919240	Barley malt Bread grapes	[75] [83]
L. fermentum, L. plantarum	Phenyllactic acid Organic acids	*A. flavus, Penicillium citrinum, Penicillium griseufulvum, A. niger* and *A. fumigatus*	Cocoa beans	[76]
L. reuteri	Reuterin	*P. chrysogenum* and *M. racemosus*	Yogurt	[84]
L. pentosus, L. plantarum, L. brevis, Lactobacillus delbrueckii, L. fermentum, Lactococcus lactis and *Lc. mesenteroides*	Hydrogen sulphide and lactic acid	*Penicillium oxalicum, Fusarium verticillioides* and *A. niger*	Fruits and vegetables	[57]
Lactobaciullus strains	Organic acids	*P. chrysogenum* and *A. favus*	Caciotta cheese	[85]
L. plantarum CECT 749	Gallic, chlorogenic, caffeic and syringic acids	*Fusarium* spp. *Penicillium* spp. and *Aspergillus* spp.	Bread	[86]
Number of LAB strains isolated from Kimchi	Lactic acid and acetic acid	*Cladosporium* sp. YS1, *Neurospora* sp. YS3, and *P. crustosum* YS2	Rice cake	[59]
Leuconostoc spp. *L. reuteri* and *L. buchneri*	Organic acids such as lactic acid, acetic acid and propionic acid	*Aspergillus, Eurotium, Penicillium, Cladosporium* and *Wallemia* spp.	Milk bread rolls	[26]
L. plantarum CH1, *L. paracasei* B20 and *Lc. mesenteroides* L1	Lactic acid and acetic acid	*M. racemosus, Penicillium commune, Yarrowia lipolytica, A. tubingensis, A. flavus* and *Paecilomyces*	Sour cream and sourdough bread	[66]

The antifungal metabolites of LAB have the potential to act in synergy. The synergistic effect between the decrease in pH resulted from the production of organic acids and other antifungal metabolites of LAB poses a more efficient final antifungal activity [7]. Peyer et al. [75] demonstrated that there are synergistic effects between organic acids and phenolic acids released by some LAB strains as antifungal metabolites against *F. culmorum*. According to Alex et al. [54], the great synergistic effect between organic acids and antifungal peptides produced by LAB allow the final biopreservation attribute to be influential in bakery products [54].

Antifungal metabolites of LAB have also shown synergy with compounds of other organisms. Ruggirello et al. [76] tested some yeast and LAB strains against six spoilage fungi belonging to *Aspergillus* and *Penicillium* genera during the cocoa fermentation process. The antifungal activity was explained by the synergic production of organic acids (from the LAB) and proteinaceous compounds (from yeasts) [76].

The understanding of the synergy mechanism between antifungal compounds could provide insight in maximizing the impact whilst altering the involved compositions of bacteria or nutrients, eventually leading to actual application in food [75].

4.3.1. Organic Acids

The production of organic acids is believed to determine an LAB strain's mycotoxigenic fungi inhibition properties; the type and quantity of the acids differ from strain to strain [87]. These acids are mainly produced by LAB as a byproduct of acidification process rather than as active synthesis of metabolic compounds aimed at restricting fungi [88].

As the main acid produced by LAB, lactic acid (2-hydroxy propionic acid) is an organic acid widely distributed in nature in two forms of L and D; L lactic acid was recognized as a safe preservative by the FDA [89]. Russo et al. [79] tested the activity of some LAB strains against *Aspergillus, Fusarium*, and *Penicillium* and reported that lactic acid was produced at a high concentration during the growth phase as the main metabolic antifungal associated with the low pH. Lactic acid was also identified as the main antifungal compound from *E. faecium, L. rhamnosus*, and *L. plantarum* [90]. In the study of Baek et al. [59], the fermentation of rice dough with some LAB isolates from kimchi greatly delayed the growth of three fungal species in rice cakes. They found lactic acid and acetic acid as the main antifungal substances [59].

The best characterized and most important antimicrobials produced by LAB are lactic acid and acetic acid, which are bioactive in the protonated form at low pH [91]. LAB can produce a variety of compounds at low concentrations and below their minimum inhibitory concentration, which are likely to act synergistically with lactic acid and acetic acid [92,93]. Acetic acid and lactic acid were also proved to display a synergistic antifungal activity in combination; however, due to higher pKa that causes a higher level of dissociation inside the cell, acetic acid has a stronger antifungal activity [94,95]. Loubiere et al. [96] suggested that lactic acid has an inhibitory effect on the metabolism and cell proliferation, which is probably due to the synergistic effect with some of the other side fermentation products such as acetic acid and formic acid [96].

Organic acids of lactic acid, oleic acid, linoleic acid, palmitic acid, 3-PLA, stearic acid, pyroglutamic acid, and 5-oxo-2-pyrrolidine-carboxylic acid were detected as antifungal compounds from *L. plantarum* K35 inhibiting the growth and aflatoxin production of *A. flavus* and *A. parasiticus* [72]. Some other carboxylic acids including benzoic, vanillic, azelaic acid, hydrocinnamic acid, and hydroxy benzoic acid were isolated as antifungal compounds from mediums of *Weissella cibaria* PS2 and three *Lactobacillus* species [97]. Hydrocinnamic acid, azelaic acid, vanillic, p-couramic, and 4-hydroxy benzoic acid were also reported from *L. reuteri* eep1 with antifungal activity [98].

The mechanism of inhibitory activity of organic acids in the growth and activity of many pathogenic and putrefactive bacteria and fungi is attributed to creating an acidic environment and reducing the pH to below the metabolic inhibition and growth range [94]. Loubiere et al. [96] suggested that the inhibitory effect of lactic acid on the metabolism and cell proliferation is probably due to the increase in osmotic pressure of the medium. Organic acids alter plasma membrane permeability and electrochemical, killing the microorganism [80]. In other words, organic acids diffuse to the fungi through the membrane and degrade the cells, thereby releasing hydrogen ions and causing a decrease in pH [99].

For organic acids to penetrate the cell wall, they must be turned to an undissociated form. The pKa of lactic, acetic, caproic acid, and 3-phenyl-L-lactic is 3.8, 4.7, 4.9, and 3.5, respectively [99]. Therefore, in an acidic environment, they will act more efficiently in inhibiting fungi. This fact was proven in the study of Cortés-Zavaleta et al. [7], where they tested the fungal inhibition of the cell-free supernatant of 13 LAB strains against four food-spoilage fungi. The results demonstrated that the inhibition properties dropped when the pH was raised to 6.5 [7].

The production of organic acids confers extra inhibition properties to LAB by activating other antifungal compounds triggered as a result of lowering the pH [99]. Furthermore, organic acids often work in synergy with other compounds, which adds to the complexity of LAB antifungal activity [54]. There is also a synergistic effect between several organic acids together produced by LAB as antifungal compounds. The antifungal activity of *L. plantarum* CCFM259 was assessed against *P. roqueforti*. Acetic acid and PLA showed better antifungal activity than other compounds, and their mixture displayed a synergistic effect [67]. The synergistic contribution of acetic acid was reported in the extension of sourdough fermented by two *Lactobacillus* strains [54].

4.3.2. Phenyllactic Acid (PLA)

PLA (2-hydroxy-3-phenyl propionic acid) is another well-studied organic acid with natural antibacterial properties derived from phenylalanine catabolism. PLA possesses a similar metabolic pathway as lactic acid and is metabolized during fermentation by the glycolytic enzyme and lactate dehydrogenase [100].

The composition of the culture medium was reported to play a great role in the quantity of PLA produced by LAB. The addition of 1.5% (*w*/*v*) phenylalanine to MRS medium of *L. reuteri* R29 significantly increased the production of PLA and, consequently, antifungal performance against *F. culmorum* [101]. The fungal inhibitory strength of PLA produced by

LAB are well-established. Lavermicocca et al. [102] reported that a 10-fold-concentrated culture supernatant of *L. plantarum* 21B inhibited *Eurotium*, *Fusarium*, *Penicillium*, *A. monilia*, and *Endomyces*. Under the same conditions, 3 mg mL^{-1} calcium propionate was not effective, while sodium benzoate performed similar to *L. plantarum* 21B. The antifungal activity of *L. plantarum* 21B was attributed to PLA and 4 OHPLA isolated from the supernatant of the bacteria [102]. In a similar study, *L. plantarum* UM55 was found to produce lactic acid, PLA, OHPLA, and indole lactic acid (ILA). The acids were individually tested against *A. flavus*, and among them, PLA showed the strongest effects with the obtained IC90 for the growth inhibition of 11.9 mg mL^{-1} [103]. PLA and 3,5-Di-O-caf- feoylquinic acids were identified as the predominant antifungal compounds in cell-free supernatant of seven LAB isolated from traditional fermented Andean products with inhibitory activity against a few spoiler fungi from *Penicillium* and *Aspergillus* genus [104]. The antifungal compounds of *L. plantarum* against *A. fumigatus* and *R. stolonifera* resembled the structure of 3PLA with the formed ligands [74].

Although promoting the metabolic pathway of PLA is likely to increase the efficiency, the antifungal properties of PLA depend on a synergistic mechanism with other metabolites [101]. Cortés-Zavaleta et al. [7] tested the fungal inhibition of the cell-free supernatant of 13 LAB strains against four food-spoilage fungi. With two exceptions, all other LAB strains produced PLA ranging from 0.021 to 0.275 mM. They concluded that even if PLA cannot be the only inhibitory compound, it very likely performs in synergy with other acidic compounds from LAB [7]. PLA and acetic acid produced from *L. plantarum* CCFM259 exhibited a synergistic inhibitory effect against *P. roqueforti* [67]. A weak synergistic inhibitory effect of PLA was also reported in combination with cyclo (L-Phe–L-Pro) produced from *L. plantarum* against *A. fumigatus* and *P. roqueforti* [105]. Acetic acid, lactic acid, and PLA produced by *L. plantarum* VE56 and *Weissella paramesenteroides* LC11 exhibited synergism against *A. tubingensis*, *A. niger*, *Candida albicans*, and *P. crustosum* [106].

4.3.3. Reuterin

Reuterin (β-hydroxy propionaldehyde) is a low-molecular-weight multi-compound system consisting of 3-HPA hydrate, 3-hydroxypropionaldehyde (3-HPA), 3-HPA dimer, and acrolein produced by the conversion of glycerol [107]. Reuterin is secreted mainly by *L. reuteri*, though some other bacterial species and genera could also secrete it [107]. Reuterin has a wide spectrum of antimicrobial properties against a range of Gram-positive and Gram-negative bacteria, bacterial spores, molds, yeasts, and protozoa [84,94].

The growth conditions and culture medium can alter the content of reuterin produced by LAB. Schaefer et al. [108] reported the optimum conditions for reuterin production from *L. reuteri* 1063 as culturing the cells for 16 h followed by suspension in 5 mL of 250 mM glycerol in distilled water and incubated for 2 h at 37 °C under anaerobic conditions. Another study reported that supplementation with 150 mM glycerol increased the antifungal activity of *L. reuteri* 5529 cultured in WFH medium. The enhanced antifungal activity of *L. reuteri* 5529 was linked to the production of reuterin. In a similar study, glycerol addition to the culture medium of *L. coryniformis* improved reuterin synthesis, consequently having an antifungal effect against yeast cells and fungal spores and conferring inhibition performance against a couple of new fungal strains [109].

The activity of LAB against fungi is mostly limited to antifungal rather than fungicidal. Reuterin, however, apart from antifungal activity, has also presented a fungicidal effect. Purified Reuterin produced by *L. reuteri* ATCC 53608 fungicidal activity by killing 99.9% of the indicator microorganisms at concentrations equal or below 15.6 mM. As an antifungal agent, it was then added to yogurt. In yogurt also, reuterin exhibited an antifungal effect at a concentration of 1.38 mM while a fungicidal effect at 6.9 mM [84].

The mechanism of action of reuterin has been reported to cause oxidative stress to fungal cells. Reuterin exposure *E. coli* increased the expression of genes regulated and expressed in response to periods of oxidative stress. It was determined that the aldehyde group of reuterin binds to thiol groups of small peptides and other molecules, leading to oxidative stress, which is hypothesized as the mechanism of inhibition [108]. Another proposed inhibition mechanism of reuterin is through the suppression of ribonuclease activity, which is the main enzyme mediating the biosynthesis of DNA [110], as cited in [68]. More recently, acrolein was reported as the main component conferring antimicrobial activity to reuterin [107].

Reuterin could be a potentially promising candidate as a food biopreservative since in vitro studies using human liver microsomes demonstrated that reuterin does not present the possibility of displaying drug interactions [63]. *P. expansum* was inhibited at concentrations of above 10 mM reuterin produced by *L. reuteri* ATCC 53608 [78]. The addition of *L. reuteri* INIA P572 with glycerol to semi-hard ewe milk cheese resulted in a lower level of 2-heptanone in cheese, which was attributed to the activity of reuterin in mold inhibition [111]. Reuterin was also found to be responsible for the antifungal performance of three *Lactobacillus* and one *Leuconostoc* strains applied in pound cake and milk bread rolls [26].

4.3.4. Peptides and Cyclic Peptides

The antimicrobial peptides are chains of 5–100 amino acid attached through peptide bonds with natural origin (held together through peptide bonds [112]. Protease enzyme treatment is usually employed to determine the peptide nature of the active compounds. The treatment of *Lactobacillus fermentum* CRL 251 supernatant with trypsin, proteinase K, and pepsin decreased the antifungal activity by 50, 4, and 3%, respectively. Further ultrafiltration analysis attributed the activity to smaller fraction of peptides (<10 kDa) [113]. In the study of Magnusson and Schnürer [71], *L. coryniformis* subsp. *coryniformis* presented a strong inhibitory activity against a number of fungi and yeast strains. The activity was attributed to the production of small (3 kDa) and heat-stable proteinaceous antifungal compounds demonstrated by the alteration of activity through the treatment with proteinase K, trypsin, and pepsin [71]. A group of peptides was purified and identified from cell-free supernatants of *L. plantarum* exhibiting inhibitory activity against *A. parasiticus* and *P. expansum* by 58% and 73%, respectively [112]. A total of 37 peptides were identified in the fraction of cell-free supernatant of *L. plantarum* TE10. Treatment of bread with the fraction resulted in slight growth and a fourfold reduction in spore formation of *A. flavus* [73].

Smaller peptides usually possess stronger antifungal activity. Low-molecular-weight peptides (<10 KDa) isolated from the supernatants of four LAB strains represented higher antifungal inhibition against six fungi in comparison with the control supernatant [73].

The mechanism of action of peptides is through binding to lipid bilayers in carpet-like and puncturing channels in it, which impairs the function. They also act through peptide-lipid interaction resulting in phase separation as well as solubilizing the membrane [114].

Low-molecular-weight peptides with high heat-stability from LAB have high potential for replacing chemical preservatives commonly used in the bakery [73]. In the study of Muhialdin, Hassan, and Saari [73], they simulated the maximum heat process of food in manufacturing (121 °C for 60 min) and exposed the supernatants of four LAB strains with antifungal activity to it. The nature of the bioactive substances secreted by bacteria was found to determine whether the activity is heat sensitive and to what extent [73].

Cyclic peptides are composed of polypeptide chains linked covalently in a circular manner. The circular structure is formed either by binding either ends of the peptide chain through an amide bond, or by lactone, thioether, ether, or disulfide bonds [115]. The cyclic dipeptide properties as antifungal agents produced by LAB have been shown in several studies and reviews. *L. plantarum* CM8, *Weissella confusa* I5, *P. pentosaceus* R47, and *W. cibaria* R16 presented inhibitory activity against *P. notatum*. Concentrated supernatants were heated to 80 °C for 1 h followed by an autoclavation step (121 °C for 15 min). No significant influence of heat was observed in the activity of the supernatants. Protease sensitivity properties of the activity implied that the bioactive substances most likely have a proteinaceous nature, perhaps (cyclic) peptides [70].

Cyclo peptide (glycyl-L-leucyl), as a compound that delays the growth of fungi *Fusarium avenaceum*, was isolated from *L. plantarum* [116]. Magnusson [109] also reported the secretion of cyclic dipeptides by *P. pentosaceus* (MiLAB 024), *L. plantarum* (MiLAB 006), and *Lactobacillus sakei* (MiLAB 091). The growth condition of the LAB strain was found to be influential in the quantity of the cyclic peptides released by them. The results obtained by Ryan et al. [88] revealed that acidification of dough fermented by *L. plantarum* FST significantly increased the quantity of cis-cyclo (LPhe-L-Pro) and cis-cyclo (L-Leu-L-Pro) as compared to nonacidified dough [88].

The mechanism of action of antimicrobial cyclic peptides is mainly attributed to the disruption of structural integrity. They target the cell envelope components, causing lysis of the membrane or inhibiting the membrane and/or cell wall biosynthesis [117].

4.3.5. Fatty Acids

Fatty acids are organic acids that possess a carboxyl group (-COOH) and a methyl group (-CH3) at either end [118]. Strong antifungal activities have been reported from fatty acids. Sjogren et al. [119] characterized 3-hydroxydodecanoic acid, 3-hydroxydecanoic acid, 3-3-hydroxy-5-cis-dodecenoic acid, and hydroxytetradecanoic acid from the supernatant of *L. plantarum* MiLAB 14. The hydroxy fatty acids displayed inhibition in the range 10 to >100 µg/mL and were reported to be much more effective than cyclic dipeptides against several molds and yeasts [119].

Fatty acids have been reported in a number of studies to be the main antifungal metabolite preserving foods fermented by LAB. Fermentation of sourdough bread and sour cream by three isolates of LAB of *L. plantarum* CH1, *L. paracasei* B20, and *Lc. mesenteroides* L1 delayed fungal growth in the final food. The main produced compounds were detected to be DL-hydroxyphenyl, 3, 3-(4-hydroxyphenyl) propionic, 4-dihydroxyhydrocinnamic, and 3-(4-hydroxy-3-methoxyphenyl) propanoic acids [66]. Black et al. [69] reported that *L. hammesii* converts linoleic acid to antifungal $C_{18:1}$ monohydroxy fatty acids. Further supplementation of linoleic acid strengthened the antifungal activity of *L. hammesii*. Hydroxylated fatty acids synthesized by the strain were found to be responsible for the extended shelf life of sourdough fermented with *L. hammesii* and the inhibition of *A. niger* and *P. roqueforti* in the bread prepared by that [69].

Little knowledge of the antifungal mechanisms of fatty acids is available so far; however, some pathways have been proposed. Detergent-like properties of the fatty acids affect the structure of cell membranes of the cells, leading to death [119]. Antifungal fatty acids disintegrate lipid bilayers of the membranes and, consequently, cause destruction of the membrane integrity, leading to the disintegration of cells and release of intracellular proteins and electrolytes [120]. Other targets of fatty acids include protein synthesis, which may be inhibited by myristic acid analogues, fatty acid metabolism, as well as topoisomerase activity, which may be inhibited by, amongst others, acetylenic fatty acids [121].

The antifungal activity of fatty acid highly depends on the structure. In the study of Black et al. [69], unsaturated monohydroxy fatty acids were antifungally active; saturated hydroxy fatty acids and unsaturated fatty acids of oleic and stearic acids, however, did not exhibit any activity. This implies the fact that for the fatty acid to function as an antifungal agent, at least one double bond as well as one hydroxyl group along a C18 aliphatic chain should be present in the structure [69].

Pathogenic fungi are less likely to become resistant to antifungal fatty acids [121]. Other antifungal compounds targeting the membrane of fungi are more susceptible to pathogen resistance, which shortens their lifespans. However, as these substances could present synergism with antifungal fatty acids, they could alternatively provide prolonged usage, reducing the required quantity of the antifungal substances [120]. An example of the synergism of fatty acids with other compounds was provided by the study of Ndagano et al. (2011). They observed that 3-hydroxylated produced by *L. plantarum* VE56 and *W. paramesenteroides* LC11 acts in synergy with other bacterial compounds secreted by the bacteria inhibiting *A. niger*, *A. tubingensis*, *C. albicans*, and *P. crustosum* [106].

5. Conclusions

Fungal growth and consequent mycotoxin release in food and feed threaten human health, which might even, in acute cases, lead to death. Addressing the consumer health concern as well as the potential negative risk of using synthetic preservatives, the substitution of LAB as a green preservative could be an alternative due to their safety, health-giving benefits, and preservation properties. LAB release antifungal metabolites against fungal species, which in many cases work in strong synergy.

The application of LAB species with antifungal properties in food can reduce the occurrence of fungal spoilage and toxicity, consequently improving its shelf life as well as causing a reduction in mycotoxins. However, case investigation is required to be carried out individually for each food candidate since the presence of LAB in food can exclusively affect its physiochemical and organoleptic properties, which may or not be desirable. On the other hand, the major population of fungi contaminating a particular food should be regarded in selecting the best LAB/combinations of LAB planned for inhibiting fungal growth in the food. The reason for that is the fact that the antifungal properties of LAB are fungal strain-specific, meaning that an LAB strain might be strongly active against a fungal strain while not causing much disturbance in the viability of another strain.

Almost all antifungal metabolites of LAB present synergy with at least one other component. This fact counts as an advantage in employing LAB as antifungal bacteria in a way to group main producers of synergic components together, thereby maximizing the final activity. The composition of the medium has also been demonstrated to be a significant factor stimulating/raising the release of antifungal compounds by LAB. Therefore, the food nutrient composition is another item to take into account when selecting LAB strains to inhibit fungal growth/mycotoxin control in food. If the formula of the food allows, supplementation with additives along the LAB strains could be a desired alternative, e.g., the addition of phenylalanine along with *L. reuteri* to food in order to increase PLA release.

For an advance in academic studies, enhancement in protection and the safety of products by LAB as probiotics and biopreservatives could be pursued. For food industrial researchers as well, the isolation, formulation, and industrialization of LAB antifungal bioactive metabolites could be of interest.

Author Contributions: Original draft preparation, A.N. and S.M.; review and editing, M.K. and P.E.J.S. All authors have read and agreed to the published version of the manuscript.

Funding: Samira Mokhtari used personal grants from Niemi Säätiö (grant number 20200071), Elintarvikkeiden Tutkimussäätiö (26.4.2021), Walter Ehlströmin Säätiö (30.3.2021).

Institutional Review Board Statement: Not applicable.

Acknowledgments: Open access funding provided by University of Helsinki.

Conflicts of Interest: There is no conflict of interest between the company and the research in this study.

References

1. Agriopoulou, S.; Stamatelopoulou, E.; Sachadyn-Król, M.; Varzakas, T. Lactic Acid bacteria as antibacterial agents to extend the shelf life of fresh and minimally processed fruits and vegetables: Quality and safety aspects. *Microorganisms* **2020**, *8*, 952. [CrossRef] [PubMed]
2. Sellamani, M.; Kalagatur, N.K.; Siddaiah, C.; Mudili, V.; Krishna, K.; Natarajan, G.; Rao Putcha, V.L. Antifungal and zearalenone inhibitory activity of *Pediococcus pentosaceus* isolated from dairy products on *Fusarium graminearum*. *Front. Microbiol.* **2016**, *7*, 890. [CrossRef] [PubMed]
3. Sevgi, E.; Tsveteslava, I.I. Antifungal activity of lactic acid bacteria, isolated from Bulgarian wheat and rye flour. *J. Life Sci.* **2015**, *9*, 1–6. [CrossRef]
4. Kowalczyk, D.; Kordowska-Wiater, M.; Zięba, E.; Baraniak, B. Effect of carboxymethylcellulose/candelilla wax coating containing potassium sorbate on microbiological and physicochemical attributes of pears. *Sci. Hortic.* **2017**, *218*, 326–333. [CrossRef]
5. Heydaryinia, A.; Veissi, M.; Sadadi, A. A comparative study of the effects of the two preservatives, sodium benzoate and potassium sorbate on *Aspergillus niger* and *Penicillium notatum*. *Jundishapur J. Microbiol.* **2011**, *4*, 301–307. Available online: www.sid.ir/en/journal/ViewPaper.aspx?id=211841 (accessed on 29 December 2021).

6. Gálvez, A.; Abriouel, H.; López, R.L.; Omar, N.B. Bacteriocin-based strategies for food biopreservation. *Int. J. Food Microbiol.* **2007**, *120*, 51–70. [CrossRef]
7. Cortés-Zavaleta, O.; López-Malo, A.; Hernández-Mendoza, A.; García, H. Antifungal activity of lactobacilli and its relationship with 3-phenyllactic acid production. *Int. J. Food Microbiol.* **2014**, *173*, 30–35. [CrossRef]
8. Yang, E.; Chang, H. Purification of a new antifungal compound produced by *Lactobacillus plantarum* AF1 isolated from kimchi. *Int. J. Food Microbiol.* **2010**, *139*, 56–63. [CrossRef]
9. Field, D.; Ross, R.P.; Hill, C. Developing bacteriocins of lactic acid bacteria into next generation biopreservatives. *Curr. Opin. Food Sci.* **2018**, *20*, 1–6. [CrossRef]
10. Pawlowska, A.M.; Zannini, E.; Coffey, A.; Arendt, E.K. Green preservatives: Combating fungi in the food and feed in-dustry by applying antifungal lactic acid bacteria. *Adv. Food Nutr. Res.* **2012**, *66*, 217. [CrossRef]
11. Liu, B.; Ge, N.; Peng, B.; Pan, S. Kinetic and isotherm studies on the adsorption of tenuazonic acid from fruit juice using inactivated LAB. *J. Food Eng.* **2018**, *224*, 45–52. [CrossRef]
12. Shetty, P.H.; Jespersen, L. *Saccharomyces cerevisiae* and lactic acid bacteria as potential mycotoxin decontaminating agents. *Trends Food Sci. Technol.* **2006**, *17*, 48–55. [CrossRef]
13. Ezat, S.W.P.; Netty, D.; Sangaran, G. Paper review of factors, surveillance and burden of food borne disease outbreak in Malaysia. *Malays. J. Public Health Med.* **2013**, *132*, 98–105. Available online: http://wprim.whocc.org.cn/admin/article/articleDetail?WPRIMID=626608&articleId=626608 (accessed on 29 December 2021).
14. CDC. Available online: www.cdc.gov/foodsafety/foodborne-germs.html (accessed on 23 September 2021).
15. WHO. Available online: www.who.int/activities/estimating-the-burden-of-foodborne-diseases (accessed on 23 September 2021).
16. Pitt, J. Toxigenic fungi: Which are important? *Med. Mycol.* **2000**, *38*, 17–22. [CrossRef] [PubMed]
17. Salaheen, S.; Peng, M.; Biswas, D. Replacement of conventional antimicrobials and preservatives in food production to improve consumer safety and enhance health benefits. In *Microbial Food Safety and Preservation Techniques*, 1st ed.; Rai, V.R., Bai, J.A., Eds.; CRC Press: Boca Raton, FL, USA, 2014; pp. 305–3010. Available online: https://www.taylorfrancis.com/chapters/edit/10.1201/b17465-20/replacement-conventional-antimicrobials-preservatives-food-production-improve-consumer-safety-enhance-health-benefits-serajus-salaheen-mengfei-peng-debabrata-biswas (accessed on 29 December 2021).
18. Shehata, M.G.; Badr, A.N.; Sohaimy, S.A.E.I.; Asker, D.; Awad, T.S. Characterization of antifungal metabolites produced by novel lactic acid bacterium and their potential application as food biopreservatives. *Ann. Agric. Sci.* **2019**, *64*, 71–78. [CrossRef]
19. Anand, S.; Sati, N. Artificial preservatives and their harmful effects: Looking toward nature for safer alternatives. *Int. J. Pharm. Sci. Res.* **2013**, *47*, 2496. [CrossRef]
20. Kumari, P.K.; Akhila, S.; Rao, Y.S.; Devi, B.R. Alternative to artificial preservatives. *Sys. Rev. Pharm.* **2019**, *10*, 99–102. Available online: www.sysrevpharm.org/fulltext/196-1568985145.pdf (accessed on 29 December 2021).
21. Dengate, S.; Ruben, A. Controlled trial of cumulative behavioural effects of a common bread preservative. *J. Paediatr. Child Health* **2002**, *38*, 373–376. [CrossRef] [PubMed]
22. Magomya, A.M.; Yebpella, G.G.; Okpaegbe, U.C.; Oko, O.J.; Gambo, S.B. Analysis and health risk assessment of sodium benzoate and potassium sorbate in selected fruit juice and soft drink brands in nigeria. *Int. J. Pharm. Chem.* **2020**, *5*, 54–59. [CrossRef]
23. Mollapour, M.; Piper, P.W. Targeted gene deletion in *Zygosaccharomyces bailii*. *Yeast* **2001**, *18*, 173–186. [CrossRef]
24. Makdesi, A.K.; Beuchat, L.R. Evaluation of media for enumerating heat-stressed, benzoate-resistant *Zygosaccharomyces bailii*. *Int. J. Food Microbiol.* **1996**, *33*, 169–181. [CrossRef]
25. Osman, N.M.; Abou Dobara, M.I.; El-Sayed, A.; Zaky, M. The inhibitory effect of some chemical food preservatives on the growth of some isolated dairy products fungi. *J. Basic. Appl. Sci.* **2021**, *2*, 253–262. [CrossRef]
26. Lay, L.C.; Mounier, J.; Vasseur, V.; Weill, A.; Le Blay, G.; Barbier, G.; Coton, E. In vitro and in situ screening of lactic acid bacteria and propionibacteria antifungal activities against bakery product spoilage molds. *Food Control* **2016**, *60*, 247–255. [CrossRef]
27. Guynot, M.; Ramos, A.; Sanchis, V.; Marín, S. Study of benzoate, propionate, and sorbate salts as mould spoilage inhibitors on intermediate moisture bakery products of low pH 4.5–5.5. *Int. J. Food Microbiol.* **2005**, *101*, 161–168. [CrossRef] [PubMed]
28. Ryan, L.; Dal Bello, F.; Arendt, E. The use of sourdough fermented by antifungal LAB to reduce the amount of calcium propionate in bread. *Int. J. Food Microbiol.* **2008**, *125*, 274–278. [CrossRef]
29. Suhr, K.I.; Nielsen, P.V. Effect of weak acid preservatives on growth of bakery product spoilage fungi at different water activities and pH values. *Int. J. Food Microbiol.* **2004**, *95*, 67–78. [CrossRef]
30. Pongsavee, M.; Mishra, R. Potassium sorbate induces oxidative stress and genotoxicity in human lymphocytes. *Indian J. Forensic Med. Toxicol.* **2021**, *15*, 2795–2803.
31. Ledenbach, L.H.; Marshall, R.T. Microbiological spoilage of dairy products. In *Compendium of the Microbiological Spoilage of Foods and Beverages*; Springer: New York, NY, USA, 2010; pp. 41–67. [CrossRef]
32. Stanojevic, D.; Comic, L.; Stefanovic, O.; Solujic-Sukdolak, S. Antimicrobial effects of sodium benzoate, sodium nitrite and potassium sorbate and their synergistic action in vitro. *Bulg. J. Agric. Sci.* **2009**, *15*, 307–311. Available online: www.agrojournal.org/15/04-05-09.pdf (accessed on 29 December 2021).
33. Piper, J.D.; Piper, P.W. Benzoate and sorbate salts: A systematic review of the potential hazards of these invaluable preservatives and the expanding spectrum of clinical uses for sodium benzoate. *Compr. Rev. Food Sci. Food Saf.* **2017**, *16*, 868–888. [CrossRef]

34. Linke, B.G.; Casagrande, T.A.; Cardoso, L.; Cardoso, A.C. Food additives and their health effects: A review on preservative sodium benzoate. *Afr. J. Biotechnol.* **2018**, *17*, 306–310. [CrossRef]
35. Tirosh, A.; Calay, E.S.; Tuncman, G.; Claiborn, K.C.; Inouye, K.E.; Eguchi, K.; Alcala, M.; Rathaus, M.; Hollander, K.S.; Ron, I.; et al. The short-chain fatty acid propionate increases glucagon and FABP4 production, impairing insulin action in mice and humans. *Sci. Transl. Med.* **2019**, *11*, eaav0120. [CrossRef] [PubMed]
36. Dehghan, P.; Mohammadi, A.; Mohammadzadeh-Aghdash, H.; Dolatabadi, J.E.N. Pharmacokinetic and toxicological aspects of potassium sorbate food additive and its constituents. *Trends Food Sci. Technol.* **2018**, *80*, 123–130. [CrossRef]
37. Lebe, E.; Baka, M.; Yavaşoğlu, A.; Aktuğ, H.; Ateş, U.; Uyanıkgil, Y. Effects of preservatives in nasal formulations on the mucosal integrity: An electron microscopic study. *Pharmacology* **2004**, *72*, 113–120. [CrossRef]
38. Sperber, W.H.; Doyle, M.P. *Compendium of the Microbiological Spoilage of Foods and Beverages*, 3rd ed.; Food microbiology and food safety; Springer Science & Business Media: Berlin/Heidelberg, Germany, 2009; p. 45. [CrossRef]
39. Stopforth, J.D.; Sofos, J.N.; Busta, F.F. Sorbic acid and sorbates. In *Antimicrobials in Food*, 2nd ed.; Davidson, P.M., Sofos, J.N., Branen, N.A., Eds.; CRC Press: New York, NY, USA, 2005; pp. 49–75. Available online: https://books.google.fi/books?hl=en&lr=&id=OU9sBgAAQBAJ&oi=fnd&pg=PA49&dq=Sorbic+acid+and+sorbates.+In+Food+Science+and+Technology&ots=hlaPNQ94qM&sig=gMY8mfCBC64bNAEcBlakPLVb2Vk&redir_esc=y#v=onepage&q=Sorbic%20acid%20and%20sorbates.%20In%20Food%20Science%20and%20Technology&f=false (accessed on 29 December 2021).
40. Ferrand, C.; Marc, F.; Fritsch, P.; Cassand, P.; Blanquat, G.D.S. Genotoxicity study of reaction products of sorbic acid. *J. Agric. Food Chem.* **2000**, *48*, 3605–3610. [CrossRef] [PubMed]
41. Wang, Y.; Deng, J.; Xu, S.; Peng, X.; Zuo, Z.; Cui, H.M.; Wang, Y.; Ren, Z.H. Effects of Zearalenone on IL-2, IL-6, and IFN-γ mRNA levels in the splenic lymphocytes of chickens. *Sci. World J.* **2012**, *2012*, 567327. [CrossRef]
42. Wiederhold, N.P. Antifungal resistance: Current trends and future strategies to combat. *Infect. Drug Resist.* **2017**, *10*, 249. [CrossRef]
43. Garnier, L.; Valence, F.A.; Pawtowski, L.; Auhustsinava-Galerne, N.; Frotté, R.; Baroncelli, F.; Déniel, E.; Coton, E.; Mounier, J. Diversity of spoilage fungi associated with various French dairy products. *Int. J. Food Microbiol.* **2017**, *241*, 191–197. [CrossRef]
44. Nielsen, P.V.; De Boer, E. Food preservatives against fungi. In *Introduction of Food-and Airborne Fungi*; Samson, R.A., Hoekstra, E.S., Frisvad, J.C., Filtenborg, O., Eds.; Centraalbureau voor Schimmelcultures: Utrecht, The Netherlands, 2000; pp. 357–363.
45. König, H.; Unden, G.; Fröhlich, J. *Biology of Microorganisms on Grapes in Must and in Wine*, 2nd ed.; Springer: Berlin/Heidelberg, Germany; New York, NY, USA, 2017; p. 4. Available online: https://link.springer.com/book/10.1007%2F978-3-319-60021-5 (accessed on 29 December 2021).
46. Beasley, S.; Tuorila, H.; Saris, P.E.J. Fermented soymilk with a monoculture of *Lactococcus lactis*. *Int. J. Food Microbiol.* **2003**, *81*, 159–162. [CrossRef]
47. García, C.; Rendueles, M.; Díaz, M. Liquid-phase food fermentations with microbial consortia involving lactic acid bacteria. *Food Res. Int.* **2019**, *119*, 207–220. [CrossRef]
48. Azizkhani, M.; Saris, P.E.J.; Baniasadi, M. An in-vitro assessment of antifungal and antibacterial activity of cow, camel, ewe, and goat milk kefir and probiotic yogurt. *J. Food Meas. Charact.* **2020**, *15*, 406–415. [CrossRef]
49. Singh, K.; Kallali, B.; Kumar, A.; Thaker, V. Probiotics: A review. *Asian Pac. J. Trop. Biomed.* **2011**, *12*, S287–S290. [CrossRef]
50. Pakdaman, M.N.; Udani, J.K.; Molina, J.P.; Shahani, M. The effects of the DDS-1 strain of lactobacillus on symptomatic re-lief for lactose intolerance-a randomized, double-blind, placebo-controlled, crossover clinical trial. *Nutr. J.* **2015**, *15*, 56. [CrossRef] [PubMed]
51. Dicks, L.; Botes, M. Probiotic lactic acid bacteria in the gastro-intestinal tract: Health benefits, safety and mode of action. *Benef. Microbes* **2009**, *1*, 11–29. [CrossRef] [PubMed]
52. Mahmoudi, M.; Khomeiri, M.; Saeidi, M.; Kashaninejad, M.; Davoodi, H. Study of potential probiotic properties of lactic acid bacteria isolated from raw and traditional fermented camel milk. *Appl. Food Biotechnol.* **2019**, *21*, 1161–1172. Available online: jast.modares.ac.ir/article-23-18423-en.pdf (accessed on 29 December 2021).
53. Mahmoudi, M.; Khomeiri, M.; Saeidi, M.; Davoodi, H. *Lactobacillus* Species from Iranian Jug Cheese: Identification and selection of probiotic based on safety and functional properties. *J. Agric. Sci.* **2021**, *8*, 47–56. [CrossRef]
54. Axel, C.; Brosnan, B.; Zannini, E.; Furey, A.; Coffey, A.; Arendt, E.K. Antifungal sourdough lactic acid bacteria as biopreservation tool in quinoa and rice bread. *Int. J. Food Microbiol.* **2016**, *23*, 86–94. [CrossRef] [PubMed]
55. Nasrollahzadeh, A.; Khomeiri, M.; Mahmoudi, M.; Sadeghi, A.; Ebrahimi, M. Identification and evaluation of the antimicrobial potential of strains derived from traditional fermented dairy products of iran as a biological preservative against *Listeria monocytogenes*, *Staphylococcus aureus*, *Salmonella enterica* and *Escherichia coli*. *Iran. J. Microbiol.* **2019**, *13*, 392–405. Available online: http://ijmm.ir/article-1-953-en.html (accessed on 29 December 2021). [CrossRef]
56. Quattrini, M.; Liang, N.; Fortina, M.G.; Xiang, S.; Curtis, J.M.; Gänzle, M. Exploiting synergies of sourdough and antifungal organic acids to delay fungal spoilage of bread. *Int. J. Food Microbiol.* **2019**, *302*, 8–14. [CrossRef]
57. Awah, J.; Ukwuru, M.; Alum, E.; Kingsley, T. Bio-preservative potential of lactic acid bacteria metabolites against fungal pathogens. *Afr. J. Microbiol. Res.* **2018**, *12*, 913–922. [CrossRef]
58. Salas, M.L.; Thierry, A.; Lemaître, M.; Garric, G.; Harel-Oger, M.; Chatel, M.; Lê, S.; Mounier, J.; Valence, F.; Coton, E. Antifungal activity of lactic acid bacteria combinations in dairy mimicking models and their potential as bioprotective cultures in pilot scale applications. *Front. Microbiol.* **2018**, *9*, 1787. [CrossRef]

59. Baek, E.; Kim, H.; Choi, H.; Yoon, S.; Kim, J. Antifungal activity of *Leuconostoc citreum* and *Weissella confusa* in rice cakes. *J. Microbiol.* **2012**, *50*, 842–848. [CrossRef] [PubMed]
60. Mandal, V.; Sen, S.K.; Mandal, N.C. Detection, isolation and partial characterization of antifungal compound is produced by *Pediococcus acidilactici* LAB 5. *Nat. Prod. Commun.* **2007**, *2*, 671–674. [CrossRef]
61. Valerio, F.; Favilla, M.; De Bellis, P.; Sisto, A.; De Candia, S.; Lavermicocca, P. Antifungal activity of strains of lactic acid bacteria isolated from a semolina ecosystem against *Penicillium roqueforti, Aspergillus niger* and *Endomyces fibuliger* contaminating bakery products. *Syst. Appl. Microbiol.* **2009**, *32*, 438–448. [CrossRef] [PubMed]
62. Lynch, K.M.; Pawlowska, A.M.; Brosnan, B.; Coffey, A.; Zannini, E.; Furey, A.; Rahman, S.U.; Chen, X.; Jiang, Y.; Zhu, D.; et al. Application of *Lactobacillus amylovorus* as an antifungal adjunct to extend the shelf-life of Cheddar cheese. *Int. Dairy J.* **2014**, *341*, 167–173. [CrossRef]
63. Fernandez, B.; Vimont, A.; Desfossés-Foucault, E.; Daga, M.; Arora, G.; Fliss, I. Antifungal activity of lactic and propionic acid bacteria and their potential as protective culture in cottage cheese. *Food Control* **2017**, *78*, 350–356. [CrossRef]
64. Cheong, E.Y.; Sandhu, A.; Jayabalan, J.; Le, T.T.; Nhiep, N.T.; Ho, H.T.M.; Zwielehner, J.; Bansal, N.; Turner, M.S. Isolation of lactic acid bacteria with antifungal activity against the common cheese spoilage mould *Penicillium commune* and their potential as biopreservatives in cheese. *Food Control* **2014**, *46*, 91–97. [CrossRef]
65. Khomeiri, M.; Esazadeh Rzelighi, S.; Nasrollahzadeh, A. Evaluation of growth inhibit of food spoilage yeast of *Lactobacillus brevis* and *Enterococcus faecium* from chal in Iranian yoghurt drink (Doogh). *Iran. J. Biosyst. Eng. IJBSE* **2017**, *47*, 643–649. Available online: https://ijbse.ut.ac.ir/m/article_60258.html?lang=en (accessed on 29 December 2021).
66. Ouiddir, M.; Bettache, G.; Salas, M.L.; Pawtowski, A.; Donot, C.; Brahimi, S.; Mabrouk, K.; Coton, E.; Mounier, J. Selection of Algerian lactic acid bacteria for use as antifungal bioprotective cultures and application in dairy and bakery products. *Food Microbiol.* **2019**, *82*, 160–170. [CrossRef]
67. Yan, B.; Zhao, J.; Fan, D.; Tian, F.; Zhang, H.; Chen, W. Antifungal activity of *Lactobacillus plantarum* against *Penicillium roqueforti* in vitro and the preservation effect on Chinese steamed bread. *J. Food Process. Preserv.* **2017**, *41*, e12969. [CrossRef]
68. Dalié, D.K.D.; Deschamps, A.M.; Richard-Forget, F. Lactic acid bacteria—Potential for control of mould growth and mycotoxins: A review. *Food Control* **2010**, *21*, 370–380. [CrossRef]
69. Black, B.A.; Zannini, E.; Curtis, J.M.; Gï, M.G. Antifungal hydroxy fatty acids produced during sourdough fermentation: Microbial and enzymatic pathways, and antifungal activity in bread. *Appl. Environ. Microbiol.* **2013**, *79*, 1866–1873. [CrossRef] [PubMed]
70. Rouse, S.; Harnett, D.; Vaughan, A.; Sinderen, D.V. Lactic acid bacteria with potential to eliminate fungal spoilage in foods. *J. Appl. Microbiol.* **2008**, *104*, 915–923. [CrossRef] [PubMed]
71. Magnusson, J.; Schnürer, J. *Lactobacillus coryniformis* subsp. *coryniformis* strain Si3 produces a broad-spectrum proteinaceous antifungal compound. *Appl. Environ. Microbiol.* **2001**, *67*, 1–5. [CrossRef] [PubMed]
72. Sangmanee, P.; Hongpattarakere, T. Inhibitory of multiple antifungal components produced by *Lactobacillus plantarum* K35 on growth, aflatoxin production and ultrastructure alterations of *Aspergillus flavus* and *Aspergillus parasiticus*. *Food Control* **2014**, *40*, 224–233. [CrossRef]
73. Muhialdin, B.J.; Hassan, Z.; Saari, N. In vitro antifungal activity of lactic acid bacteria low molecular peptides against spoilage fungi of bakery products. *Ann. Microbiol.* **2018**, *68*, 557–567. [CrossRef]
74. Prema, P.; Smila, D.; Palavesam, A.; Immanuel, G. Production and characterization of an antifungal compound 3-phenyllactic acid produced by *Lactobacillus plantarum* strain. *Food Bioprocess Technol.* **2010**, *33*, 379–386. [CrossRef]
75. Peyer, L.C.; Axel, C.; Lynch, K.M.; Zannini, E.; Jacob, F.; Arendt, E.K. Inhibition of *Fusarium culmorum* by carboxylic acids released from lactic acid bacteria in a barley malt substrate. *Food Control* **2016**, *69*, 227–236. [CrossRef]
76. Ruggirello, M.; Nucera, D.; Cannoni, M.; Peraino, A.; Rosso, F.; Fontana, M.; Cocolin, L.; Dolci, P. Antifungal activity of yeasts and lactic acid bacteria isolated from cocoa bean fermentations. *Food Res. Int.* **2019**, *115*, 519–525. [CrossRef]
77. Nasrollahzadeh, A.; Khomeiri, M.; Sadeghi, A. Molecular identification of lactic acid bacteria strains isolated from the Chal in Golestan province and study antifungal activity of *Lactobacillus brevis* and *Enterococcus faecium* isolated against *Penicillium chrysogenum*. *J. Appl. Microbiol. Food Ind.* **2016**, *2*, 56–69. Available online: https://www.sid.ir/en/Journal/ViewPaper.aspx?ID=546149 (accessed on 29 December 2021).
78. Ortiz-Rivera, Y.; Sánchez-Vega, R.; Gutiérrez-Méndez, N.; León-Félix, J.; Acosta-Muñiz, C.; Sepulveda, D.R. Production of reuterin in a fermented milk product by *Lactobacillus reuteri*: Inhibition of pathogens, spoilage microorganisms, and lactic acid bacteria. *J. Dairy Sci.* **2017**, *100*, 4258–4268. [CrossRef]
79. Russo, P.; Arena, M.P.; Fiocco, D.; Capozzi, V.; Drider, D.; Spano, G. *Lactobacillus plantarum* with broad antifungal activity: A promising approach to increase safety and shelf-life of cereal-based products. *Int. J. Food Microbiol.* **2017**, *247*, 48–54. [CrossRef] [PubMed]
80. Oliveira, P.M.; Zannini, E.; Arendt, E.K. Cereal fungal infection, mycotoxins, and lactic acid bacteria mediated bioprotection: From crop farming to cereal products. *Food Microbiol.* **2014**, *37*, 78–95. [CrossRef] [PubMed]
81. Muhialdin, B.J.; Hassan, Z.; Sadon, S.K.H. Antifungal activity of *Lactobacillus fermentum* Te007, *Pediococcus pentosaceus* Te010, *Lactobacillus pentosus* G004, and *L. paracasi* D5 on selected foods. *J. Food Sci.* **2011**, *76*, M493–M499. [CrossRef] [PubMed]
82. Gupta, R.; Srivastava, S. Antifungal effect of antimicrobial peptides (AMPs LR14) derived from *Lactobacillus plantarum* strain LR/14 and their applications in prevention of grain spoilage. *Food Microbiol.* **2014**, *42*, 1–7. [CrossRef] [PubMed]

83. Oirdi, E.S.; Lakhlifi, T.; Bahar, A.A.; Yatim, M.; Rachid, Z.; Belhaj, A. Isolation and identification of *Lactobacillus plantarum* 4F, a strain with high antifungal activity, fungicidal effect, and biopreservation properties of food. *J. Food Process. Preserv.* **2020**, *6*, e15517. [CrossRef]
84. Vimont, A.; Fernandez, B.; Ahmed, G.; Fortin, H.P.; Fliss, I. Quantitative antifungal activity of reuterin against food isolates of yeasts and moulds and its potential application in yogurt. *Int. J. Food Microbiol.* **2019**, *28*, 182–188. [CrossRef] [PubMed]
85. Cosentino, S.; Viale, S.; Deplano, M.; Fadda, M.E.; Pisano, M.B. Application of autochthonous *Lactobacillus* strains as biopreservatives to control fungal spoilage in Caciotta cheese. *BioMed Res. Int.* **2018**, *18*, 381–393. [CrossRef]
86. Luz, C.; D'Opazo, V.; Mañes, J.; Meca, G. Antifungal activity and shelf life extension of loaf bread produced with sourdough fermented by Lactobacillus strains. *J. Food Process. Preserv.* **2019**, *43*, e14126. [CrossRef]
87. Guimarães, A.; Venancio, A.; Abrunhosa, L. Antifungal effect of organic acids from lactic acid bacteria on *Penicillium nordicum*. *Food Addit. Contam.* **2018**, *359*, 1803–1818. [CrossRef]
88. Ryan, L.A.; Dal Bello, F.; Arendt, E.K.; Koehler, P. Detection and quantitation of 2, 5-diketopiperazines in wheat sourdough and bread. *J. Agric. Food Chem.* **2009**, *57*, 9563–9568. [CrossRef]
89. Martinez, F.A.C.; Balciunas, E.M.; Salgado, J.M.; González, J.M.D.; Converti, A.; Oliveira, R.P.S. Lactic acid properties, applications and production: A review. *Trends Food Sci. Technol.* **2013**, *30*, 70–83. [CrossRef]
90. Nasrollahzadeh, A.; Khomeiri, M.; Mahmoudi, M.; Sadeghi, A.; Ebrahimi, M. Antifungal activity of lactic acid bacteria isolated from masske, camel dough, and local yoghurt against *Aspergillus flavus* and *Aspergillus niger*. *J. Food Hyg.* **2020**, *4*, 1–11. Available online: https://www.sid.ir/en/Journal/ViewPaper.aspx?ID=747257 (accessed on 29 December 2021).
91. Arena, M.P.; Silvain, A.; Normanno, G.; Grieco, F.; Drider, D.; Spano, G.; Fiocco, D. Use of *Lactobacillus plantarum* strains as a bio-control strategy against food-borne pathogenic microorganisms. *Front. Microbiol.* **2016**, *7*, 464. [CrossRef] [PubMed]
92. Honoré, A.H.; Aunsbjerg, S.D.; Ebrahimi, P.; Thorsen, M.; Benfeldt, C.; Knøchel, S.; Skov, T. Metabolic footprinting for investigation of antifungal properties of *Lactobacillus paracasei*. *Anal. Bioanal. Chem.* **2016**, *408*, 83–96. [CrossRef] [PubMed]
93. Salas, M.L.; Mounier, J.; Maillard, M.B.; Valence, F.; Coton, E.; Thierry, A. Identification and quantification of natural compounds produced by antifungal bioprotective cultures in dairy products. *Food Chem.* **2019**, *301*, 125260. [CrossRef] [PubMed]
94. Crowley, S.; Mahony, J.; Sinderen, D.V. Current perspectives on antifungal lactic acid bacteria as natural bio-preservatives. *Trends Food Sci. Technol.* **2013**, *33*, 93–109. [CrossRef]
95. Dagnas, S.; Gauvry, E.; Onno, B.; Membre, J.M. Quantifying effect of lactic, acetic, and propionic acids on growth of molds isolated from spoiled bakery products. *J. Food Prot.* **2015**, *78*, 1689–1698. [CrossRef]
96. Loubiere, P.; Cocaign-Bousquet, M.; Matos, J.; Goma, G.; Lindley, N. Influence of end-products inhibition and nutrient limitations on the growth of *Lactococcus lactis* subsp. *Lactis*. *J. Appl. Microbiol.* **1997**, *82*, 95–100. [CrossRef]
97. Brosnan, B.; Coffey, A.; Arendt, E.K.; Furey, A. Rapid identification, by use of the LTQ Orbitrap hybrid FT mass spectrometer, of antifungal compounds produced by lactic acid bacteria. *Anal. Bioanal. Chem.* **2012**, *403*, 2983–2995. [CrossRef]
98. Guo, J.; Brosnan, B.; Furey, A.; Arendt, E.; Murphy, P.; Coffey, A. Antifungal activity of *Lactobacillus* against *Microsporum canis*, *Microsporum gypseum* and *Epidermophyton floccosum*. *Bioeng. Bugs* **2012**, *32*, 104–113. [CrossRef]
99. Muynck, C.; Leroy, A.I.D.; Maeseneire, S.D.; Arnaut, F.; Soetaert, W.; Vandamme, E.J. Potential of selected lactic acid bacteria to produce food compatible antifungal metabolites. *Microbiol. Res.* **2004**, *159*, 339–346. [CrossRef] [PubMed]
100. Jung, S.; Hwang, H.; Lee, J.H. Effect of lactic acid bacteria on phenyllactic acid production in kimchi. *Food Control* **2019**, *106*, 106701. [CrossRef]
101. Schmidt, M.; Lynch, K.M.; Zannini, E.; Arendt, E.K. Fundamental study on the improvement of the antifungal activity of *Lactobacillus reuteri* R29 through increased production of phenyllactic acid and reuterin. *Food Control* **2018**, *88*, 139–148. [CrossRef]
102. Lavermicocca, P.; Valerio, F.; Evidente, A.; Lazzaroni, S.; Corsetti, A.; Gobbetti, M. Purification and characterization of novel antifungal compounds from the sourdough *Lactobacillus plantarum* strain 21B. *Appl. Environ. Microbiol.* **2000**, *66*, 4084–4090. [CrossRef] [PubMed]
103. Guimarães, A.; Santiago, A.; Teixeira, J.A.; Venâncio, A.; Abrunhosa, L. Anti-aflatoxigenic effect of organic acids produced by *Lactobacillus plantarum*. *Int. J. Food Microbiol.* **2018**, *264*, 31–38. [CrossRef] [PubMed]
104. Yépez, A.; Luz, C.; Meca, G.; Vignolo, G.; Mañes, J.; Aznar, R. Biopreservation potential of lactic acid bacteria from Andean fermented food of vegetal origin. *Food Control* **2017**, *78*, 393–400. [CrossRef]
105. Ström, K.; Sjögren, J.; Broberg, A.; Schnürer, J. *Lactobacillus plantarum* MiLAB 393 produces the antifungal cyclic dipeptides cyclo(L-Phe-L-Pro) and cyclo(L-Phe-trans-4-OH-L-Pro) and 3-phenyllactic acid. *Appl. Environ. Microbiol.* **2002**, *68*, 4322–4327. [CrossRef]
106. Ndagano, D.; Lamoureux, T.; Dortu, C.; Vandermoten, S.; Thonart, P. Antifungal activity of 2 lactic acid bacteria of the *Weissella* genus isolated from food. *J. Food Sci.* **2011**, *76*, M305–M311. [CrossRef]
107. Engels, C.; Schwab, C.; Zhang, J.; Stevens, M.J.; Bieri, C.; Ebert, M.O.; McNeill, K.; Sturla, S.J.; Lacroix, C. Acrolein contributes strongly to antimicrobial and heterocyclic amine transformation activities of reuterin. *Sci. Rep.* **2016**, *6*, 36246. [CrossRef]
108. Schaefer, L.; Auchtung, T.A.; Hermans, K.E.; Whitehead, D.; Borhan, B.; Britton, R.A. The antimicrobial compound reuterin 3-hydroxypropionaldehyde induces oxidative stress via interaction with thiol groups. *Microbiology* **2010**, *156*, 1589–1599. [CrossRef]
109. Magnusson, J. Antifungal activity of lactic acid bacteria. Ph.D. Thesis, Swedish University of Agricultural Sciences, Uppsala, Sweden, 2003. Available online: https://pub.epsilon.slu.se/247/ (accessed on 29 December 2021).

110. Talarico, T.L.; Dobrogosz, W.J. Chemical characterization of an antimicrobial substance produced by *Lactobacillus reuteri*. *Antimicrob. Agents Chemother.* **1989**, *33*, 674–679. [CrossRef] [PubMed]
111. Gómez-Torres, N.; Ávila, M.; Delgado, D.; Garde, S. Effect of reuterin-producing *Lactobacillus reuteri* coupled with glycerol on the volatile fraction, odour and aroma of semi-hard ewe milk cheese. *Int. J. Food Microbiol.* **2016**, *232*, 103–110. [CrossRef] [PubMed]
112. Luz, C.; Saladino, F.; Luciano, F.; Mañes, J.; Meca, G. In vitro antifungal activity of bioactive peptides produced by *Lactobacillus plantarum* against *Aspergillus parasiticus* and *Penicillium expansum*. *LWT* **2017**, *81*, 128–135. [CrossRef]
113. Gerez, C.L.; Torres, M.J.; Valdez, G.F.D.; Rollán, G. Control of spoilage fungi by lactic acid bacteria. *Biol. Control* **2013**, *64*, 231–237. [CrossRef]
114. Guillén, G.; López Caballero, M.E.; Alemán, A.; López de Lacey, A.; Giménez, B.; Montero García, P. Antioxidant and antimicrobial peptide fractions from squid and tuna skin gelatin. In *Sea By-Products as Real Material: New Ways of Application*; Bihan, E., Ed.; Transworld Research Network: Trivandrum, India, 2010; pp. 98–115.
115. Joo, S.H. Cyclic peptides as therapeutic agents and biochemical tools. *Biomol. Ther.* **2012**, *20*, 19–26. [CrossRef]
116. Niku-Paavola, M.L.; Laitila, A.; Mattila-Sandholm, T.; Haikara, A. New types of antimicrobial compounds produced by *Lactobacillus plantarum*. *J. Appl. Microbiol.* **1999**, *861*, 29–35. [CrossRef]
117. Shai, Y. Molecular recognition between membrane-spanning polypeptides. *Trends Biochem. Sci.* **1995**, *20*, 460–464. [CrossRef]
118. Insel, P.M. *Discovering Nutrition*, 4th ed.; Jones and Bartlett: Burlington, MA, USA, 2013. Available online: https://books.google.com/books/about/Discovering_Nutrition.html?id=an0vCrQxaB8C (accessed on 29 December 2021).
119. Sjögren, J.; Magnusson, J.; Broberg, A.; Schnürer, J.; Kenne, L. Antifungal 3-hydroxy fatty acids from *Lactobacillus plantarum* MiLAB 14. *Appl. Environ. Microbiol.* **2003**, *69*, 7554–7557. [CrossRef]
120. Avis, T.J.; Bélanger, R.R. Specificity and mode of action of the antifungal fatty acid cis-9-heptadecenoic acid produced by *Pseudozyma flocculosa*. *Appl. Environ. Microbiol.* **2001**, *672*, 956–960. [CrossRef]
121. Pohl, C.H.; Kock, J.L.F.; Thibane, V.S. Antifungal free fatty acids: A review. In *Science against Microbial Pathogens: Communicating Current Research and Technological Advances*; Formatex Research Center: Guadalajara, Mexico, 2011; Volume 1, pp. 61–71. Available online: www.researchgate.net/profile/Carolina_Pohl/publication/266463207_Antifungal_free_fatty_acids_A_Review/links/54daf14b0cf261ce15ce9643.pdf (accessed on 29 December 2021).

MDPI
St. Alban-Anlage 66
4052 Basel
Switzerland
www.mdpi.com

Foods Editorial Office
E-mail: foods@mdpi.com
www.mdpi.com/journal/foods

www.ingramcontent.com/pod-product-compliance
Lightning Source LLC
Chambersburg PA
CBHW061513210926
51850CB00008B/1906
* 9 7 8 3 0 3 6 5 9 9 6 9 4 *